Minimax and Applications

Nonconvex Optimization and Its Applications

Volume 5

The titles published in this series are listed at the end of this volume.

"You cannot step twice into the same river, for other waters are continually flowing on"

"Time is a child moving counters in a game; the royal power is a child's"

- Heraclitus

Contents

Preface

Techniques and principles of minimax theory play a key role in many areas of research, including game theory, optimization, and computational complexity. In general, a minimax problem can be formulated as

$$\min_{x \in X} \max_{y \in Y} f(x, y) \tag{1}$$

where $f(x, y)$ is a function defined on the product of X and Y spaces. There are two basic issues regarding minimax problems:

The first issue concerns the establishment of sufficient and necessary conditions for equality

$$\min_{x \in X} \max_{y \in Y} f(x, y) = \max_{y \in Y} \min_{x \in X} f(x, y). \tag{2}$$

The classical minimax theorem of von Neumann is a result of this type. Duality theory in linear and convex quadratic programming interprets minimax theory in a different way.

The second issue concerns the establishment of sufficient and necessary conditions for values of the variables x and y that achieve the global minimax function value

$$f(x^*, y^*) = \min_{x \in X} \max_{y \in Y} f(x, y). \tag{3}$$

There are two developments in minimax theory that we would like to mention. First, it has been shown that some minimax problems (which are NP-hard in general) are polynomially solvable when condition (2) holds [see e.g., Andras Recski, Minimax Results and Polynomial Algorithms in VLSI Routing, in Proceedings of the Fourth Czechoslovakian Symposium on Combinatorics, Graphs, and Complexity (Prachatice, Czechoslovakia, June 1990) North-Holland, 1992 (Editors M. Fiedler and J. Nesetril), pp. 261-273]. Secondly, a long-standing open problem on minimum networks, the Gilbert-Pollak conjecture on the Steiner ratio, was solved by using a minimax approach [see e.g., D.-Z. Du and F.K. Hwang, An approach for proving lower bounds: solution of Gilbert-Pollak's conjecture on the Steiner ratio, In Proceedings of 31th FOCS Conference (1990), pp. 76-85]. Central to this approach is a result regarding the characterization of global optima of (3). These developments indicate that minimax theory will continue to be an important tool for solving difficult and interesting problems. In addition, minimax methods provide a paradigm for investigating analogous problems. An exciting future with new unified theories may be expected.

A remark on terminology may be necessary. Many researchers reserve the term "minimax" for results of von Neumann's type. However, for study of the second issue, the term "min-max" or "minmax" is used. Since the two types of problems discussed above are of the same nature, we suggest that the term "minimax" is used.

The collection of papers in this book covers a diverse number of topics and provides a good picture of recent research in minimax theory. Most of the papers in the first group present results on classical minimax theorems and optimization problems using duality theory. The papers in the second group are mainly concerned with optimality and approximate algorithms of minimax problems. Instead of a postface, the final paper provides a brief survey and open questions on minimax problems in combinatorial optimization.

The book will be a valuable source of information to faculty, students and researchers in optimization, computer sciences, and related areas. We would like to take the opportunity to thank the authors of the papers, the referees, and the publisher for helping us to produce this book.

Ding-Zhu Du and Panos M. Pardalos

University of Minnesota and University of Florida
March 1995

MINIMAX THEOREMS AND THEIR PROOFS

STEPHEN SIMONS
Department of Mathematics, University of California, Santa Barbara, CA 93106-3080.

1. Introduction

We suppose that X and Y are nonempty sets and $f : X \times Y \to \mathbb{R}$. A *minimax theorem* is a theorem which asserts that, under certain conditions,

$$\min_Y \max_X f = \max_X \min_Y f.$$

The original motivation for the study of minimax theorems was, of course, Von Neumann's work on games of strategy. After a lapse of nearly ten years, generalizations of Von Neumann's original result for matrices started appearing. As time went on, these generalizations became progressively more remote from game theory, and minimax theorems started becoming objects of study in their own right. In this article, we will trace the development of minimax theorems starting from Von Neumann's original result. We will discuss infinite dimensional *bilinear* results and their connection with weak compactness. We will discuss the results for *concave-convex* functions, and their generalizations to *quasiconcave-quasiconvex* functions. We will discuss various minimax theorems in which X and Y are not assumed to be subsets of vector space. These fall naturally into three classes, *topological minimax theorems* in which various connectedness hypotheses are assumed for X, Y and f, *quantitative minimax theorems* in which no special properties are assumed for X and Y, but various quantitative properties are assumed for f and, finally, *mixed minimax theorems* in which the quantitative and the topological properties are mixed. Recent developments have included *unifying metaminimax theorems*, theorems which imply simultaneously the minimax theorems of all the above three types. These latter results would tend to indicate that our initial classification of minimax theorems is too rigid. We have kept it, however, for historical reasons. We will also discuss *minimax inequalities for two or more functions.*

To a certain extent, a survey like this will always reflect the interests (prejudices) of the author. For instance, we will not discuss the kind of "local minimax theorem" that uses arguments related to the Palais-Smale condition and Ekeland's variational principle. We will also not discuss computational methods for solving games – we refer the reader to the 1974 survey article [131] by Yanovskaya. That article also contains a discussion of various infinite dimensional games. Yanovskaya's survey work was continued by Irle [34] in 1981. We would like to acknowledge that we have used both [131] and [34], as well as the many papers of Kindler, as sources

D.-Z. Du and P. M. Pardalos (eds.), Minimax and Applications, 1–23.

for the history of the development of the subject. While on the subject of general references, we should mention that, in addition to his penetrating study of optimal decision rules, Aubin has a very complete bibliography in his book [1], and that there is a section on minimax theorems in the book [3] by Barbu-Precupanu.

Finally, we would like to express our sincere thanks to Heinz König for some very insightful comments on a preliminary version of this paper.

For simplicity, we shall assume that all topological spaces are Hausdorff.

2. The First Minimax Theorem

The first minimax theorem was proved by von Neumann in 1928 using topological arguments:

Theorem 1 ([124]) Let A be an $m \times n$ matrix, and X and Y be the sets of nonnegative row and column vectors with unit sum. Let $f(x, y) := xAy$. Then

$$\min_{Y} \max_{X} f = \max_{X} \min_{Y} f.$$

In this case f is a jointly continuous function of x and y, X and Y are finite dimensional simplexes, and f is bilinear. In 1938, Ville [123] gave the first elementary proof of Theorem 1, using the *Theorem of the alternative* for matrices. It is Ville's proof that von Neumann-Morgenstern expounded in [126]. Another elementary proof of Theorem 1 was given by Weyl [129] in 1950. Karlin gave an extensive analysis of the matrix case in [44]. Finally, Berge gave a proof of Theorem 1 in [5] using his theory of *regular nonlinear convexity*.

3. Infinite Dimensional Bilinear Results

Ville [123] in 1938 and Wald [127] in 1945 considered generalizations of Theorem 1 in which the function f is still bilinear, but defined on infinite-dimensional sets. (Wald's interest was motivated by his work on statistical decision functions - see [128].) This line of investigation was pursued further by Karlin [42]-[43] ([43] also contains some results on non-bilinear games), Dulmage-Peck [12] and Tjoe-Tie [120]. In 1952, Kneser [62] gave a very simple proof of a bilinear result in which X and Y are nonempty convex subsets of vector spaces, X does not even have to be topologized, Y is required to be compact and the only topological condition that is needed is that f be lower semicontinuous on Y. Kneser's result was extended by by Peck-Dulmage [97]. Wald's result was extended by Young [132], with an axiomatic theory of games. This discussion is continued in the sections **Quantitative minimax theorems** and **Minimax theorems and weak compactness**.

4. Minimax Theorems When X and Y Are More General Convex Sets

It will be convenient at this point to give a more compact notation for the various "level sets" associated with f. If $\lambda \in \mathbb{R}$ and $W \subset X$ we define

$$LE(W, \lambda) := \bigcap_{x \in W} \{y : y \in Y, f(x, y) \leq \lambda\}$$

and

$$LT(W,\lambda) := \bigcap_{x\in W}\{y : y \in Y, f(x,y) < \lambda\}.$$

If $x \in X$ we write (by abuse of notation)

$$LE(x,\lambda) = \{y : y \in Y, f(x,y) \le \lambda\}$$

and

$$LT(x,\lambda) = \{y : y \in Y, f(x,y) < \lambda\}.$$

If $\lambda \in \mathbb{R}$ and $V \subset Y$ we define

$$GE(\lambda,V) := \bigcap_{y\in V}\{x : x \in X, f(x,y) \ge \lambda\}$$

and

$$GT(\lambda,V) := \bigcap_{y\in V}\{x : x \in X, f(x,y) > \lambda\}.$$

If $y \in Y$ we write (by abuse of notation)

$$GE(\lambda,y) = \{x : x \in X, f(x,y) \ge \lambda\}$$

and

$$GT(\lambda,y) = \{x : x \in X, f(x,y) > \lambda\}.$$

In 1937, the result of [123] was extended by von Neumann as follows:

Theorem 2 ([125]) Let X and Y be nonempty compact, convex subsets of Euclidean space, and f be jointly continuous. Suppose that f is *quasiconcave* on X, that is to say,

$$\text{for all } y \in Y \text{ and } \lambda \in \mathbb{R}, GE(\lambda,y) \text{ is convex}$$

and f is *quasiconvex* on Y, that is to say,

$$\text{for all } x \in X \text{ and } \lambda \in \mathbb{R}, LE(x,\lambda) \text{ is convex.}$$

Then

$$\min_Y \max_X f = \max_X \min_Y f.$$

In 1941, Kakutani [41] analyzed von Neumann's proof and, as a result, discovered the fixed-point theorem that bears his name. In 1952, Fan [13] generalized Theorem 2 to the case when X and Y are compact, convex subsets of (infinite dimensional) locally convex spaces and the quasiconcave and quasiconvex conditions are somewhat relaxed, while Nikaidô [92], using Brouwer's fixed-point theorem directly, generalized the same result to the case when X and Y are nonempty compact, convex subsets of (not necessarily locally convex) topological vector spaces and f is only required to be separately continuous. Nikaidô also showed in [93] that, if we replace the words *quasiconcave* and *quasiconvex* by *concave* and *convex*, then it is possible to give a proof of the minimax theorem by elementary calculus. Likewise, Moreau [87] showed that it is possible to give a proof using Fenchel duality. In 1980, Joó [37] gave a proof based on the properties of level sets, and then pointed out in [38] the connections between the level set technique and the Hahn-Banach theorem. In fact, the techniques of [37] and [38] really belonged in a much more general context. See the section **Topological minimax theorems**.

Motivated by problems in optimization theory and the theory of Lagrangians, Rockafellar [101] developed in 1970 a calculus of (possibly infinite valued) concave-convex functions in finite dimensional situations, which he extended in [102] to the infinite dimensional case. In these two paper, Rockafellar considered the concepts usually associated with convex analysis such as subdifferentials, duality and monotone operators.

5. Minimax Theorems for Separately Semicontinuous Functions

Sion, using the lemma of Knaster, Kuratowski and Mazurkiewicz on closed subsets of a finite dimensional simplex, proved the following *quasiconcave-quasiconvex u.s.c-l.s.c* result in 1958:

Theorem 3 ([114]) Let X be a convex subset of a linear topological space, Y be a compact convex subset of a linear topological space, and $f : X \times Y \to \mathbb{R}$ be upper semicontinuous on X and lower semicontinuous on Y. Suppose that,

$$\text{for all } y \in Y \text{ and } \lambda \in \mathbb{R},\ GE(\lambda, y) \text{ is convex}$$

and,

$$\text{for all } x \in X \text{ and } \lambda \in \mathbb{R},\ LE(x, \lambda) \text{ is convex.}$$

Then

$$\min_Y \sup_X f = \sup_X \min_Y f.$$

By abuse of terminology, we will continue to describe this kind of result as a *minimax theorem*. The importance of Sion's weakening of continuity to semicontinuity was that it indicated that many kind of minimax problems had equivalent formulations in terms of subsets of $X \times Y$, and led to Fan's 1972 work [16] on sets with convex sections and minimax inequalities, which have since found many applications in economic theory. (Now we would phrase some of these results in terms of *coincidence theorems* for multifunctions.) Ha [29] proved a result similar to Theorem 3, assuming instead that f was lower semicontinuous on $X \times Y$. Brézis-Nirenberg-Stampacchia [8], Kindler [46], Ha [30] and Hartung [32] showed that it was possible to relax somewhat the compactness condition in Theorem 3. It was believed for some time that Brouwer's fixed-point theorem (or the related Knaster-Kuratowski-Mazurkiewicz lemma) was required to order to prove Theorem 3. In 1966, Ghouila-Houri [22] showed that one could prove Theorem 3 using a simple combinatorial property of convex sets in finite dimensional space. Another very simple proof of Theorem 3 was given in 1988 by Komiya [63]. Using the concept of a *quarter continuous* multifunction, Komiya [65] subsequently showed that, despite the fact that the semicontinuities in Sion's result and Ha's result are different, there is a result that simultaneously generalizes both of them. See the section **Unifying**

metaminimax theorems for more recent and abstract applications of the concept of a quarter continuous multifunction.

In [84], McLinden investigated minimax results in which X and Y are not compact and f takes infinite values, using the concept of *closed saddle function* introduced by Rockafellar in [101] and [102], and also the concept of an *ε-minimax solution*. McLinden also investigated the connection between ε-minimax theorems and Ekeland's variational principle in [83]. The investigation of this type of minimax theorem appropriate for the discussion of Lagrangians was continued by Gwinner-Jeyakumar in [27] and Gwinner-Oettli in [28]. In [86], Mertens investigates the general problem of minimax theorems for separately semicontinuous functions. In this paper, he also discusses some measure-theoretic results. There is also a discussion of related results in the paper [99] by Pomerol, which has a very complete bibliography.

6. Topological Minimax Theorems

It was also believed at a more general level that proofs of minimax theorems required either the machinery of algebraic topology, or the machinery of convexity. However, in 1959, Wu, motivated by Sion's result, had already initiated research in a new direction by proving the first minimax theorem in which the conditions of convexity were totally replaced by conditions related to connectedness. Wu's paper, which appeared only a year after Sion's, evidently did not received very wide circulation in the west.

Theorem 4 ([130]) Let X be a topological space, Y be a compact separable topological space, and $f : X \times Y \to \mathbb{R}$ be separately continuous. Suppose that, for all $x_0, x_1 \in X$, there exists a continuous map $h : [0,1] \to X$ such that $h(0) = x_0, h(1) = x_1$ and,

$$\text{for all } y \in Y \text{ and } \lambda \in \mathbb{R}, \{t : t \in [0,1], f(h(t), y\} \geq \lambda\} \text{ is connected in } [0,1].$$

Suppose also that,

$$\text{for all nonempty finite subsets } W \text{ of } X \text{ and } \lambda \in \mathbb{R},\ LT(W, \lambda) \text{ is connected in } Y.$$

Then

$$\min_Y \sup_X f = \sup_X \min_Y f.$$

In 1974, Tuy removed the restrictive topological assumptions in Theorem 4, and proved the following result that generalized both Theorem 3 and Theorem 4.

Theorem 5 ([121] and [122]) Let X be a topological space, Y be a compact topological space, and $f : X \times Y \to \mathbb{R}$ be upper semicontinuous on X and lower semicontinuous on Y. Suppose that there exists a sequence $\{\lambda_s\}_{s \geq 1}$ such that $\lambda_s \downarrow \sup_X \min_Y f$ and, for all $s \geq 1$ and $x_0, x_1 \in X$, there exists a continuous map $h : [0,1] \to X$ such that $h(0) = x_0, h(1) = x_1$,

$$(b \in [a,c] \subset [0,1], y \in Y \text{ and } f(h(b), y) \leq \lambda_s) \Longrightarrow (f(h(a), y) \leq \lambda_s \text{ or } f(h(c), y) \leq \lambda_s).$$

Suppose also that,

for all $s \geq 1$ and nonempty finite subsets W of X, $LE(W, \lambda_s)$ is connected in Y.

Then

$$\min_Y \sup_X f = \sup_X \min_Y f.$$

A different result that generalizes both Theorem 3 and Theorem 4 was given in 1984 by Geraghty-Lin:

Theorem 6 ([19]) Let X be a topological space, Y be a compact topological space, and $f : X \times Y \to \mathbb{R}$ be lower semicontinuous on X and also lower semicontinuous on Y. Suppose that, for all $x_0, x_1 \in X$, there exists a continuous map $h : [0,1] \to X$ such that $h(0) = x_0, h(1) = x_1$, and

$$b \in [a,c] \subset [0,1] \Longrightarrow f(h(b),.) \geq f(h(a),.) \wedge f(h(c),.) \text{ on } Y,$$

where "$\wedge$" stands for "minimum". Suppose also that,

for all nonempty finite subsets W of X and $\lambda \in \mathbb{R}$, $LE(W, \lambda)$ is connected in Y.

Then

$$\min_Y \sup_X f = \sup_X \min_Y f.$$

After analyzing the paper [37] of Joó already mentioned, Stachó introduced in 1980 the concept of an *interval space*, a topological space X such that, for all $x_1, x_2 \in X$, there exists a connected subset $[x_1, x_2]$ of X such that $[x_1, x_2] = [x_2, x_1] \supset \{x_1, x_2\}$. A subset C of X is called *interval convex* if $x_1, x_2 \in C \Longrightarrow [x_1, x_2] \subset C$. Stachó then established the following result:

Theorem 7 ([115]) Let X and Y be compact interval spaces, and $f : X \times Y \to \mathbb{R}$ be continuous. Suppose that,

for all $y \in Y$ and $\lambda \in \mathbb{R}$, $GE(\lambda, y)$ is interval convex

and,

for all $x \in X$ and $\lambda \in \mathbb{R}$, $LE(x, \lambda)$ is interval convex.

Then

$$\min_Y \max_X f = \max_X \min_Y f.$$

Komornik subsequently proved in [66] a minimax theorem for interval spaces which generalized both Theorem 7 and the result [29] of Ha already mentioned.

Stachó also introduced the concept of a *Dedekind complete* interval space and proved a second minimax theorem, which generalizes Theorem 3. Here is a slightly simplified (see below) version of this second result:

Theorem 8 ([115]) Let X be a Hausdorff, Dedekind complete interval space, Y be a compact interval space, and $f : X \times Y \to \mathbb{R}$ be upper semicontinuous on X and lower semicontinuous on Y. Suppose that,

$$\text{for all } y \in Y \text{ and } \lambda \in \mathbb{R},\ GE(\lambda, y) \text{ is interval convex}$$

and,

$$\text{for all } x \in X \text{ and } \lambda \in \mathbb{R},\ LE(x, \lambda) \text{ is interval convex.}$$

Then

$$\min_Y \sup_X f = \sup_X \min_Y f.$$

In 1989, Kindler-Trost established a very general topological minimax theorem. Here is a slightly simplified (see below) version of their result, which contains both Theorem 5 and Theorem 8.

Theorem 9 ([61]) Let X be an interval space, Y be a compact topological space, and $f : X \times Y \to \mathbb{R}$ be upper semicontinuous on X and lower semicontinuous on Y. Suppose that there exists a sequence $\{\lambda_s\}_{s \geq 1}$ such that $\lambda_s > \sup_X \min_Y f$, $\lambda_s \to \sup_X \min_Y f$ and,

$$\text{for all } s \geq 1 \text{ and } y \in Y,\ GT(\lambda_s, y) \text{ is interval convex.}$$

Suppose also that,

$$\text{for all } s \geq 1 \text{ and nonempty finite subsets } W \text{ of } X,\ LE(W, \lambda_s) \text{ is connected in } Y.$$

Then

$$\min_Y \sup_X f = \sup_X \min_Y f.$$

In fact, the semicontinuity conditions and the compactness of Y assumed in Theorem 8 and Theorem 9 are stronger than the topological conditions actually assumed in [115] and [61]. We have adopted these simplifications so as not to overburden the reader with too many technicalities, and also to achieve a certain unity of presentation.

The above results are all subsumed by the following general topological minimax theorem established by König. Again, we have simplified the statement somewhat.

Theorem 10 ([72]) Let X be a connected topological space, Y be a compact connected topological space, and $f : X \times Y \to \mathbb{R}$ be upper semicontinuous on X and lower semicontinuous on Y. Suppose that, for all $\lambda > \sup_X \min_Y f$ **either**

$$\text{for all nonempty subsets } V \text{ of } Y,\ GT(\lambda, V) \text{ is connected in } X, \text{ and}$$
$$\text{for all nonempty finite subsets } W \text{ of } X,\ LE(W, \lambda) \text{ is connected in } Y,$$

or

$$\text{for all nonempty subsets } V \text{ of } Y,\ GT(\lambda, V) \text{ is connected in } X, \text{ and}$$
$$\text{for all nonempty finite subsets } W \text{ of } X,\ LT(W, \lambda) \text{ is connected in } Y,$$

or

$$\text{for all nonempty subsets } V \text{ of } Y,\ GE(\lambda, V) \text{ is connected in } X, \text{ and}$$
$$\text{for all nonempty finite subsets } W \text{ of } X,\ LT(W, \lambda) \text{ is connected in } Y,$$

or

$$\text{for all nonempty subsets } V \text{ of } Y,\ GE(\lambda, V) \text{ is connected in } X, \text{ and}$$
$$\text{for all nonempty finite subsets } W \text{ of } X,\ LE(W, \lambda) \text{ is connected in } Y.$$

Then

$$\min_Y \sup_X f = \sup_X \min_Y f.$$

There is also in [72] a result similar to Theorem 10 with different semicontinuity assumptions. We note the basic asymmetry in the above results: in one variable we allow arbitrary intersections, while in the other variable we only allow finite intersections. König [73] has recently given an example showing the failure of the "symmetric" theorem in which we allow only finite intersections in both variables.

In [33], Horvath proved a result similar to Theorem 6 only with X a convex set in some vector space, and the topology of X replaced by the natural topology of all the line segments in X. More results in the direction of Theorem 9 were proved by Ricceri in [100].

This discussion will be continued in the section **Unifying metaminimax theorems**.

7. Quantitative Minimax Theorems

In 1953, Fan was the first person to take the theory of minimax theorems out of the context of convex subsets of vector spaces when he established the following result generalizing [62]:

Theorem 11 ([14]) Let X be a nonempty set and Y be a nonempty compact topological space. Let $f : X \times Y \to \mathbb{R}$ be lower semicontinuous on Y. Suppose that f is *concavelike* on X and *convexlike* on Y, that is to say:

$$\text{for all } x_1, x_2 \in X \text{ and } \alpha \in [0,1], \text{ there exists } x_3 \in X \text{ such that}$$
$$f(x_3, .) \geq \alpha f(x_1, .) + (1-\alpha) f(x_2, .) \text{ on } Y,$$

and

$$\text{for all } y_1, y_2 \in Y \text{ and } \beta \in [0,1], \text{ there exists } y_3 \in Y \text{ such that}$$
$$f(., y_3) \leq \beta f(., y_1) + (1-\beta) f(., y_2) \text{ on } X.$$

Then

$$\min_Y \sup_X f = \sup_X \min_Y f.$$

See Parthasarathy [96] for further developments in this direction. In [14], Fan also proved a minimax theorem for almost periodic functions of two variables, which was subsequently generalized by Tjoe-The [120] and Parthasarathy [95]. Aubin ([1] and [2]) proved results related to Theorem 11, in terms of the concepts of γ-vexity. In [6], Borwein and Zhuang give a very short proof of Theorem 11 using the Eidelheit separation theorem. König in 1968, and then Simons in 1971, proved the following result generalizing Theorem 11:

Theorem 12 ([67] and [105]) Let X be a nonempty set and Y be a nonempty compact topological space. Let $f : X \times Y \to \mathbb{R}$ be lower semicontinuous on Y. Suppose that:

$$\text{for all } x_1, x_2 \in X, \text{ there exists } x_3 \in X \text{ such that } f(x_3, .) \geq \frac{f(x_1, .) + f(x_2, .)}{2} \text{ on } Y,$$

and,

$$\text{for all } y_1, y_2 \in Y, \text{ there exists } y_3 \in Y \text{ such that } f(., y_3) \leq \frac{f(., y_1) + f(., y_2)}{2} \text{ on } X.$$

Then

$$\min_Y \sup_X f = \sup_X \min_Y f.$$

At first sight, the difference between Theorem 11 and Theorem 12 is not very striking. The proofs in [67] and [105] both used a version of the Hahn-Banach theorem due to Mazur-Orlicz. However, both proofs followed the same pattern as that of [14], replacing the convexity of the the sets X and Y by statements about the convexity of the functional values of f. It turned out subsequently that the difference between Theorem 11 and Theorem 12 was quite significant, and led eventually, *via* the steps that will be described below, to the *unifying metaminimax theorems* to be discussed in a later section.

Since the Mazur-Orlicz theorem is itself a very special kind of minimax theorem, and is not as well known as it deserves to be, it seems appropriate for us to take a small digression and give a few additional details. First, here is a statement of the result itself:

Theorem 13 ([85]) Let E be a real vector space, $S : E \to \mathbb{R}$ be sublinear and C be a nonempty convex subset of E. Then there exists a linear functional L on E such that

$$L \leq S \text{ on } E \text{ and } \inf_C L = \inf_C S.$$

Other applications of the Mazur-Orlicz theorem and related *sandwich theorems* are discussed by König in [68], [69] and [70], and Neumann in [89], [90] and [91]. These include not only numerous applications to measure theory and Hardy algebra theory, but also to the theory of flows in infinite networks. See also the paper [99] by Pomerol already mentioned in a previous section. We now return to our historical narrative. In 1977, Neumann proved the following generalization of Theorem 12:

Theorem 14 ([88] and [74]) Let X be a nonempty set and Y be a nonempty compact topological space. Let $f : X \times Y \to \mathbb{R}$ be lower semicontinuous on Y. Suppose that there exists $\alpha \in (0,1)$ such that,

$$\text{for all } x_1, x_2 \in X, \text{ there exists } x_3 \in X \text{ such that}$$

$$f(x_3,.) \geq \alpha f(x_1,.) + (1-\alpha) f(x_2,.) \text{ on } Y,$$

and there exists $\beta \in (0,1)$ such that,

$$\text{for all } y_1, y_2 \in Y, \text{ there exists } y_3 \in Y \text{ such that}$$

$$f(.,y_3) \leq \beta f(.,y_1) + (1-\beta) f(.,y_2) \text{ on } X.$$

Then

$$\min_Y \sup_X f = \sup_X \min_Y f.$$

Actually, Neumann proved a result that was more general that this by a factor of ε, but we will not discuss these technicalities in this article. The result of [88] was subsequently extended by Fuchssteiner-König [17], but their results do not qualify as minimax theorems in our sense. In [48], Kindler investigated the connections between minimax theorems and the representation of integrals. Then there was a hiatus of several years in this line of research.

Activity in this area resumed with a sequence of papers in which minimax theorems were proved without recourse to arguments based ultimately on convexity. We cite the 1989 paper by Lin-Quan, who generalized Theorem 14 with the following result:

Theorem 15 ([76]) Let X be a nonempty set and Y be a nonempty compact topological space. Let $f : X \times Y \to \mathbb{R}$ be lower semicontinuous on Y. Suppose there exists $\alpha \in (0,1)$ such that,

$$\left.\begin{array}{c}\text{for all } x_1, x_2 \in X, \text{ there exists } x_3 \in X \text{ such that} \\ f(x_3,.) \geq \alpha[f(x_1,.) \vee f(x_2,.)] + (1-\alpha)[f(x_1,.) \wedge f(x_2,.)] \text{ on } Y,\end{array}\right\} \tag{15.1}$$

and there exists $\beta \in (0,1)$ such that,

$$\text{for all } y_1, y_2 \in Y, \text{ there exists } y_3 \in Y \text{ such that}$$
$$f(.,y_3) \leq \beta[f(.,y_1) \vee f(.,y_2)] + (1-\beta)[f(.,y_1) \wedge f(.,y_2)] \text{ on } X,$$

where "$\vee$" stands for "maximum". Then

$$\min_Y \sup_X f = \sup_X \min_Y f.$$

There was actually a slightly earlier (1985) more general result by Irle which, unfortunately, did not receive much circulation. Here we present a slightly simplified version of Irle's result. Irle defines an *averaging function* to be a continuous function $\phi : \mathbb{R}^2 \to \mathbb{R}$ such that ϕ is nondecreasing in each variable, $\phi(\lambda, \lambda) = \lambda$,

$$\text{if } \lambda \neq \mu \text{ then } \lambda \wedge \mu < \phi(\lambda, \mu) < \lambda \vee \mu,$$

and then proves:

Theorem 16 ([35]) Let X be a nonempty set and Y be a nonempty compact topological space. Let $f : X \times Y \to \mathbb{R}$ be lower semicontinuous on Y. Suppose there exist averaging functions ϕ and ψ such that,

$$\text{for all } x_1, x_2 \in X, \text{ there exists } x_3 \in X \text{ such that}$$

$$f(x_3, .) \geq \phi(f(x_1, .), f(x_2, .)) \text{ on } Y,$$

and,

$$\text{for all } y_1, y_2 \in Y, \text{ there exists } y_3 \in Y \text{ such that}$$

$$f(., y_3) \leq \psi(f(., y_1), f(., y_2)) \text{ on } X.$$

Then

$$\min_Y \sup_X f = \sup_X \min_Y f.$$

Irle [36] has given an application of the above result to *hide-and-seek games*. In 1990, Kindler [54] gave a generalization of Theorem 16 using the concept of a *mean function*, which is too complicated to describe in detail here. Averaging functions in the sense of Irle are mean functions, but mean functions do not have the restriction of being continuous that averaging functions have. Kindler's paper [54] has a deeper significance which we will return to in the section **Unifying metaminimax theorems**. In 1990, Simons gave another result that extends Theorem 15:

Theorem 17 ([109]) Let X be a nonempty set and Y be a nonempty compact topological space. Let $f : X \times Y \to \mathbb{R}$ be lower semicontinuous on Y. Suppose that, for all $\varepsilon > 0$, there exists $\delta > 0$ such that,

$$\text{for all } x_1, x_2 \in X, \text{ there exists } x_3 \in X \text{ such that } f(x_3, .) \geq f(x_1, .) \wedge f(x_2, .) \text{ on } Y \tag{17.1}$$

and

$$y \in Y \text{ and } |f(x_1, y) - f(x_2, y)| \geq \varepsilon \Longrightarrow f(x_3, y) \geq f(x_1, y) \wedge f(x_2, y) + \delta, \tag{17.2}$$

and

$$\text{for all } y_1, y_2 \in Y, \text{ there exists } y_3 \in Y \text{ such that } f(., y_3) \leq f(., y_1) \vee f(., y_2) \text{ on } X \tag{17.3}$$

and

$$x \in X \text{ and } |f(x, y_1) - f(x, y_2)| \geq \varepsilon \Longrightarrow f(x, y_3) \leq f(x, y_1) \vee f(x, y_2) - \delta.$$

Then

$$\min_Y \sup_X f = \sup_X \min_Y f.$$

At this point, it would be in order to explain the motivation behind the series of results discussed above. It is easy to see that, even given the strongest topological conditions, (17.1) and (17.3) are *not* sufficient to force the minimax relation to hold. Theorem 12, Theorem 14, Theorem 15 and Theorem 16 had successively weaker hypotheses, which *did* force the minimax relation to hold. Theorem 17 was another result with hypotheses weaker than Theorem 15 which also forced the minimax relation to hold. The hypotheses of *all* the results mentioned above imply (17.1) and (17.3). Theorem 16 and its generalization by Kindler both use external functions ϕ and ψ, while Theorem 17 does not. These two kinds of results were unified by Simons in [111] using the concept of a *staircase*, which is quite technical and too complicated to go into here. A deep and very penetrating study of this kind of problem was also made by König-Zartmann in [75]. Nevertheless, there is in fact a very simple combinatorial principle behind all these results, which we will discuss in the section **Unifying metaminimax theorems**. We should mention finally that Kindler has recently incorporated some of the techniques described above to obtain extensions of the results in [48] to statistical decision theory and the theory of convex metric spaces – see [59] and [60], and also the paper [98] on minimax risk by Pinelis.

8. Mixed Minimax Theorems

In 1972, Terkelsen proved the first *mixed* minimax theorem. Specifically, one of the conditions in the following result is taken from the topological Theorem 6 and the other from the quantitative Theorem 12:

Theorem 18 ([119]) Let X be a nonempty set and Y be a nonempty compact topological space. Let $f : X \times Y \to \mathbb{R}$ be lower semicontinuous on Y. Suppose that,

$$\left.\begin{array}{c}\text{for all } x_1, x_2 \in X, \text{ there exists } x_3 \in X \text{ such that} \\ f(x_3, .) \geq [f(x_1, .) + f(x_2, .)]/2 \text{ on } Y.\end{array}\right\} \tag{18.1}$$

Suppose also that,

$$\text{for all nonempty finite subsets } W \text{ of } X \text{ and } \lambda \in \mathbb{R},\ LE(W, \lambda) \text{ is connected in } Y. \tag{18.2}$$

Then

$$\min_Y \sup_X f = \sup_X \min_Y f.$$

This result was subsequently generalized by Geraghty-Lin [18] who proved in 1983 that (18.1) can be weakened to: there exists $\alpha \in (0, 1)$ such that,

$$\begin{array}{c}\text{for all } x_1, x_2 \in X, \text{ there exists } x_3 \in X \text{ such that} \\ f(x_3, .) \geq \alpha[f(x_1, .) \vee f(x_2, .)] + (1 - \alpha)[f(x_1, .) \wedge f(x_2, .)] \text{ on } Y,\end{array}$$

that is to say, (15.1) is satisfied. Simons [110] proved that this condition can be further weakened to: for all $\varepsilon > 0$, there exists $\delta > 0$ such that,

$$\text{for all } x_1, x_2 \in X, \text{ there exists } x_3 \in X \text{ such that } f(x_3, .) \geq f(x_1, .) \wedge f(x_2, .) \text{ on } Y$$

and

$$y \in Y \text{ and } |f(x_1, y) - f(x_2, y)| \geq \varepsilon \Longrightarrow f(x_3, y) \geq f(x_1, y) \wedge f(x_2, y) + \delta,$$

that is to say, (17.1) and (17.2) are satisfied. Kindler, in the paper [54] already mentioned gave another generalization of Theorem 18 using his concept of a *mean function*.

Takahashi showed in [117] that (18.2) could also be weakened somewhat. Furthermore, Takahashi-Takahashi have applied Terkelsen's methods in [118] to obtain results on fuzzy sets. Finally, Stefanescu in [116] proved a minimax theorem similar to Theorem 18 in which (18.2) was replaced by the appropriate set-theoretic assumption.

9. Unifying Metaminimax Theorems

The first hint that our classification of general minimax theorems into *topological*, *quantitative* and *mixed* might be too rigid, was probably provided in 1982 by Joó-Stachó, who showed in [40] that Theorem 11, which we have classified as *quantitative*, could be deduced using Radon measures from the result [8] of Brézis-Nirenberg-Stampacchia already mentioned which could, in turn, be deduced from Theorem 8, which we have classified as *topological*. A second hint was provided in 1985 and 1986 by Geraghty-Lin, who investigated in [20] and [21] a continuum of minimax theorems joining Theorem 4, which we have classified as *topological*, and Theorem 18, which we have classified as *mixed*. A third hint was provided in 1989 by Komiya, who proved in [64] a result which contained both Theorem 11 and also [19], which we have classified as *topological*. It was Kindler in [54] who first realized in 1990 that some concept akin to connectedness might be involved in minimax theorems where the topological condition of connectedness was not explicitly assumed. This idea was pursued by Simons with the introduction in 1992 of the concept of *pseudoconnectedness*. We say that sets H_0 and H_1 are *joined* by a set H if

$$H \subset H_0 \cup H_1, H \cap H_0 \neq \emptyset \text{ and } H \cap H_1 \neq \emptyset.$$

We say that a family $\mathcal{H}$ of sets is *pseudoconnected* if:

$$\text{if } H_0, H_1, H \in \mathcal{H} \text{ and } H_0 \text{ and } H_1 \text{are joined by } H \text{ then } H_0 \cap H_1 \neq \emptyset.$$

Any family of closed connected subsets of a topological space is pseudoconnected. So also is any family of open connected subsets. Our next result is an improvement suggested by some comments of Heinz König of a result from [112] and [113].

We shall say that a subset W of X is *good* if W is finite and,

$$\text{for all } x \in X,\ LE(x, \sup_X \inf_Y f) \cap LE(W, \sup_X \inf_Y f) \neq \emptyset.$$

Theorem 19 Let Y be a topological space, and Λ be a nonempty subset of $\mathbb{R}$ such that $\inf \Lambda = \sup_X \inf_Y f$. Suppose that, for all $\lambda \in \Lambda$ and good subsets W of X,

$$\text{for all } x \in X, LE(x, \lambda) \text{ is closed and compact,} \tag{19.1}$$

$$\{LE(x, \lambda) \cap LE(W, \lambda)\}_{x \in X} \text{ is pseudoconnected} \tag{19.2}$$

and, for all $x_0, x_1 \in X$, there exists $x \in X$ such that

$$LE(x_0, \lambda) \text{ and } LE(x_1, \lambda) \text{ are joined by } LE(x, \lambda) \cap LE(W, \lambda). \tag{19.3}$$

Then

$$\min_Y \sup_X f = \sup_X \min_Y f.$$

Proof Let $x \in X$. If $\mu \in \mathbb{R}$ and $\mu > \sup_X \min_Y f$ then $\mu > \min f(x, Y)$, from which $LE(x, \mu) \neq \emptyset$. From (19.1) and the finite intersection property, $LE(x, \sup_X \min_Y f) \neq \emptyset$. Thus $\emptyset$ is good. We now prove by induction that all finite subsets of X are good. So suppose that $n \geq 1$ and

$$W \subset X \text{ and } \operatorname{card} W \leq n-1 \Longrightarrow W \text{ is good.} \tag{19.4}$$

Let $V \subset X$ and $\operatorname{card} V = n$. Let $x_0 \in V$ and set $W := V \backslash \{x_0\}$. From the induction hypothesis (19.4), W is good. Let $x_1 \in X$ be arbitrary. Let $\lambda \in \Lambda$ be arbitrary. From (19.3), there exists $x \in X$ such that $LE(x_0, \lambda)$ and $LE(x_1, \lambda)$ are joined by $LE(x, \lambda) \cap LE(W, \lambda)$. Equivalently,

$$LE(x_0, \lambda) \cap LE(W, \lambda) \text{ and } LE(x_1, \lambda) \cap LE(W, \lambda) \text{ are joined by } LE(x, \lambda) \cap LE(W, \lambda).$$

From (19.2), $LE(x_0, \lambda) \cap LE(x_1, \lambda) \cap LE(W, \lambda) \neq \emptyset$, that is to say, $LE(x_1, \lambda) \cap LE(V, \lambda) \neq \emptyset$. Since this holds for all $\lambda \in \Lambda$, from (19.1) and the finite intersection property again,

$$LE(x_1, \sup_X \min_Y f) \cap LE(V, \sup_X \min_Y f) \neq \emptyset.$$

Since this is valid for all $x_1 \in X$, V is good. This completes the inductive step of the proof that all finite subsets of X are good. It now follows from (19.1) and the finite intersection property for a third time that $LE(X, \sup_X \min_Y f) \neq \emptyset$. This completes the proof of Theorem 19. Given the obvious topological motivation behind the concept of pseudoconnectedness, it is hardly surprising that Theorem 19 implies all the results mentioned in the section **Topological minimax theorems** (except for some of the parts of Theorem 10). What is more unexpected is that Theorem 19 implies all the results mentioned in the sections **Quantitative minimax theorems** and **Mixed minimax theorems** (except possibly that of [116]), as well as some new results. On the other hand, as we have seen above the proof of Theorem 19 is certainly not profound — the real work is done in proving that the conditions (19.2) and (19.3) are satisfied in any of the particular cases. That is why we prefer to describe Theorem 19 as a **metaminimax theorem** rather than a **minimax theorem**, it is really a device for obtaining minimax theorems rather than a minimax theorem in its own right. Another way of looking at Theorem 19 is as a "decomposition of the minimax property". This avenue is pursued quite profoundly in the paper [75] by König-Zartmann (already mentioned in the section **Quantitative minimax theorems**).

The remainder of this section is devoted to some results which are at the interface between minimax theory and abstract set theory. Since many of them are quite technical, we will not go into them in great detail. Most of results discussed below were motivated by Theorem 19, [75], and Theorem 10.

Suppose that $\{C_x\}_{x\in X}$ is a family of subsets of Y. If $y \in Y$, define the *conjugate set* by

$$C_y^* := \{x : x \in X,\ y \notin C_x\}$$

Kindler proved in [55] that the sets $\{C_x\}_{x\in X}$ have the finite intersection property if, and *only if*, there exist there exist topologies on X and Y such that

Y is compact,

all the sets C_x are closed in Y,

for each closed subset F of Y, $\bigcup\{C_y^* : y \in F\}$ is open in X,

all finite intersections of the sets C_x are connected in Y,

and

all intersections of the sets C_y^* are connected in X.

Kindler deduced a number of minimax theorems from this observation. In [56], he gave necessary *and sufficient* conditions in terms of these and allied concepts that the minimax relation hold. In [57], motivated by the *interval spaces* introduced by Stachó in [115], Kindler considered a *midset space*, which is simply a set S and a function $S \times S \to 2^S$, and went on in [58] to use this concept and the concept of *quarter continuous multifunction* introduced by Komiya in [65] to prove a generalization of Theorem 3.

10. Connections with Weak Compactness

In 1971, Simons proved that there are limitations on the extent to which one can generalize the minimax theorem. Specifically:

Theorem 20 ([105]) Suppose that X is a nonempty bounded, convex, complete subset of a locally convex space E with dual space E^*, and

$$\inf_{y\in Y}\sup_{x\in X}\langle x,y\rangle = \sup_{x\in X}\inf_{y\in Y}\langle x,y\rangle$$

whenever Y is a nonempty convex, equicontinuous, subset of E^*. Then

X is weakly compact.

Coupled with the following result, one can obtain a proof of R. C. James's *sup theorem*:

Theorem 21 ([106]) . If X is a nonempty bounded, convex subset of a locally convex space E such that every element of the dual space E^* attains its supremum on X, and Y is any nonempty convex equicontinuous subset of E^*, then

$$\inf_{y\in Y}\sup_{x\in X}\langle x,y\rangle = \sup_{x\in X}\inf_{y\in Y}\langle x,y\rangle.$$

In 1974, using the concept of *ordered iterated limits*, De Wilde [10] simplified and extended the result of Theorem 21. Further work on this topic was also done by Ha in [31]. In [103] and [104], Rodé introduced the related theory of *superconvexity*, an axiomatic theory of infinite convex combinations. See the articles [68], [69] and [71] by König for later developments in this direction.

One of the aspects of von Neumann's original minimax theorem that we have not mentioned explicitly is the idea of extending a game from a finite set of *pure* strategies to a convex set of *mixed* strategies. What Theorem 1 showed is that, even if there is no saddle point for the original game, there is always one for the extended game. Indeed, many of the results mentioned in the section **Infinite dimensional bilinear results** were motivated by the problems involved in extending a game. Generalizing a result from the paper [132] of Young already mentioned, Kindler established the following in 1976:

Theorem 22 ([45]) Let X, Y be nonempty sets and $a : X \times Y \to \mathbb{R}$ be bounded. Suppose that

$$\lim_{n\to\infty}\lim_{m\to\infty} a(x_m, y_n) \leq \lim_{m\to\infty}\lim_{n\to\infty} a(x_m, y_n)$$

whenever $\{x_m\}_{m\geq 1}$ and $\{y_n\}_{n\geq 1}$ are sequences in X and Y, respectively, such that the iterated limits exist. Then

$$\inf_{\nu\in P(Y)}\sup_{\mu\in P(X)}\int_X\int_Y a\, d\nu d\mu = \sup_{\mu\in P(X)}\inf_{\nu\in P(Y)}\int_X\int_Y a\, d\nu d\mu,$$

where $P(S)$ is the set of all probability measures on S with finite support.

The hypothesis on a in Theorem 22 is exactly the condition on *ordered iterated limits* introduced by De Wilde in [10]. In [49], Kindler combined Theorem 12 and Theorem 22 to obtain results on the extension of games, which led to simple proofs of the Krein-Smulian and Eberlein-Smulian theorems. This shows again the close connection between minimax theorems and weak compactness. In [50] and [51], Kindler considers generalizations of Theorem 22 to the case when μ and ν are allowed to be more general finitely additive measures, and the connections with other concepts related to weak compactness, such as Fubini's theorem for finitely additive measures, and Pták's combinatorial lemma. Kindler then showed that the connection between ordered iterated limits and minimax theorems was even tighter with a two-function generalization of Theorem 22. Here is a slightly simplified version of Kindler's result:

Theorem 23 ([52]) Let X, Y be nonempty sets and $a, b : X \times Y \to \mathbb{R}$ be bounded. Then:

$$\lim_{n\to\infty}\lim_{m\to\infty} a(x_m, y_n) \leq \lim_{m\to\infty}\lim_{n\to\infty} b(x_m, y_n)$$

whenever $\{x_m\}_{m\geq 1}$ and $\{y_n\}_{n\geq 1}$ are sequences in X and Y, respectively, such that the iterated limits exist,

if, and only if,

$$\inf_{\nu\in P(T)}\sup_{\mu\in P(S)}\int_S\int_T a\, d\nu d\mu \leq \sup_{\mu\in P(S)}\inf_{\nu\in P(T)}\int_S\int_T b\, d\nu d\mu$$

whenever S is a nonempty subset of X and T is a nonempty subset of Y.

11. Minimax Inequalities for Two or More Functions

Motivated by Nash equilibrium and the theory of non-cooperative games, Fan generalized Theorem 3 in 1964 to the case of more than one function. In particular, he proved the following two-function minimax inequality:

Theorem 24 ([15]) Let X and Y be nonempty compact, convex subsets of topological vector spaces and $f, g : X \times Y \to \mathbb{R}$. Suppose that f is lower semicontinuous on Y,

$$\text{for all } y \in Y \text{ and } \lambda \in \mathbb{R}, \{x : x \in X, f(x,y) \geq \lambda\} \text{ is convex,}$$

$$\text{for all } x \in X \text{ and } \lambda \in \mathbb{R}, \{y : y \in Y, g(x,y) \leq \lambda\} \text{ is convex,}$$

g is upper semicontinuous on X, and

$$f \leq g \text{ on } X \times Y.$$

Then

$$\min_Y \sup_X f \leq \sup_X \inf_Y g.$$

Liu [81] observed that Theorem 24 is true even if X is not assumed to be compact, and that Theorem 24 actually unifies the theory of minimax theorems and the theory of variational inequalities. Theorem 24 was extended even further by Ben-El-Mechaiekh, Deguire and Granas who proved:

Theorem 25 ([4]) Let X and Y be nonempty compact, convex subsets of topological vector spaces and $f, s, t, g : X \times Y \to \mathbb{R}$. Suppose that f is lower semicontinuous on Y,

$$\text{for all } y \in Y \text{ and } \lambda \in \mathbb{R}, \{x : x \in X, s(x,y) \geq \lambda\} \text{ is convex,}$$

$$\text{for all } x \in X \text{ and } \lambda \in \mathbb{R}, \{y : y \in Y, t(x,y) \leq \lambda\} \text{ is convex,}$$

g is upper semicontinuous on X, and

$$f \leq s \leq t \leq g \text{ on } X \times Y.$$

Then

$$\inf_Y \sup_X f \leq \sup_X \inf_Y g.$$

In [23], Granas-Liu extended Ha's result of [30] to two and three functions In 1981, Fan (unpublished) and Simons generalized Theorem 12 by proving the following two-function minimax inequality:

Theorem 26 ([107]) Let X be a nonempty set, Y be a compact topological space and $f, g : X \times Y \to \mathbb{R}$. Suppose that f is lower semicontinuous on Y,

$$\text{for all } y_1, y_2 \in Y, \text{ there exists } y_3 \in Y \text{ such that } f(.,y_3) \leq \frac{f(.,y_1) + f(.,y_2)}{2} \text{ on } X,$$

$$\text{for all } x_1, x_2 \in X, \text{ there exists } x_3 \in X \text{ such that } g(x_3,.) \geq \frac{g(x_1,.) + g(x_2,.)}{2} \text{ on } Y,$$

and

$$f \leq g \text{ on } X \times Y.$$

Then

$$\min_Y \sup_X f \leq \sup_X \inf_Y g.$$

Theorem 26 also unifies the theory of minimax theorems and the theory of variational inequalities. The curious feature about Theorem 24 and Theorem 26 is that they have *opposite geometric pictures.* This question is discussed in [107] and [108]. The relationship between Theorem 24 and Brouwer's fixed-point theorem is quite interesting. As we have already pointed out, Sion's minimax theorem, Theorem 3, can be proved in an elementary fashion without recourse to fixed-point related concepts. On the other hand, Theorem 24, which is a generalization of Theorem 3 can, in fact, be used to *prove* Tychonoff's fixed-point theorem. (See [15] for more details.)

In [39], Joó-Kassay defined a *pseudoconvex space* to be a topological space X with an appropriate family of continuous maps from finite dimensional simplices into certain subsets of X. They then proved that Theorem 24 can be generalized to this more abstract situation. In the same paper, they gave a counterexample showing that the "obvious" generalization of Theorem 17 to two functions fails.

On the other hand, Lin-Quan gave the following two function generalization of Theorem 15:

Theorem 27 ([77]) Let X be a nonempty set and Y be a nonempty compact topological space. Let $f, g : X \times Y \to \mathbb{R}$ be lower semicontinuous on Y. Suppose there exists $\alpha \in (0,1)$ such that,

$$\text{for all } x_1, x_2 \in X, \text{ there exists } x_3 \in X \text{ such that}$$
$$f(x_3, .) \geq \alpha[f(x_1, .) \vee g(x_2, .)] + (1-\alpha)[f(x_1, .) \wedge g(x_2, .)] \text{ on } Y,$$

and there exists $\beta \in (0,1)$ such that,

$$\text{for all } y_1, y_2 \in Y, \text{ there exists } y_3 \in Y \text{ such that}$$
$$g(., y_3) \leq \beta[f(., y_1) \vee g(., y_2)] + (1-\beta)[f(., y_1) \wedge g(., y_2)] \text{ on } X,$$

and

$$f \leq g \text{ on } X \times Y.$$

Then

$$\min_Y \sup_X f \leq \sup_X \min_Y g.$$

By contrast with the counterexample of Joó-Kassay mentioned above, the conditions in Theorem 27 "mix up" the functional values of f and g. In [78], Lin-Quan generalize Theorem 27 using the concept of a *staircase* introduced in [111]. In [79], they gave a generalization of Komiya's topological result of [65] to two functions, while in [80] they show that the compactness condition in [79] can be relaxed.

Granas-Liu have proved a number of minimax inequalities for two or more functions. In [24], they extended Fan's result of [14] to three and four functions, and in [26] they extended the result of Theorem 26 to four functions with a weakening of the conditions of semiconvexlike and semiconcavelike, but under the assumption that X is also a compact topological space.

In [53], Kindler consolidated the results of his papers [45]-[52] and considered approximate two-function minimax inequalities.

12. Coincidence Theorems

A *coincidence theorem* is a theorem that asserts that if $S : X \to 2^Y$ and $T : Y \to 2^X$ have nonempty values and satisfy certain other conditions then there exist $x_0 \in X$ and $y_0 \in Y$ such that $y_0 \in Sx_0$ and $x_0 \in Ty_0$. The connection with minimax theorems is as follows: suppose that $\inf_Y \sup_X f \neq \sup_X \inf_Y f$. Then there exists $\lambda \in \mathbb{R}$ such that

$$\sup_X \inf_Y f < \lambda < \inf_Y \sup_X f.$$

Hence,

$$\text{for all } x \in X, \text{ there exists } y \in Y \text{ such that } f(x,y) < \lambda$$

and,

$$\text{for all } y \in Y, \text{ there exists } x \in X \text{ such that } f(x,y) > \lambda.$$

Define $S : X \to 2^Y$ and $T : Y \to 2^X$ by

$$Sx := \{y : y \in Y, f(x,y) < \lambda\} \text{ and } Tx := \{x : x \in X, f(x,y) > \lambda\}.$$

Then the values of the multifunctions S and T are nonempty. If S and T were to satisfy a coincidence theorem then we would have $x_0 \in X$ and $y_0 \in Y$ such that

$$f(x_0, y_0) < \lambda \text{ and } f(x_0, y_0) > \lambda,$$

which is clearly impossible. Thus this coincidence theorem would imply that

$$\inf_Y \sup_X f = \sup_X \inf_Y f.$$

The coincidence theorems known in algebraic topology consequently give rise to corresponding minimax theorems. The first person to have used this idea seems to have been Debreu [11] in 1952, and this line of investigation was pursued by Bourgin [7] and McClendon [82]. See also the paper [25] by Granas-Liu for analogous results for minimax inequalities involving more than one function. There is a very extensive literature of coincidence theorems. We refer the reader to the paper [94] by Park for some further pointers in this direction.

References

1. J.-P. Aubin, *Mathematical methods of game and economic theory*, North Holland, Amsterdam-New York-Oxford(1979).
2. J.-P. Aubin, *Théorème du minimax pour une classe de functions*, C. R. Acad. Sci. Paris **274**(1972), 455-458.
3. V. Barbu and T. Precupanu, *Convexity and optimization in Banach space*, D. Reidel Publishing Company, Dordrecht-Boston-Lancaster(1986).

4. H. Ben-El-Mechaiekh, P. Deguire and A. Granas, *Points fixes et coïncidences pour les fonctions multivoques II (Applications de type ϕ and ϕ^*)*, C. R. Acad. Sci. Paris **295**(1982), 381-384.
5. C. Berge, *Sur une convexité regulière non linéaire et ses applications à la théorie des jeux*, Bull Soc. Math. France **82**(1954), 301-319.
6. J. M. Borwein and D. Zhuang, *On Fan's minimax theorem*, Math. Programming **34**(1986), 232-234.
7. [7]D. G. Bourgin, *Fixed point and min-max theorems*, Pac. J. Math. **45**(1973), 403-412.
8. H. Brézis, L. Nirenberg and G. Stampacchia, *A remark on Ky Fan's minimax principle*, Boll. Uni. Mat. Ital. (4)**6**(1972), 293-300.
9. B. C. Cuóng, *Some remarks on minimax theorems*, Acta Math. Vietnam **1**(1976), 67-74.
10. M. De Wilde, *Doubles limites ordonées et théorèmes de minimax*, Ann. Inst. Fourier **24**(1974), 181-188.
11. G. Debreu, *A social equilibrium existence theorem*, Proc. Nat. Acad. Sci. U.S.A. **38**(1952), 886-893.
12. A. L. Dulmage and J. E. Peck, *Certain infinite zero-sum two-person games*, Canadian J. Math. **8**(1956), 412-416.
13. K. Fan, *Fixed-point and minimax theorems in locally convex topological linear spaces*, Proc. Nat. Acad. Sci. U.S.A. **38**(1952), 121-126.
14. K. Fan, *Minimax theorems*, Proc. Nat. Acad. Sci. U.S.A. **39**(1953), 42-47.
15. K. Fan, *Sur un théorème minimax*, C. R. Acad. Sci. Paris **259**(1964), 3925-3928.
16. K. Fan, *A minimax inequality and its applications*, Inequalities III, O. Shisha Ed., Academic Press(1972), 103-113.
17. B. Fuchssteiner and H. König, *New versions of the Hahn-Banach theorem*, General inequalities 2, E. F. Beckenbach Ed., ISNM 47, Birkhaüser, Basel(1980), 255-266.
18. M. A. Geraghty and B.-L. Lin, *On a minimax theorem of Terkelsen*, Bull. Inst. Math. Acad. Sinica **11**(1983), 343-347.
19. M. A. Geraghty and B.-L. Lin, *Topological minimax theorems*, Proc. Amer. Math. Soc. **91**(1984), 377-380.
20. M. A. Geraghty and B.-L. Lin, *Minimax theorems without linear structure*, Linear and Multilinear Algebra **17**(1985), 171-180.
21. M. A. Geraghty and B.-L. Lin, *Minimax theorems without convexity*, Contemporary Mathematics **52**(1986), 102-108.
22. M. A. Ghouila-Houri, *Le théorème minimax de Sion*, Theory of games, Engl. Univ. Press, London (1966), 123-129.
23. A. Granas and F. C. Liu, *Théorèmes de minimax*, C. R. Acad. Sci. Paris **298**(1984), 329-332.
24. A. Granas and F. C. Liu, *Quelques théorèmes du minimax sans convexité*, C. R. Acad. Sci. Paris **300**(1985), 347-350.
25. A. Granas and F. C. Liu, *Coincidences for set-valued maps and minimax inequalities*, J. de Math. Pures et Appliquées **65**(1986), 119-148.
26. A. Granas and F. C. Liu, *Some minimax theorems without convexity*, Nonlinear and convex analysis, B.-L. Lin and S. Simons, Eds., Marcel Dekker, New York-Basel(1987), 61-75.
27. J. Gwinner and V. Jeyakumar, *Stable minimax on noncompact sets*, Fixed point theory and applications, M. A. Théra and J.-B. Baillon eds., Pitman research notes **252**(1991), 215-220.
28. J. Gwinner and W. Oettli, *Theorems of the alternative and duality for inf-sup problems*, Preprint, 1993.
29. C. W. Ha, *Minimax and fixed point theorems*, Math. Ann. **248**(1980), 73-77.
30. C. W. Ha, *A non-compact minimax theorem*, Pac. J. Math. **97**(1981), 115-117.
31. C. W. Ha, *Weak compactness and the minimax equality*, Nonlinear and convex analysis, B.-L. Lin and S. Simons, Eds., Marcel Dekker, New York-Basel(1987), 77-82.
32. J. Hartung, *An extension of Sion's minimax theorem, with an application to a method for constrained games*, Pac. J. Math. **103**(1982), 401-408.
33. C. Horvath, *Quelques théorèmes en théorie des mini-max*, C. R. Acad. Sci. Paris **310** (1990), 269-272.
34. I. Irle, *Minimax theorems in convex situations*, Game theory and mathematical economics, O. Moeschlin and D. Pallaschke, Eds., North Holland, Amsterdam-New York-Oxford (1981), 321-331.
35. I. Irle, *A General minimax theorem*, Zeitschrift für Operations Research **29**(1985), 229-247.
36. I. Irle, *On minimax theorems for hide-and-seek games*, Methods Oper. Res. **54**(1986), 373-383.

37. I. Joó, *A simple proof for von Neumann's minimax theorem*, Acta Sci. Math. **42**(1980), 91-94.
38. I. Joó, *Note on my paper "A simple proof for von Neumann's minimax theorem"*, Acta. Math. **44**(1984), 363-365.
39. I. Joó and G. Kassay, *Convexity, minimax theorems and their applications*, Preprint.
40. I. Joó and L. L. Stachó, *A note on Ky Fan's minimax theorem*, Acta. Math. **39**(1982), 401-407.
41. S. Kakutani, *A generalization of Brouwer's fixed-point theorem*, Duke Math. J. **8**(1941), 457-459.
42. S. Karlin, *Operator treatment of minmax principle*, Contributions to the theory of games 1, Princeton. Univ. Press(1950), 133-154.
43. S. Karlin, *The theory of infinite games*, Ann. Math. **58**(1953), 371-401.
44. S. Karlin, *Mathematical methods and theory in games, programming and economics*, Addison Wesley, Reading, Mass. (1959).
45. J. Kindler, *Über ein Minimaxtheorem von Young*, Math. Operationsforsch. Statist. **7** (1976), 477-480.
46. J. Kindler, *Über Spiele auf konvexen Mengen*, Methods Oper. Res. **26**(1977), 695-704.
47. J. Kindler, *Schwach definite Spiele*, Math. Operationsforsch. Statist. Ser. Optimization **8**(1977), 199-205.
48. J. Kindler, *Minimaxtheoreme und das Integraldarstellungsproblem*, Manuscripta Math. **29**(1979), 277-294.
49. J. Kindler, *Minimaxtheoreme für die diskrete gemischte Ereiterung von Spielen und ein Approximationssatz*, Math. Operationsforsch. Statist. Ser. Optimization **11**(1980), 473-485.
50. J. Kindler, *Some consequences of a double limit condition*, Game theory and mathematical economics, O. Moeschlin and D. Pallaschke, Eds., North Holland, Amsterdam-New York-Oxford(1981), 73-82.
51. J. Kindler, *A general solution concept for two-person, zero-sum games*, J. Opt. Th. Appl. **40**(1983), 105-119.
52. J. Kindler, *A minimax version of Pták's combinatorial lemma*, J. Math. Anal. Appl. **94**(1983), 454-459.
53. J. Kindler, *Equilibrium point theorems for two-person games*, Siam. J. Control. Opt. **22**(1984), 671-683.
54. J. Kindler, *On a minimax theorem of Terkelsen's*, Arch Math. **55**(1990),573-583.
55. J. Kindler, *Topological intersection theorems*, Proc. Amer. Math. Soc. **117**(1993),1003-1011.
56. J. Kindler, *Intersection theorems and minimax theorems based on connectedness*, J. Math. Anal. Appl. **178**(1993), 529-546.
57. J. Kindler, *Intersecting sets in midset spaces. I*, Arch. Math. **62**(1994), 49-57.
58. J. Kindler, *Intersecting sets in midset spaces. II*, Arch. Math. **62**(1994), 168-176.
59. J. Kindler, *Minimax theorems with one-sided randomization*, Preprint, 1993.
60. J. Kindler, *Minimax theorems with applications to convex metric spaces*, Preprint, 1993.
61. J. Kindler and R. Trost, *Minimax theorems for interval spaces*, Acta Math. Hung. **54** (1989).
62. H. Kneser, *Sur un théorème fondamental de la théorie des jeux*, C. R. Acad. Sci. Paris **234**(1952), 2418-2420.
63. H. Komiya, *Elementary proof for Sion's minimax theorem*, Kodai Math. J. **11**(1988),5-7.
64. H. Komiya, *On minimax theorems*, Bull. Inst. Math. Acad. Sinica, **17**(1989), 171-178.
65. H. Komiya, *On minimax theorems without linear structure*, Hiyoshi Review of Natural Science **8**(1990), 74-78.
66. V. Komornik, *Minimax theorems for upper semicontinuous functions*, Acta Math. Acad. Sci. Hungar. **40**(1982), 159-163.
67. H. König, *Über das Von Neumannsche Minimax-Theorem*, Arch. Math. **19**(1968), 482-487.
68. H. König, *On certain applications of the Hahn-Banach and minimax theorems*, Arch. Math. **21**(1970), 583-591.
69. H. König, *Neue Methoden und Resultate aus Funktionalanalysis und konvexe Analysis*, Oper. Res. Verf. **28**(1978), 6-16.
70. H. König, *On some basic theorems in convex analysis*, Modern applied mathematics - optimization and operations research, B. Korte Ed., North Holland, Amsterdam-New York-Oxford (1982), 108-144.
71. H. König, *Theory and applications of superconvex spaces*, Aspects of positivity in functional analysis, R. Nagel, U. Schlotterbeck and M. P. H. Wolff, Eds., Elsevier Science Publishers

(North-Holland) (1986), 79-117.
72. H. König, *A general minimax theorem based on connectedness*, Arch. Math. **59**(1992), 55-64.
73. H. König, *A note on the general topological minimax theorem*, Preprint, 1994.
74. H. König and M. Neumann, *Mathematische Wirtschaftstheorie - mit einer Einführing in die konvexe Analysis*, Mathematical Systems in Economics 100, A. Hain (1986).
75. H. König and F. Zartmann, *New versions of the minimax theorem*, Preprint, 1992.
76. B.-L. Lin and X.-C. Quan, *A symmetric minimax theorem without linear structure*, Arch. Math. **52**(1989), 367-370.
77. B.-L. Lin and X.-C. Quan, *A two functions symmetric nonlinear minimax theorem*, Arch. Math. **57**(1991), 75-79.
78. B.-L. Lin and X.-C. Quan, *Two functions minimax theorem with staircase*, Bull. Inst. Math. Acad. Sinica **19**(1991), 279-287.
79. B.-L. Lin and X.-C. Quan, *A two functions nonlinear minimax theorem*, Fixed point theory and applications, M. A. Théra and J.-B. Baillon eds., Pitman research notes **252**(1991), 321-325.
80. B.-L. Lin and X.-C. Quan, *A noncompact topological minimax theorem*, J. Math. Anal. Appl. **161**(1991), 587-590.
81. F. C. Liu, *A note on the von Neumann-Sion minimax principle*, Bull. Inst. Math. Acad. Sinica **6**(1978), 517-524.
82. J.-F. Mertens, *The minimax theorem for u.s.c-l.s.c payoff functions*, Int. J. of Game Theory **15**, 237-250.
83. J. F. McClendon, *Minimax theorems for ANR's*, Proc. Amer. Math. Soc. **90**(1984), 149-154.
84. L. McLinden, *An application of Ekeland's theorem to minimax problems*, J. Nonlinear Anal.: Theory, Methods and Appl. **6**(1982), 189-196.
85. L. McLinden, *A minimax theorem*, Math. Oper. Res. **9**(1984), 576-591.
86. S. Mazur and W. Orlicz, *Sur les espaces métriques linéaires II*, Studia Math. **13**(1953), 137-179.
87. J. Moreau, *Théorèmes "inf-sup"*, C. R. Acad. Sci. Paris **258**(1964), 2720-2722.
88. M. Neumann, *Bemerkungen zum von Neumannschen Minimaxtheorem*, Arch. Math. **29** (1977), 96-105.
89. M. Neumann, *Some unexpected applicatons of the sandwich theorem*, Proceedings of the conference on optimization and convex analysis, University of Mississippi, 1989.
90. M. Neumann, *On the Mazur-Orlicz theorem*, Czechoslovak Mathematical J. **41**(1991), 104-109.
91. M. Neumann, *Generalized convexity and the Mazur-Orlicz theorem*, Proceedings of the Orlicz memorial conference, University of Mississippi, 1991.
92. H. Nikaidô, *On von Neumann's minimax theorem*, Pac. J. Math. **4**(1954), 65-72.
93. H. Nikaidô, *On a method of proof for the minimax theorem*, Proc. Amer. Math. Soc. **10**(1959), 205-212.
94. S. Park, *Some coincidence theorems on acyclic multifunctions and applications to KKM theory*, Proceedings of the Second International Conference on Fixed Point Theory and Applications, Halifax, Nova Scotia, Canada, K.-K. Tan. ed., World Scientific, River Edge, NJ, 1992.
95. T. Parthasarathy, *A note on a minimax theorem of T. T. Tie*, Sankhyā, Series A, **27**(1965), 407-408.
96. T. Parthasarathy, *On a general minimax theorem*, Math. Student. **34**(1966), 195-196.
97. J. E. Peck and A. L. Dulmage, *Games on a compact set*, Canadian J. Math. **9**(1957), 450-458.
98. I. F. Pinelis, *On minimax risk*, Th. Prob. Appl. **35**(1990), 104-109.
99. J-Ch. Pomerol, *Inequality systems and minimax theorems*, J. Math. Anal. Appl. **103** (1984), 263-292.
100. B. Ricceri, *Some topological minimax theorems via an alternative principle for multifunctions*, Preprint.
101. R. T. Rockafellar, *Convex analysis*, Princeton. Univ. Press(1970).
102. R. T. Rockafellar, *Saddle-points and convex analysis*, Differential games and related topics, H. W. Kuhn and G. P. Szegö Eds., North Holland (American Elsevier), Amsterdam-London-New York(1971), 109-127.
103. G Rodé, *Superkonvexe Analysis*, Arch. Math. **34**(1980), 452-462.
104. G Rodé, *Superkonvexität und schwache Kompaktheit*, Arch. Math. **36**(1981), 62-72.

105. S. Simons, *Critères de faible compacité en termes du théorème de minimax*, Seminaire Choquet 1970/1971, no. 23, 8 pages.
106. S. Simons, *Maximinimax, minimax, and antiminimax theorems and a result of R. C. James*, Pac. J. Math. **40**(1972), 709-718.
107. S. Simons, *Minimax and variational inequalities, are they or fixed point or Hahn-Banach type?*, Game theory and mathematical economics, O. Moeschlin and D. Pallaschke, Eds., North Holland, Amsterdam-New York-Oxford(1981), 379-388.
108. S. Simons, *Two-function minimax theorems and variational inequalities for functions on compact and noncompact sets, with some comments on fixed-points theorems*, Proc. Symp. in Pure Math. **45**(1986), 377-392.
109. S. Simons, *An upward-downward minimax theorem*, Arch. Math. **55**(1990), 275-279.
110. S. Simons, *On Terkelsen's minimax theorem*, Bull. Inst. Math. Acad. Sinica **18**(1990), 35-39.
111. S. Simons, *Minimax theorems with staircases*, Arch. Math. **57**(1991), 169-179.
112. S. Simons, *A general framework for minimax theorems*, Nonlinear Problems in Engineering and Science, Shutie Xiao and Xian-Cheng Hu eds., Science Press, Beijing and New York (1992), 129-136.
113. S. Simons, *A flexible minimax theorem*, Acta. Mathematica. Hungarica **63**(1994), 119-132.
114. M. Sion, *On general minimax theorems*, Pac. J. Math. **8**(1958), 171-176.
115. L. L. Stachó, *Minimax theorems beyond topological vector spaces*, Acta Sci. Math. **42** (1980), 157-164.
116. A. Stefanescu, *A general min-max theorem*, Optimization **16**(1985), 497-504.
117. W. Takahashi, *Nonlinear variational inequalities and fixed point theorems*, J. Math. Soc. Japan **28**(1976), 168-181.
118. M. Takahashi and W. Takahashi, *Separation theorems and minimax theorems for fuzzy sets*, J. Optimization Theory and Applications **31**(1980), 177-194.
119. F. Terkelsen, *Some minimax theorems*, Math. Scand. **31**(1972), 405-413.
120. T. Tjoe-Tie, *Minimax theorems on conditionally compact sets*, Ann Math. Stat. **34**(1963), 1536-1540.
121. [121]H. Tuy, *On a general minimax theorem*, Soviet. Math. Dokl. **15**(1974), 1689-1693.
122. H. Tuy, *On the general minimax theorem*, Colloquium Math. **33**(1975), 145-158.
123. J. Ville, *Sur la théorie générale des jeux où interviennent l'habilité des joueurs*, Traité du calcul des probabilités et de ses applications **2**(1938), 105-113.
124. J. von Neumann, *Zur Theorie der Gesellschaftspiele*, Math. Ann. **100**(1928), 295-320. For an English translation see: *On the theory of games of strategy*, Contributions to the theory of games 4, Princeton. Univ. Press(1959), 13-42.
125. J. von Neumann, *Über ein ökonomisches Gleichungssystem und eine Verallgemeinerung des Brouwerschen Fixpunktsatzes*, Ergebn. Math. Kolloq. Wein **8**(1937), 73-83.
126. J. von Neumann and O. Morgenstern, *Theory of games and economic behaviour*, 3rd edition, Princeton. Univ. Press(1953).
127. A. Wald, *Generalization of a theorem of von Neumann concerning zero sum two person games*, Annals of Math. **46**(1945), 281-286.
128. A. Wald, *Statistical decision functions*, Chelsea Pub. Co., Bronx, N.Y., (1971).
129. H. Weyl, *Elementary proof of a minimax theorem due to von Neumann*, Contributions to the theory of games 1, Princeton. Univ. Press(1950), 19-25.
130. Wu Wen-Tsün, *A remark on the fundamental theorem in the theory of games*, Sci. Rec., New. Ser. **3**(1959), 229-233.
131. E. B. Yanovskaya, *Infinite zero-sum two-person games*, J. Soviet. Math. **2**(1974), 520-541.
132. N. J. Young, *Admixtures of two-person games*, Proc. London Math. Soc. Ser. III **25**(1972), 736-750.

A SURVEY ON MINIMAX TREES AND ASSOCIATED ALGORITHMS

CLAUDE G. DIDERICH* and MARC GENGLER
Computer Science Department, Swiss Federal Institute of Technology – Lausanne, CH-1015 Lausanne, Switzerland

Abstract. This paper surveys theoretical results about minimax game trees and the algorithms used to explore them. The notion of game tree is formally introduced and its relation with game playing described. The first part of the survey outlines major theoretical results about minimax game trees, their size and the structure of their subtrees. In the second part of this paper, we survey the various sequential algorithms that have been developed to explore minimax trees. The last part of this paper tries to give a succinct view on the state of the art in parallel minimax game tree searching.

1. Introduction

With the introduction of computers, also started the interest in having machines play games. Programming a computer such that it could play, for example chess, was seen as giving it some kind of intelligence. Starting in the mid fifties, a theory on how to play two player zero sum perfect information games, like chess or go, was developed. This theory is essentially based on traversing a tree called minimax or game tree, compute its value and find a path from the root node to one of its leaves. An edge in the tree represents a move by either of the players and a node a configuration of the game.

A large number of theoretical work has been done to more closely understand minimax game trees, their size and their structure. Two major algorithms have emerged to compute the best sequence of moves in such a minimax tree. On one hand, we have the alpha-beta algorithm suggested around 1956 by John McCarthy and first published in Slagle and Dixon [52]. On the other hand, Stockman [55] introduced the SSS* algorithm. Both methods try to minimize the number of nodes explored in the game tree using special traversal strategies and cut conditions. During recent years many researchers have investigated the parallel exploration of minimax trees. This has led to a more or less large group of parallel algorithms. A complete bibliography on minimax trees and related work may be found in [18].

At this point we would like to alert the reader that, in this paper, we are not investigating techniques that are specific to particular games, or other methods used to make game playing programs more efficient, like heuristics, hash tables, hardware move generators or specific data representation. Information on these topics may be found in [10, 11, 24, 26, 30, 36, 44, 49, 51], to name just a few.

This paper is organized as follows. In Sec. 2 we describe various theoretical results about minimax trees. We also present how minimax trees are related to games

* Supported by Swiss National Science Foundation grant SPP-IF 5003-034349.

D.-Z. Du and P. M. Pardalos (eds.), Minimax and Applications, 25–54.

like chess, go and others. Section 3 is completely devoted to sequential minimax algorithms and their analysis. In Sec. 4 we present some of the most interesting and new parallel algorithms that have been developed to explore minimax trees. Finally, in Sec. 5 we state some open problems before concluding.

2. Minimax Trees and the Theory Behind Them

The theory on minimax or game trees has been introduced to allow and simplify reasoning about two player, zero sum, perfect information games. A *two player, zero sum, perfect information game*, also called *minimax game*, is a game which involves exactly two players who alternatively make moves. No information is hidden from the adversary. No coins are tossed, that is, the game is completely deterministic, and there is a perfect symmetry in the quality of the moves allowed.

Go, checkers and chess are such minimax games whereas backgammon (the outcome of a die determines the moves available) or card games (cards are hidden from the adversary) are not.

2.1. Definition of minimax trees

In this section we are going to introduce some definitions about minimax trees, their structure and their evaluation. We usually start by giving a formal definition and only show in second place its intention and use. This is especially true for the most important concept in this paper, minimax or game trees.

Definition 2.1 (Minimax tree) A *minimax tree* (*game tree*) T is a tree such that
- each node of the tree is either of type min or of type max,
- each node has zero or more sons,
- there exists a unique special node called root which has no father,
- the sons of a node of type max (type min) are of type min (type max),
- a node which has no sons is called a *leaf* node.

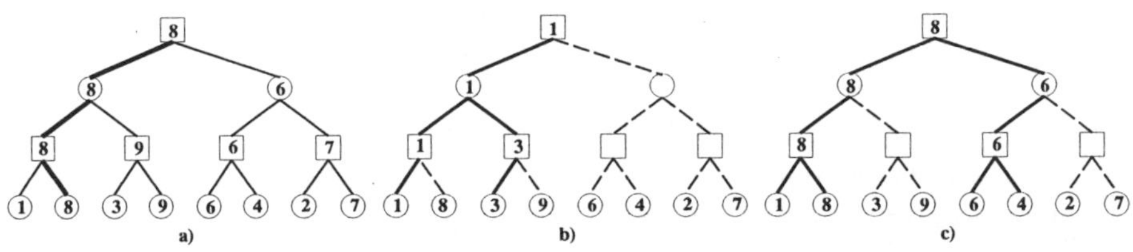

Fig. 1. a) A minimax game tree T, in bold a solution path of T, b) an associated α tree, c) an associated β tree.

Usually a node of type max is represented by a square box and a node of type min by a circle. Some papers refer to nodes of type max and min as AND and OR nodes[1]. Figure 1.a) represents such a tree. A minimax tree is an explicit representation of all possible board configurations in a minimax game. A node of type max represents

[1] In fact, if, according to Def. 2.2, one defines the set E as being $E = \{\text{false}, \text{true}\}$, then the max (resp. min) operation may bee seen as a boolean OR (resp. AND) operation.

a board configuration where the next move will be a personal move whereas a min node represents a board configuration where the next move will be a move of the adversary. Therefore the two players are sometimes called min and max. The goal of each game is to find a winning sequence of moves, given that the opponent always plays his best move.

The quality of a node n in the minimax game tree, representing a configuration, is given by its value $e(n)$ which is defined as follows.

Definition 2.2 (Value of a minimax tree) Let T be a minimax tree, E a totally ordered and topologically open set and $f: L(T) \longrightarrow E$ a function called *leaf evaluation function* where $L(T)$ is the set of leaf nodes in T. The value $e(t)$ of a node $t \in T$ is defined recursively as

- if $t \in L(T)$, then $e(t) = f(t)$,
- if $t \notin L(T)$ and t is of type min, then $e(t) = \min_{x \in \mathsf{sons}(t)} e(x)$,
- if $t \notin L(T)$ and t is of type max, then $e(t) = \max_{x \in \mathsf{sons}(t)} e(x)$.

Usually the set E is identified with $\mathbf{Z}$ or $\mathbf{R}$. Figure 1.a) represents a minimax tree and the values associated with each of its nodes. As an alternative one may define the value of a minimax tree T by

$$e(t) = \begin{cases} f(t) & \text{if } t \in L(T) \text{ and of type max} \\ -f(t) & \text{if } t \in L(T) \text{ and of type min} \\ \max_{x \in \mathsf{sons}(t)} -e(x) & \text{otherwise} \end{cases}$$

This definition scheme is known under the name of *negamax tree evaluation*. Such a scheme simplifies the notation and the description of some algorithms but partially hides the information about whose move one is reasoning about. The basic idea is to reverse the sign of all nodes of type max, that is, use the identity $\max_i e_i = -\min_i(-e_i)$. Without loss of generality, we will thus assume in the rest of this paper that the root node of any minimax tree is of type max.

If the considered minimax tree T represents a complete game, that is, all possible board configurations, where $r = \text{root}(T)$, the function f may be defined as follows:

$$f(t) = \begin{cases} +1 & \text{if the moves from } r \text{ to } t \text{ lead to a winning position} \\ 0 & \text{if the moves from } r \text{ to } t \text{ lead to a tie position} \\ -1 & \text{if the moves from } r \text{ to } t \text{ lead to a losing position} \end{cases} \tag{1}$$

The value of the root node of a minimax tree is called the value of that tree. Therefore we usually write $e(T)$ for $e(\text{root}(T))$.

Formally, a *solution path* in a minimax tree T is defined as a sequence of nodes $s_1, \ldots, s_h$ such that $s_1 = \text{root}(T)$, $\forall i < h: s_{i+1} \in \mathsf{sons}(s_i)$ and $e(s_i) = e(s_j)$ for all i, j such that $1 \leq i, j \leq h$. It is not necessarily unique. A solution path of tree T in Fig. 1.a) is shown in bold. It represents a sequence of moves by players max and min alternatively such that at each stage the best move is played, from the players point of view.

Definition 2.3 (Best-case and worst-case minimax trees) A minimax tree T is called a *best-case* (resp. *worst-case*) *minimax tree* if and only if for all $n \in T - L(T)$

$$\text{node } n \text{ is of type max (resp. min)} \Leftrightarrow e(s_1) \geq e(s_2) \geq \cdots \geq e(s_b)$$

$$\text{node } n \text{ is of type } \mathsf{min} \text{ (resp. } \mathsf{max}) \Leftrightarrow e(s_1) \leq e(s_2) \leq \cdots \leq e(s_b)$$

where $s_1, \ldots, s_b$ are the b sons of node n.

Definition 2.4 (α and β trees) Let T be a minimax tree. An α tree (resp. β tree) is defined as a subtree T_α (resp. T_β) of the tree T obtained by including all the sons of any non leaf node of type min (resp. type max) and exactly one son, no matter which one, of any non leaf node of type max (resp. type min).

The notions of α and β trees correspond to the trees T^- and T^+ in [44]. Sometimes these trees are also called *solution trees*. An α tree may be seen as a strategy for player max as it indicates, for any board position and for any move of the adversary (not necessarily his best move), the best move to play. A strategy is a winning one if and only if whatever move the opponent plays, the strategy leads to a winning position.

Figure 1.b) (resp. 1.c)) represents an α tree (resp. β tree) associated with the tree T in Fig. 1.a). As can easily be seen, α and β trees are not unique. Furthermore the inequality $e(T_\alpha) \leq e(T) \leq e(T_\beta)$ obviously always holds for any T_α and T_β associated with T.

2.2. The relation between minimax trees and games

Up to now we have introduced the most important notions about minimax trees and illustrated their relation to minimax games. Table I summarizes the relations that exist between minimax trees and minimax games. In Fig. 2 we show a portion of the minimax tree associated with the tic, tac, toe game. Shaded nodes represent leaves with their associated value, as defined in (1). The bold path shows a possible solution path.

Minimax tree notion	Minimax game notion
Minimax tree	All possible board configurations
Node in the tree	Board configuration
Edge from a max node to a min node	Personal move (move by player max)
Edge from a min node to a max node	Adversary move (move by player min)
Node value	Quality of a given board position
Leaf node	Outcome of a game (either win, tie or lose)
Solution path	Sequence of moves leading to the best outcome
α (resp. β) tree	Strategy for player max (resp. min)

TABLE I
Relation between minimax trees and minimax games.

2.3. Some statistic models of minimax game trees

In this section we will describe the most important models of minimax trees suggested in the literature. As analyzing minimax trees with variable number of sons

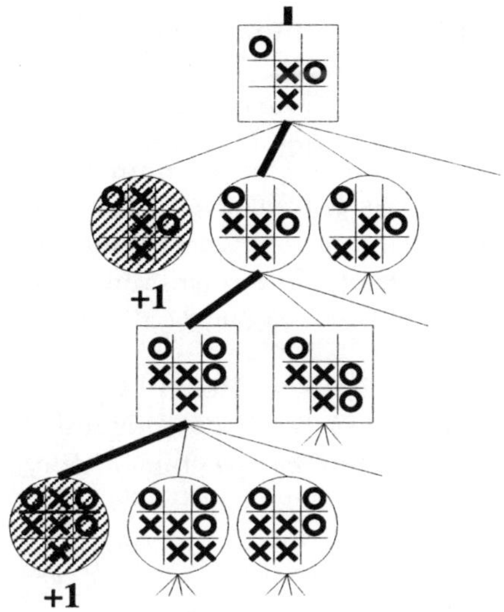

Fig. 2. Portion of a minimax tree associated with the minimax game tic, tac, toe.

and variable depths is difficult, most research has focused on uniform game trees. In some sense, this approach may be oversimplified. Indeed, there is no evidence that results obtained on regular trees may be verified on irregular trees. A *uniform minimax tree* is a minimax tree T in which each non leaf node has exactly b sons and all leaf nodes are at depth or height h (or at distance h from the root node) and in which the type of the root node is arbitrarily fixed to max. Such a tree is denoted by $T(b,h)$.

Definition 2.5 We call *win-loss tree*, and write $T(b,h,p)$, any uniform minimax tree for which the leaf evaluation function is defined by

$$f(t) = \begin{cases} +1 & \text{with probability } p \\ -1 & \text{with probability } 1-p \end{cases}$$

Win-loss trees are very simply structured and allow an easy study of their expected value. In fact, Pearl has shown that, for sufficiently deep win-loss trees, its value can be determined, with high probability, in advance.

Theorem 2.1 (Pearl [43]) *Let $T(b,h,p)$ be a win-loss minimax tree. Then*

$$\lim_{h\to\infty} \Pr\Big\{e\Big(\text{root}\big(T(b,h,p)\big)\Big) = +1\Big\} = \begin{cases} 1 & \text{if } p > \xi_b \\ 0 & \text{if } p \leq \xi_b \end{cases}$$

where ξ_b is the positive solution of the equation $x^b + x - 1 = 0$.

For a proof of any of the theorems, lemmas and corollaries in this paper, the reader is referred to the original paper mentioned in the theorem's header.

Definition 2.6 (Pearl [43]) A uniform minimax tree $T(b,h)$ in which the values assigned to the leaf nodes (via the function f) are independent identically distributed random variables is called a *random uniform minimax tree* or *rumtree*. It is denoted by $T(b,h,F)$, where F is the distribution function.

It is easy to see that the values of all nodes at the same depth in a rumtree are also independently and identically distributed random variables because the value of a node only depends on the value of its sons. But what is surprising is that the value of any sufficiently large rumtree may be predicted quite accurately.

Theorem 2.2 (Minimax convergence, Pearl [43]) *The root value of a rumtree $T(b,h,F)$ with continuous strictly increasing terminal distribution F converges (in probability), as h tends to infinity, to the $(1-\xi_b)$-quantile of F, were ξ_b is the positive solution of $x^b + x - 1 = 0$.*

If the terminal values are discrete ($v_1 < \ldots < v_m$), then the root value converges to a definite limit if $1 - \xi_b \neq F(v_i)$ for all i, in which case the limit is the smallest v_i satisfying $F(v_{i-1}) < 1 - \xi_b < F(v_i)$.

As Pearl mentioned, the remarkable feature of Thm. 2.2 is that it holds for any distribution. Thus, for a ternary tree ($b = 3$), with terminal values uniformly distributed in $]0,1[$, the value of its root node would converge to 0.3177 while the depth h tends to infinity.

In his PhD thesis Michon [41] shows some results about the branching factor, the probability to win, the size and the finiteness of a class of minimax trees called impartial minimax trees and the associated games. We won't go into further details as the rules for winning impartial minimax games differ slightly from the ones of minimax games as shown up to now.

2.4. On the number of nodes to explore to compute the value of a minimax tree

When reasoning about minimax trees one important question which arises is: "What is the minimal number of nodes one has to explore to compute the value of a minimax tree?" The following results have been proved in a less general framework by Slagle and Dixon [52] and Knuth and Moore [34]. The formulation presented below is due to Diderich and Gengler [20]

Lemma 2.1 *Let T be a minimax tree, $v \in E$ and $n \in T$. To show that $v \leq e(n)$ (resp. $v \geq e(n)$) it is necessary to explore a subtree T_α (resp. T_β) rooted at node n.*

Theorem 2.3 *Let T be a minimax tree. There exist two trees T_α and T_β (not necessarily unique), rooted at* root(T), *associated with T such that to compute the value $e(T)$ a necessary and sufficient condition is to explore the tree $T_\alpha \cup T_\beta$.*

Furthermore, for the α and β trees of Thm. 2.3,we have $e(T_\alpha) = e(T_\beta) = e(T)$.

If the set E is topologically closed, that is, $E = [a,b]$, then the lower bound of Thm. 2.3 does no longer hold. In fact, for any tree T such that $e(T) = b$, there is no need to explore any T_β tree, because showing that $\exists T_\alpha \in T$ such that $e(T_\alpha) = b$ is sufficient to show that $e(T) = b$.

Corollary 2.1 $|L(T_\alpha \cup T_\beta)| = |L(T_\alpha)| + |L(T_\beta)| - 1$

For uniform trees of depth h and of constant branching b we deduce that $|L(T_\alpha \cup T_\beta)| = b^{\lfloor h/2 \rfloor} + b^{\lceil h/2 \rceil} - 1$, a well known result first proved by Slagle and Dixon [52]. This gives us a lower bound on the number of nodes a deterministic algorithm needs to explore in order to compute the value of a minimax tree over a topologically open set E.

We will write $N_A(T)$ for the number of leaf nodes evaluated by algorithm A when traversing a minimax tree T in order to compute its value. We will now define a cost measure for studying the efficiency of any minimax algorithm[2].

Definition 2.7 (Branching factor) Let $T(b,h)$ be a uniform minimax tree of branching b and depth h and A a game tree evaluation algorithm. We define the *branching factor* $\mathcal{R}_A(T(b,h))$ corresponding to the algorithm A by

$$\mathcal{R}_A\big(T(b,h)\big) = \lim_{h\to\infty} N_A\big(T(b,h)\big)^{1/h}$$

The notion of *branching factor* has been first introduced by Knuth and Moore [34]. It indicates the average number of sons of a node one has to explore in order to compute is minimax value.

Theorem 2.4 (Pearl [43]) *The optimal branching factor for any minimax algorithm A which evaluates a rumtree $T(b,h,F)$ having discrete leaf values is*

$$\mathcal{R}_A\big(T(b,h,F)\big) = \begin{cases} \sqrt{b} & \text{if } \forall v \in L(T)\colon F(v) \neq 1-\xi_b \\ \xi_b/(1-\xi_b) & \text{otherwise} \end{cases}$$

and

$$\mathcal{R}_A\big(T(b,h,F)\big) = \frac{\xi_b}{1-\xi_b} = \Theta\left(\frac{b}{\log b}\right)$$

if $T(b,h,F)$ is a rumtree having continuous leaf values, where ξ_b is the positive solution of $x^b + x - 1 = 0$.

In Thm. 2.4 we have given the optimal value for the branching factor for any minimax algorithm. Any algorithm achieving this branching factor, can be said asymptotically optimal.

3. Sequential Minimax Game Tree Algorithms

In order so simplify the description of the various sequential and parallel minimax algorithms, we will introduce some general functions. Let T be a minimax tree and $t \in T$ a node. Then the function *first_son*(t) returns the first son node s_1 of t and *next_son*(s_i, t) returns the $i+1$-th son of node t. The function *no_more_sons*(s,t) returns true of s is the last son of t. Otherwise it returns false. The ordering of the sons introduced by these functions is arbitrary. In practice it is given by some

[2] A minimax algorithm is an algorithm A that computes the minimax value $e(T)$ of a minimax tree T.

heuristic function. The function *father*(t) returns the father node of t, the function *is_leaf*(t) returns whether or not $t \in L(T)$ and *node_type*(t) return the type of node t, either max or min.

3.1. The Minimax algorithm

The most basic minimax algorithm is called the *Minimax* algorithms. It systematically traverses, in a depth first, left to right fashion, the whole minimax tree. All nodes are visited exactly once.

---

function *Minimax*(n) **return** E **is**
begin
  if *is_leaf*(n) **then return** $f(n)$
  if *node_type*$(n) =$ max **then**
    return $\max_{s \in \mathsf{sons}(n)} Minimax(s)$
  else
    return $\min_{s \in \mathsf{sons}(n)} Minimax(s)$
  end if
end *Minimax*

As one can easily see, the branching factor is $\mathcal{R}_{Minimax}(T(b,h)) = b$, which is not optimal.

3.2. The alpha-beta algorithm

The first non trivial algorithm introduced to compute the minimax value of a game tree was the alpha-beta algorithm. The early history is somewhat obscure, because it is based on undocumented recollections. According to Knuth and Moore, McCarthy's comments at the Dartmouth summer research conference on artificial intelligence led to the use of alpha-beta pruning[3] in game playing programs since the late 1950s. The first published discussion of an algorithm for minimax tree pruning appeared in a paper by Newell, Shaw and Simon in 1958 (see [21], page 56). But the algorithm was not given any specific name. In fact the name of alpha-beta algorithm is attributed to McCarthy. Two early extensive studies of the algorithm may be found in Slagle and Dixon [52] and in Knuth and Moore [34].

The idea behind the alpha-beta algorithm is to traverse the minimax tree in a depth first, left to right fashion. In this point it is equivalent to the minimax algorithm of Sec. 3.1. Furthermore it tries to prune subtrees that can not influence the minimax value of the tree. The conditions used to prune subtrees are called cut conditions. The idea behind the suggested cut conditions is to associate to each node a lower and an upper bound. These bounds are usually called α and β bounds. Sometimes one also talks about the α-β window. The bounds of a node are passed to its sons and tightened during the execution of the algorithm. It is easy to see that the following two cut conditions are correct.

- If the lower bound of a node t of type max is larger than its upper bound, that is, $\alpha \geq_E \beta$, then all unvisited sons of node t can be pruned.

[3] Pruning a minimax tree means removing subtrees that need not be visited to evaluate the tree.

– If the upper bound of a node t of type min is smaller than its lower bound, that is, $\beta \leq_E \alpha$, then all unvisited sons of node t can be pruned.

It has formally been proved in [34] that the alpha-beta algorithm correctly computes the minimax value of a tree. In fact it speeds up the computation of the minimax value without losing any information. The following pseudo-code describes the alpha-beta algorithm.

```
function AlphaBeta(n, α, β) return E is
begin
   if is_leaf(n) then return f(n)
   s ← first_son(n)
   if node_type(n) = max then
      loop
         α ← max{α, AlphaBeta(s, α, β)}
         if α ≥_E β then return β
         exit loop when no_more_sons(n, s)
         s ← next_son(n, s)
      end loop
      return α
   else
      loop
         β ← min{β, AlphaBeta(s, α, β)}
         if β ≤_E α then return α
         exit loop when no_more_sons(n, s)
         s ← next_son(n, s)
      end loop
      return β
   end if
end AlphaBeta
```

The minimax value of a tree T may be computed as follows.

$$e(\text{root}(T)) \leftarrow AlphaBeta(\text{root}(T), -\infty_E, +\infty_E)$$

In Fig. 3 we illustrate the execution of the alpha-beta algorithm on a given minimax tree. Dashed edges indicate subtrees that have been pruned by the alpha-beta cut conditions. The value contained inside the various nodes are either their actual value or one of their bounds, if they are preceded by a less than or greater than sign.

Fuller, Gaschnig and Gillogy [28] have shown that the number of leaf nodes evaluated by the alpha-beta algorithm on a uniform minimax tree $T(b, h)$ for small values of b and h may be reasonably approximated by the formula

$$N_{AlphaBeta}(T(b, h)) = c(h) \cdot b^{0.72h+0.277}$$

where $1 \leq c(h) \leq 2$ (at least for the range of values they considered).

The order in which the sons of a node are visited plays an important role. In fact, the number of possible prunings depends on it. Knuth and Moore [34] have shown that there always exists an ordering of the sons of each node such that the number of nodes visited by the alpha-beta algorithm is optimal.

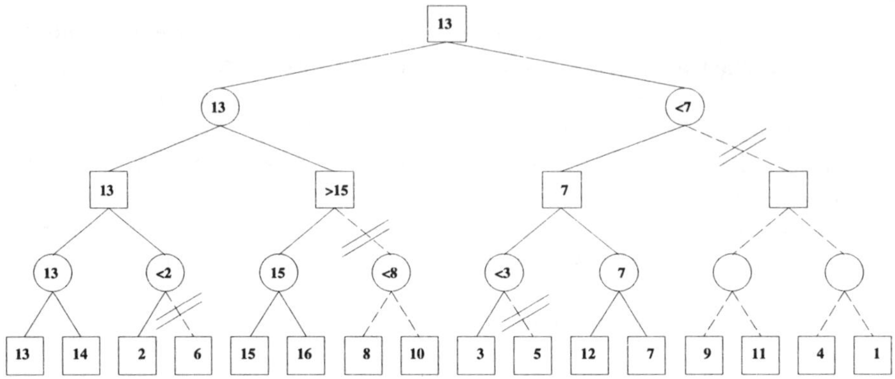

Fig. 3. Execution of the alpha-beta algorithm on a minimax tree T.

Theorem 3.1 (Knuth and Moore [34]) *It is always possible to arrange the leaves of a minimax tree T in such a way that the alpha-beta algorithm exactly searches $|L(T_\alpha \cup T_\beta)|$ leaf nodes.*

Various results about the efficiency of the alpha-beta algorithm have been proved. They are summarized by the following theorems.

Theorem 3.2 (Pearl [43]) *Let $T(b, h, F)$ be a rumtree with either continuous leaf values or distinct discrete leaf values. Then the branching factor of the alpha-beta algorithm is given by*

$$\mathcal{R}_{AlphaBeta}(T(b,h,F)) = \frac{\xi_b}{1-\xi_b} = \Theta\left(\frac{b}{\log b}\right)$$

where ξ_b is the positive solution of $x^b + x - 1 = 0$.

Corollary 3.1 (Pearl [43]) *The alpha-beta algorithm is asymptotically optimal over any minimax algorithm.*

In fact the branching factor of the alpha-beta algorithm (Thm. 3.2) achieves the optimal value given in Thm. 2.4.

Theorem 3.3 (Knuth and Moore [34]) *Let $T(b, 2, F)$ be a rumtree of depth 2. Then the expected number of nodes explored by the alpha-beta algorithm is given by*

$$E\Big[N_{AlphaBeta}(T(b, h=2, F))\Big] = \Theta\left(\frac{b^2}{\log b}\right)$$

Darwish [16] showed in 1983 a similar result. He proved that the expected number of nodes visited by the alpha-beta algorithm when traversing a rumtree of depth 2 is bound below by the quantity

$$E\Big[N_{AlphaBeta}(T(b, h=2, F))\Big] = O\left(b \cdot \left(1 + \frac{1}{2} + \frac{1}{3} + \cdots + \frac{1}{b}\right)\right)$$

Baudet, in [9], deduced the general value for the expected number of nodes explored by the alpha-beta algorithm when traversing a rumtree.

3.3. The optimal state space search algorithm SSS*

The optimal state space search algorithm SSS* has been introduced by G. C. Stockman in 1979. It originates not in game playing but in systematic pattern recognition. In fact the SSS* algorithm was developed to produce ambiguous parses of waveforms by a context-free grammar. The interested reader may find further details in [54]. The algorithm was first analyzed and criticized in a paper called "A Minimax Algorithm Better than Alpha-Beta? Yes and No" by Roizen and Pearl [48] in 1983.

The idea behind the SSS* algorithm is to use a tree traversal strategy that is, better than the depth first and left to right strategy found in the alpha-beta algorithm. The criteria used to order the nodes yet to visit is an upper bound of their value. Nodes are stored in non increasing order of their upper bound, also sometimes called merit, in a list called OPEN[4].

The SSS* algorithm first traverses the minimax tree from top to bottom. Nodes whose sons have not yet been visited and which cannot yet be pruned are marked LIVE. Nodes marked SOLVED have already been visited once and have therefore their best upper bound associated.

The operation *purge*(t, OPEN) removes all the nodes from the OPEN list for which the node t is an ancestor. Due to the fact that the nodes in the OPEN list are sorted in non increasing order of their associated upper bound, the pruning operation only eliminates nodes that need no further consideration. A formal proof may be found in [55].

Unfortunately, the SSS* algorithm is very consuming in memory. It requires at least $O(b^{\lceil h/2 \rceil})$ memory cells to store the OPEN list whereas algorithms like alpha-beta only require $O(h)$ stack space when traversing a uniform minimax tree $T(b, h)$.

The SSS* algorithm is described by the following pseudo-code.

---

function *SSS**(T) **return** E **is**
begin
  OPEN $\leftarrow \emptyset$
  insert$((\text{root}(T), \text{LIVE}, +\infty_E), \text{OPEN})$
  loop
    $(s, t, m) \leftarrow$ *remove*(OPEN)
    if $s = \text{root}(T) \wedge t = \text{SOLVED}$ **then return** m
    ⟨Apply the Γ operator to node s⟩
  end loop
end *SSS**

To each node s extracted from the OPEN list is applied an operator called $\Gamma(s)$.

⟨Apply the Γ operator to node s⟩ ≡

[4] The name OPEN came from the fact that this list contains all nodes that are still open for evaluation.

if $t = \text{LIVE} \wedge node\_type = \mathsf{max} \wedge \neg is\_leaf(t)$ **then**
  $s \leftarrow first\_son(t)$
  loop
    $insert((s, \text{LIVE}, m), \text{OPEN})$
    exit loop when $no\_more\_sons(s, t)$
    $s \leftarrow next\_son(s, t)$
  end loop
end if
if $t = \text{LIVE} \wedge node\_type = \mathsf{min} \wedge \neg is\_leaf(t)$ **then**
  $insert((first\_son(t), \text{LIVE}, m), \text{OPEN})$
end if
if $t = \text{LIVE} \wedge is\_leaf(t)$ **then**
  $insert((t, \text{SOLVED}, \min\{f(t), m\}), \text{OPEN})$
end if
if $t = \text{SOLVED} \wedge node\_type = \mathsf{max} \wedge \neg no\_more\_sons(t, father(t))$ **then**
  $insert((next\_son(t, father(t)), \text{LIVE}, m), \text{OPEN})$
end if
if $t = \text{SOLVED} \wedge node\_type = \mathsf{max} \wedge no\_more\_sons(t, father(t))$ **then**
  $insert((father(t), \text{SOLVED}, m), \text{OPEN})$
end if
if $t = \text{SOLVED} \wedge node\_type = \mathsf{min}$ **then**
  $insert((father(t), \text{SOLVED}, m), \text{OPEN})$
  $purge(father(t), \text{OPEN})$
end if

If the SSS* algorithm is applied to the minimax tree in Fig. 3, the same set of nodes is explored as is by the alpha-beta algorithm.

As one may notice, the SSS* only considers upper bounds. It is possible to define a dual version of the SSS*, which may be called SSS*-dual, in which the computation of upper bounds is replaced by the computation of lower bounds. The SSS*-dual algorithm has been suggested by Marsland *et al.* [39].

The theoretical optimality of the SSS* algorithm is proved by the following theorem.

Theorem 3.4 (Stockman [55]) *If the SSS* algorithm explores a node t, then this node is also explored by the alpha-beta algorithm.*

In fact, the alpha-beta algorithm loses efficiency (in the number of nodes visited) against the SSS* algorithm when the value of the minimax tree is found towards the right of the tree. If there exists an accurate ordering of the sons of each node, this loss of efficiency will only very seldomly occur. According to Stockman, to obtain an efficient algorithm, the alpha-beta algorithm and the SSS* procedure should be combined. Furthermore, if the SSS* algorithm is applied to win-lose trees then it visits exactly the same nodes in the same order as would the alpha-beta algorithm. In this case, the SSS* algorithm may start with the upper bound m of the root node being set to $+1$ instead of $+\infty_E$.

Roizen and Pearl [48] have studied the average number of leafs evaluated by the SSS* algorithm. The formulas deduced for the general case are quite complicated and therefore not reproduced herein. As can be seen from Thm. 3.5, the branching factor of the SSS* algorithm is optimal and therefore it is asymptotically optimal over any minimax algorithm.

Theorem 3.5 (Roizen and Pearl [48]) *Let $T(b, h, F)$ be a rumtree with either continuous leaf values or distinct discrete leaf values. Then the branching factor of the SSS* algorithm is given by*

$$\mathcal{R}_{SSS^*}(T(b,h,F)) = \frac{\xi_b}{1-\xi_b} = \Theta\left(\frac{b}{\log b}\right)$$

where ξ_b is the positive solution of $x^b + x - 1 = 0$.

3.4. SCOUT — A MINIMAX ALGORITHM OF THEORETICAL INTEREST

In the three previous sections, we have described the most common minimax algorithms. While trying to show the optimality of the alpha-beta algorithm, Pearl [43] introduced the SCOUT algorithm. His idea was to show that the SCOUT algorithm is dominated by the alpha-beta algorithm and to prove that SCOUT achieves an optimal performance. But counterexamples showed that the alpha-beta algorithm does not dominate the SCOUT algorithm because the conservative testing approach of the SCOUT algorithm may sometimes cut off nodes that would have been explored by the alpha-beta algorithm.

The idea behind the SCOUT algorithm is to only verify the value of all but the first son of a node, by constructing α trees (resp. β trees) and applying lemma 2.1.

The test function $Test(n, v, R)$ verifies whether or not $e(n)\ R\ v$ is true by visiting a minimal number of nodes.

```
function Test(n, v, R) return B is
begin
   if is_leaf(n) then return f(n) R v
   s ← first_son(n)
   loop
      exit loop when no_more_sons(n, s)
      s ← next_son(n, s)
      if node_type(n) = max then if Test(s, v, R) then return true
      if node_type(n) = min then if Test(s, v, ¬R) then return false
   end loop
   if node_type(n) = max then return false else return true
end Test
```

The SCOUT algorithm itself recursively computes the value of the first of its sons. Then it tests to see if the value of the first son is better that the value of the other sons. In case of a negative result, the son that failed the test is completely evaluated by recursively calling SCOUT.

```
function SCOUT(n) return E is
begin
   if is_leaf(n) then return f(n)
```

```
    s ← first_son(n), v ← SCOUT(s)
    loop
        exit loop when no_more_sons(n, s)
        s ← next_son(n, s)
        if node_type(n) = max then if Test(s, v, <) then v ← SCOUT(s)
        if node_type(n) = min then if Test(s, v, ≥) then v ← SCOUT(s)
    end loop
    return v
end SCOUT
```

Pearl [43] has derived the following two results characterizing the quality of the SCOUT algorithm in theory on rumtrees.

Theorem 3.6 (Pearl [43]) *The expected number of leaf nodes examined by the* SCOUT *algorithm in the evaluation of a $T(b, h, F)$ rumtree with continuous leaf values has a branching factor of*

$$\mathcal{R}_{SCOUT}(T(b, h)) = \frac{\xi_b}{1 - \xi_b} = \Theta\left(\frac{b}{\log b}\right)$$

Theorem 3.7 (Pearl [43]) *For any rumtree $T(b, h, F)$ with discrete leaf values satisfying $\forall v \in L(T)$: $F(v) \neq \xi_b$, the* SCOUT *algorithm is asymptotically optimal over all minimax algorithms.*

Although the SCOUT algorithm is more of theoretical interest, there are some problem instances where it outperforms all other minimax algorithms. A last advantage of the SCOUT algorithm versus one of its major competitors, the SSS* algorithm, is that its storage requirements are similar to those of the alpha-beta algorithm.

3.5. GSEARCH — A GENERALIZED GAME TREE SEARCH ALGORITHM

In 1986, Ibaraki [31] proposed a generalization of the previously known algorithms to compute the minimax value of a game tree. His idea was to use a branch and bound[5] like approach. Nodes of the considered tree which have not yet been evaluated are stored in a list called OPEN list. This list is ordered according to a given criteria. Different orderings give different traversal strategies. Furthermore, as found in the branch and bound algorithm, he suggested to associate a lower and upper bound to each node. These bounds generalize the α and β values found in the alpha-beta algorithm.

In the same paper Ibaraki also introduced the model of *informed minimax tree*. In such a tree not only is it possible to evaluate a leaf node, but some nodes have a lower and an upper bound of their value associated. These nodes are called *tip nodes* because they give some kind of hint to guide the search. The bounds, for any tip node $n \in T$ are written $l(n)$ and $u(n)$ respectively.

The main idea is to associate to each node $n \in T$ a lower and an upper bound ($l_b(n)$ and $u_b(n)$ resp.). They are computed using the equations (2) and (3).

[5] Branch and bound is the name of an algorithm to exactly solve combinatorial optimization problems. A fairly well description can be found in [42].

$$l_b(n) = \begin{cases} f(n) & \text{if } n \text{ is a leaf node} \\ \max\{l(n), l(father(n))\} & \text{if } n \text{ is a tip node} \\ \max_{s \in \mathsf{sons}(n)} l_b(s) & \text{if } n \text{ is of type } \mathsf{max} \\ \min_{s \in \mathsf{sons}(n)} l_b(s) & \text{if } n \text{ is of type } \mathsf{min} \end{cases} \tag{2}$$

$$u_b(n) = \begin{cases} f(n) & \text{if } n \text{ is a leaf node} \\ \max\{u(n), u(father(n))\} & \text{if } n \text{ is a tip node} \\ \max_{s \in \mathsf{sons}(n)} u_b(s) & \text{if } n \text{ is of type } \mathsf{max} \\ \min_{s \in \mathsf{sons}(n)} u_b(s) & \text{if } n \text{ is of type } \mathsf{min} \end{cases} \tag{3}$$

It is easy to see that $l(n) \leq l_b(n) \leq e(n) \leq u_b(n) \leq u(n)$ for any tip node n in any minimax tree T. Given a minimax tree T, it is sometimes possible to conclude that there is no need to expand a tip node n further. Let AMAX(n) (resp. AMIN(n)) be the set of max (resp. min) nodes in T which are proper ancestors of n. Then we define $l_t(n)$ and $u_t(n)$ by

$$l_t(n) = \max_{v \in \mathrm{AMAX}(n)} l_b(v) \tag{4}$$

$$u_t(n) = \min_{v \in \mathrm{AMIN}(n)} u_b(v) \tag{5}$$

This allows us to state the following theorem.

Theorem 3.8 (Ibaraki [31]) *Let T be a minimax tree and $n \in T$ a node. Then the node n needs no further exploration to compute the minimax value, that is, can be cut off, if and only if*

$$\max\{l_t(n), l_b(n)\} \geq \min\{u_t(n), u_b(n)\}$$

Now we are able to describe the generalized minimax tree search algorithm, called GSEARCH, as introduced by Ibaraki.

---

function *GSEARCH*(n) **return** E **is**
begin
  insert($(n, -\infty_E, +\infty_E)$, OPEN)
  loop
    exit loop when $l_b(n) = u_b(n)$
    $(s, l_b(s), u_b(s)) \leftarrow$ *remove*(OPEN)
    if $l_t(s) < u_t(s)$ **then**
      compute $l(s)$ and $u(s)$ and update l_b, u_b, l_t and u_t according to equations (2), (3), (4) and (5)
      if $\max\{l_t(n), l_b(n)\} < \min\{u_t(n), u_b(n)\}$ **then**
        $s \leftarrow$ *first_son*(t)
        loop
          insert($(s, -\infty_E, +\infty_E)$, OPEN)
          exit loop when *no_more_sons*(s, t)
          $s \leftarrow$ *next_son*(s, t)
        end loop

```
            end if
        end if
    end loop
    return l(n)
end GSEARCH
```

Finally Ibaraki showed how the algorithm GSEARCH is related to other minimax algorithms like alpha-beta or SSS*, and proved that his algorithm always surpasses the alpha-beta algorithm.

To conclude this section, we would like to note that Pijls and de Bruin have proposed in [46] a variant of their SSS-2 algorithm for the informed minimax tree model of Ibaraki.

3.6. SSS-2 — A RECURSIVE STATE SPACE SEARCH ALGORITHM

The SSS-2 algorithm has been proposed by Pijls and de Bruin [45]. It is based on the idea of computing an upper bound for the root node and then repeatedly transforming this upper bound in a tighter one.

An upper bound on a minimax tree T can be found by simply computing the value of a β tree. To allow refining the so constructed β tree, the node selected as a son of a node of type min is chosen according to some total ordering introduced on the tree. Once a β tree has been constructed it is refined by using the following fact. If a node n in the β tree is of type max, then we can obtain a better upper bound for it only if for all its sons s_i there exist β trees rooted at s_i such that $u(n) < u(s_i)$[6] for all s_i. If, on the other hand, node n is of type min, then there are more possibilities to generate a better upper bound. First, one can try to obtain (recursively) a better upper bound for the current son c of n. But it is also possible to select another son s' of n, and to try to establish an upper bound for the node c' by building a new β tree rooted at c' such that the new upper bound for c' is smaller than the old one.

It is easy to see that, in the same way as the SSS* algorithm, the SSS-2 algorithm admits a dual version in which lower bounds and α trees are considered instead of upper bounds and β trees.

A more detailed description of the SSS-2 algorithm can be found in [45]. It has also been adapted to searching informed minimax trees[7] (see [46]).

Pijls and de Bruin [45] have shown that the SSS-2 algorithm exactly expands the same nodes as those to which the SSS* algorithm applies the Γ operator. Therefore the following result, whose proof is based on the computation and refinement of upper bounds using β trees, is not surprising.

Theorem 3.9 (Pijls and de Bruin [45]) *The SSS-2 algorithm surpasses the alpha-beta algorithm in the sense that the SSS-2 algorithm visits a subset of the nodes visited by the alpha-beta algorithm.*

[6] We denote by $u(n)$ the upper bound associated with node $n \in T$ during the executing of the algorithm.

[7] See Sec. 3.5 for a formal definition of the notion of informed minimax trees.

3.7. Some variations on the subject

In this section we will briefly describe some techniques and algorithms that have been proposed to enhance the efficiency of the more classical algorithms like alpha-beta or SSS*.

3.7.1. *Aspiration search*

Computing the minimax value of a game tree may be seen as aspiring the solution value from a leaf node through the whole tree up to the root node. While moving closer to the root node, more and more useless subtrees will be eliminated, as we have already stated for the alpha-beta algorithm. The better the α and β bounds, the more subtrees may be pruned.

If, for instance, one knows that the minimax value will, with high probability, be found in the subset $]a,b[$[8] of E, then it may be worth calling the alpha-beta algorithm as

$$e \leftarrow \mathit{AlphaBeta}\big(\mathrm{root}(T), a, b\big)$$

If, indeed, the minimax value $e(\mathrm{root}(T))$ belongs to the set $]a,b[$, the algorithm will correctly return that value. If the minimax value does not belong to the set $]a,b[$, then the value returned will be either a or b, depending on whether the minimax value belongs to $]-\infty_e, a]$ or $[b, +\infty_E[$. We then say that the alpha-beta algorithm failed low, resp. high. In the case where the algorithm failed low, the call

$$e \leftarrow \mathit{AlphaBeta}\big(\mathrm{root}(T), -\infty_E, a+1\big)$$

will return the correct value. But it would also be possible to reiterate this procedure on a subset $]a_1, a+1[$.

The technique of limiting the interval in which the solution may be found is called *aspiration search*. If the minimax value belongs to the specified interval, then a much larger number of cut conditions are verified and the tree actually traversed is much smaller than the one traversed by the alpha-beta algorithm without initial alpha and beta bounds. If the minimax value of a tree T is e, then the call $\mathit{AlphaBeta}(\mathrm{root}(T), e-1, e+1)$[9] will exactly explore the union $T_\alpha \cup T_\beta$ of an alpha and a beta tree. Kaindl et al. [32] as well as Baudet [8] and Marsland et al. [39] have extensively studied the efficiency in practice of the technique of aspiration search.

Furthermore it is interesting to note that aspiration search is at the bases of a technique called *iterative deepening* which is used in many game playing programs. Without going into more details, the technique of iterative deepening consists in iteratively increasing the depth of the tree searched and then using the solution path of the previous depth as traversal strategy for the next one. An in depth description of the iterative deepening technique may be found in [35].

3.7.2. *Other interesting results*

Althöfer [6] has suggested an incremental negamax algorithm which uses estimates of all nodes in the minimax tree, rather than only those of the leaf nodes, to determine

[8] $]a,b[$ denotes the open set ranging from a to b, but not including these two values.

[9] This is called a null window search.

the value of the root node. This algorithm is useful when dealing with erroneous leaf evaluation functions. Under the assumption of independently occurring and sufficiently small errors, the proposed algorithm is shown to have exponentially reduced error probabilities with respect to the depth of the tree.

Rivest [47] proposed an algorithm for searching minimax trees based on the idea of approximating the min and the max operators by generalized mean-value operators. The approximation is used to guide the selection of the next leaf node to expand, since the approximation allows to select efficiently that leaf node upon whose value the minimax value most highly depends. Ballard [7] proposed a similar algorithm where the value of some nodes (the chance node as he calls them) is a, possibly weighted, average of the values of its sons. In fact he considers one additional type of nodes called chance nodes.

Conspiracy numbers have bee introduced by McAllester in [40] as a measurement of the accuracy of the minimax value of an incomplete tree. They measure the number of leaf nodes whose value must change in order to change the minimax value of the root node by a given amount. A study of conspiracy numbers may be found in [51].

4. Parallel Minimax Tree Algorithms

In this section we will describe some of the most important parallel algorithms for traversing minimax trees and evaluate their efficiencies. In parallel computing one distinguishes between two large classes of machines: first, machines that execute one instruction per cycle but on multiple data streams — single instruction multiple data stream machines (SIMD) and second, machines on which each processor executes its own instructions on its own data stream — multiple instructions multiple data stream machines (MIMD). This terminology has been introduced by Flynn in 1966. As minimax trees are irregular structures, all the algorithms we will describe have been developed for the second class of parallel machines. Within this category one further distinguishes between shared memory machines, where each processor shares all the memory with all the other processors and distributed memory machines where each processor has its own local memory and communication between processors is explicit through some kind of communication network.

Minimax algorithms were seen very soon as good candidates for parallelization. Indeed the huge size of computations makes them ideal candidates.

Parallelizing the minimax algorithm of Sec. 3.1 is trivial over uniform trees. Even on irregular trees, the parallelization remains easy. The only additional problem arises from the fact that the size of the subtrees to explore may now vary. Different processors will be attributed problems of varying computational volume. All what is needed then to achieve excellent speedups, is a load-balancing scheme, that is, a mechanism by means of which processors may, during run-time, exchange problems so as to keep all processors busy all the time.

The parallelization of the alpha-beta and the SSS* algorithms are of course much more interesting than the more theoretical minimax algorithm. There exist basically two approaches or techniques to parallelize the alpha-beta algorithm. In the first approach, which has been one of the first techniques used, all processors explore the entire tree but using different search-intervals. This approach is at the basic of the algorithm called parallel aspiration search by Baudet (see Sec. 4.3). The second one consists in exploring simultaneously different parts of the minimax tree.

4.1. A SIMPLE WAY TO PARALLELIZE THE EXPLORATION OF MINIMAX TREES

Exploring a minimax tree in parallel can very simply be obtained by generating the sons of the root node, and their sons and so on up to the point where one has as many son nodes waiting to be explored as there are processors. At this point, each processor will explore the subtree rooted at one of these nodes, using any given sequential minimax algorithm. When all processors have completed their exploration, the solution for the entire tree is computed by using the partial results obtained from each of the processors. This principle is illustrated in Fig. 4.a) for a binary tree and four processors, processor P_i being in charge of the subtree rooted at node s_i.

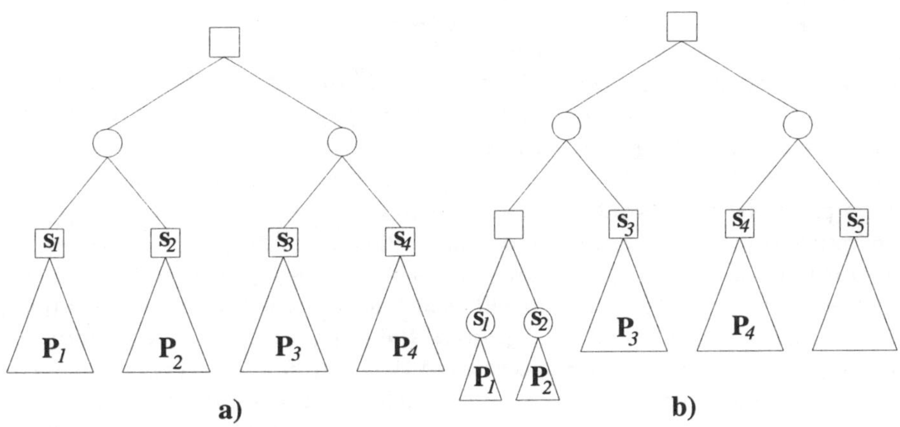

Fig. 4. a) A processor allocation scheme for parallel minimax tree searching. b) Another processor allocation scheme for parallel minimax tree searching.

Such a work distribution may be non optimal. Indeed, if we suppose the evaluations of both s_1 and s_2 to be no smaller than the one of node s_3, then the alpha-beta algorithm would prune the subtree rooted at node s_4 from search. Thus the work given to processor P_4 is completely useless. The problem comes from the fact that we do not known in advance whether or not the subtree rooted at node s_4, or any other node s_i, will be pruned.

In the absence of any additional information, it is not possible to use another allocation scheme than the one shown in Fig. 4.a) which would be superior in most situations. In practice however, the sons of a node may be ordered in such a way that any son has a probability of yielding the locally optimal path that is no smaller than the corresponding probabilities for its right neighbours. The probability to find the optimum in the subtree rooted at a given son then always decreases when traversing the sons in a left to right order. Such ordering information is generally available in game-playing programs, the ordering function being a heuristic function based on the knowledge of the game to be played.

With such an assumption, the allocation shown in Fig. 4.a) is not very good as we use the same computation resources for exploring the subtree rooted at node s_1, which will be, with high probability, on the solution path, as for exploring the subtree rooted at node s_4, which will be, with high probability, the not considered at all. We should therefore rather allocate the processors has shown in Fig. 4.b).

The tree in Fig. 4.b) contains five subtrees rooted at s_1 to s_5. The processors $P_1, \ldots, P_4$ are allocated to the nodes $s_1, \ldots, s_4$ and node s_5 is not allocated à computing resource at the beginning of the exploration. If the nodes ordering function correctly guessed the ordering of nodes s_1 to s_5, then all processors will participate in useful work in the sense that nodes s_1 to s_4 need do be explored. Note that this does not mean that the processors explore those nodes in the optimal way.

4.2. A MANDATORY WORK FIRST ALGORITHM

The algorithm proposed by Hewett and Krishnamurthy [29] is a straightforward illustration of the parallelization described in Sec. 4.1. It is designed for shared memory machines. In [29], they have shown that an efficiency of roughly 50% for an number of processors in the range of 2 to 25 is feasible.

All the nodes that still need to be explored are maintained in a list called OPEN list. This list is ordered with respect to the lexicographical order[10] of **d**, where **d** is a vector of values indicating how the node n has been reached. A node a has a higher priority than an node b if $(a_1, \ldots, a_k)$ is lexicographically smaller than $(b_1, \ldots, b_l)$.

More precisely, the algorithm maintains two lists called OPEN and CLOSED, and a tree called CUT. The OPEN list contains all the nodes yet to be explored, the CLOSED list contains the expanded nodes not yet pruned and the CUT tree contains the pruned nodes. The OPEN list initially contains only the root node. All processors fetch nodes from the OPEN list and process them if they cannot be discarded, that is, they do not have any of their ancestors in the CUT tree. Leaf nodes are evaluated and their result is returned to the parent which may update its value and check for possible prunings by traversing the CUT tree up to the root node applying the usual alpha and beta cutoffs. If the node selected is not a leaf node, it is expanded and its sons are inserted into the OPEN list and itself into the CLOSED list.

The algorithm proposed by Akl *et al.* [1, 2] uses the same approach for exploring the minimax tree. Their priority function is computed as

$$\text{priority}(n_i) = \text{priority}(\mathit{father}(n_i)) - (b_{n_i} + 1 - i) \cdot 10^{(h-f-1)}$$

where n_i is the i-th son of node $\mathit{father}(n_i)$, b_{n_i} the branching of node $\mathit{father}(n_i)$, h the search depth (the maximal depth of the minimax tree) and f the depth of node $\mathit{father}(n_i)$ in the minimax tree.

Almquist *et al.* [3] also developed an algorithm based on the idea of having two categories of unexplored nodes which are ordered according to a given priority function. Furthermore they add to this concept parallel aspiration search as well as a novel scheduling algorithm.

In the same direction, Cung and Roucairol [15] have proposed a shared memory parallel minimax algorithm which distinguishes between critical and non critical

---

[10] A vector $(a_1, \ldots, a_k)$ is lexicographically smaller than a vector $(b_1, \ldots, b_k)$ if and only if for $1 \leq i \leq m$ we have $a_i = b_i$ and $a_m < b_m$ for some m.

nodes. In their algorithm one processor is assigned to each node.

In the algorithm by Steinberg and Solomon [53], which is also a mandatory work first type algorithm, the list containing the speculative work or non critical nodes is dynamically ordered.

4.3. Aspiration search

The parallel algorithm called ***aspiration search*** has been introduced by Baudet in 1978 [8]. In this algorithm the search interval $]-\infty,+\infty[$ used by the sequential alpha-beta algorithm is divided into a certain number of subintervals that cover the entire range $]-\infty,+\infty[$. Now, every processor explores the entire minimax tree using one subinterval, different processors being assigned different intervals. Any processor searching an interval $]a_i, a_{i+1}]$ may either fail low or high. The principle is the same as in the sequential version of the algorithm. Exactly one processor will neither fail low, nor fail high. The value computed by this processor is the value of the minimax tree to explore.

The aspiration search algorithm may be described by the following pseudo-code.

$\langle$On processor $1 \le i \le p$ execute the following code$\rangle \equiv$
  $e \leftarrow AlphaBeta(\text{root}(T), a_i, a_{i+1}+\varepsilon)$
  if $e \in]a_i, a_{i+1}]$ **then return** the minimax value is e

---

where ε is a infinitesimal quantity and $-\infty < a_1 < \cdots < a_{p-1} < +\infty$ a subdivision of the interval $]-\infty,+\infty[$ in p subintervals.

The implementation of the aspiration search algorithm is really simple. Furthermore, there is no information exchange needed between processors. If the nodes in the to explore minimax tree T are ordered in such a way that the alpha-beta algorithm has to explore the whole tree, then the speedup obtained by using the aspiration search algorithm on p processors equals p. But, when the aspiration search algorithm is applied to randomly generated trees, that is, rumtrees, then Baudet has shown that the speedup is limited to about six and is independent of the number of processors used. This limited speedup is astonishing at first glance. Nevertheless, it can somehow be explained. First of all, the parallel aspiration search algorithm is not faster than the alpha-beta algorithm when executed on a best-first minimax tree. Indeed, the processor that finds the solution within its search window has to perform a complete alpha-beta search. On the other hand, given the fact that the alpha-beta algorithm is asymptotically optimal, it will perform almost as well as the alpha-beta algorithm on best-first trees on a large minimax tree, even if it starts with a search window of $]-\infty,+\infty[$. Thus the parallel aspiration search yields in fact an asymptotic speedup of one.

4.4. The tree-splitting algorithm

Among the early parallel minimax algorithms we find the tree-splitting algorithm by Finkel and Fishburn [27]. This algorithm is based on the idea to look at the available processors as a tree of processors. Each processor, except for the ones representing leaves in the processor tree, have a fixed number p_b of son or slave processors. During the execution of the algorithm a non leaf processor associated with a node n in the minimax tree to explore spawns the exploration of the sons s_i of n to its p_b slaves. As soon as one slave returns the next unexplored son s_j is spawned to that slave or the current value is returned to the father processor if the cut condition is satisfied. If all the sons of a node have been spawned to its slaves, the father processor waits for the results of all its slaves. Leaf processors simply compute the value of their associated node using the sequential alpha-beta algorithm.

An important advantage of the tree-splitting algorithm over other more elaborated algorithms is that it may be simply implemented as well on a shared memory parallel machine as on a distributed memories parallel machine. To simplify the notation, we will describe the tree-splitting algorithm using the negamax notation.

```
function TreeSplit(n, α, β) return E is
begin
   if ⟨I am a leaf processor⟩ then return AlphaBeta(n, α, β)
   for s ∈ sons(n) loop in parallel
      ⟨Wait until a slave node is idle⟩
      v_i ← TreeSplit(s, −β, −α)
      if v_i > α then
         α ← v_i
         ⟨Update the bounds according to α on all slaves⟩
      end if
      if α > β then
         ⟨Terminate all slave processors⟩
         return α
      end if
   end loop
end TreeSplit
```

The notation **in parallel** means that the executing processor will logically create a micro-task for each body of the loop. In practice this construction would be implemented by using a **select** statement.

The code sequence ⟨Update the bounds according to α on all slaves⟩ means that the new bound should be sent to all the descendents of the executing node. On a distributed memories machines this operation may be implemented by each node asynchronously updating its α or β bounds and transmitting the new bounds to all its sons or slaves. On a shared memory machine the α and β bounds may be stored in a global array, one entry for each level of the tree depth, which is accessed under the mutual exclusion principle.

Finkel and Fishburn have shown that the speedup of their algorithm when executed on a p processor machine is at least $\sqrt{p}$ and even p in some special cases.

Theorem 4.1 *On a worst-first minimax tree, the speedup obtained by the tree-splitting algorithm, when executed on p processors, is p. On a best-first minimax tree, the speedup obtained by the tree-splitting algorithm, when executed on p processors, is* $\sqrt{p}$.

The tree-splitting algorithm has been implemented and its execution has been simulated. On a 27 processor simulated machine, in which each processor has tree slave sons associated, the average speedup was 5.31 for trees of depth 8 and a branching of 3.

To conclude this section, we would like to mention that the parallel minimax algorithm known under the name *palphabeta* in the literature is a simplified version of the tree-splitting algorithm

4.5. The principal variation splitting algorithm – PVSPLIT

We will now discuss another very interesting parallel minimax algorithm. The principal variation splitting algorithm (PVSPLIT) has been proposed by Marsland and Campbell in [37] and is by far the most often implemented algorithm, especially in chess playing programs. The algorithm is based on the structure of the sequential alpha-beta algorithm. The idea is to first explore in a sequential fashion a path from the root node to its leftmost leaf. This traversal is done in order to obtain alpha and beta bounds. This path is called the principal variation path. If the minimax tree to explore is of type best first, then the explored principal variation path represents the solution path. In a second phase, for each level of the minimax tree all the yet to be visited sons are explored in parallel by using the bounds computed during the principal variation path computation and the traversal of the lower levels of the minimax tree.

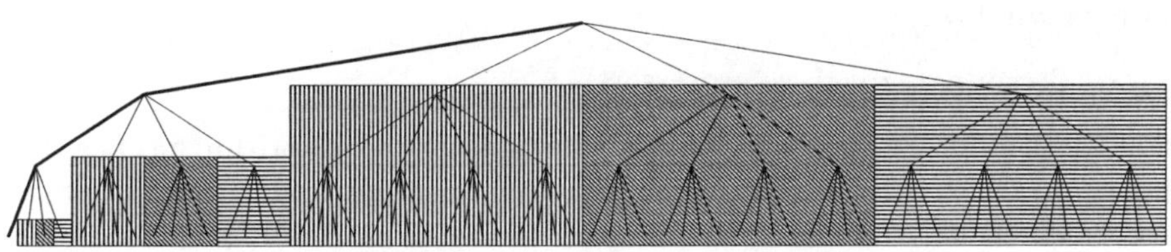

Fig. 5. Execution of the principal variation splitting algorithm. The bold path represents the principal variation path. Shaded subtrees are searched by different processors. Subtrees boxed with the same size boxes are traversed in parallel.

In Fig. 5 we show the execution of the PVSPLIT algorithm when applied to a uniform minimax tree of branching 4 and depth 3. The PVSPLIT algorithm is completely described by the following pseudo-code using the negamax notation.

---

function $PVSPLIT(n, \alpha, \beta)$ **return** E **is**
begin
  if $is\_leaf(n)$ **then return** $f(n)$
  $s \leftarrow first\_son(n)$
  $\alpha \leftarrow -PVSPLIT(s, -\beta, -\alpha)$;
  if $\alpha \geq \beta$ **then return** α
  for $s' \in \mathsf{sons}(n) - \{s\}$ **loop in parallel**
    ⟨Wait until a slave node is idle⟩
    $v_i \leftarrow TreeSplit(s', -\beta, -\alpha)$
    if $v_i > \alpha$ **then** $\alpha \leftarrow v_i$
    if $\alpha > \beta$ **then**

```
        ⟨Terminate all slave processors⟩
        return α
      end if
    end loop
    return α
end PVSPLIT
```

The PVSPLIT algorithm has been implemented by Marsland and Popowich [38] on a network of Sun workstations. An acceleration of 3.06 has been measured on 4 processors when traversing minimax trees representing real chess games. The main problem of the PVSPLIT algorithm is that, during the second phase, the subtrees explored in parallel are not necessarily of the same size. Some processors will have to wait for their brothers to terminate.

The PVSPLIT algorithm is the most efficient when the iterative deepening technique is used, because with each iteration is is increasingly likely that the first move tried, that is, the one on the principal variation path, is the best one.

The following variant may speedup somehow the executing of the PVSPLIT algorithm. Once a candidate solution has been found, by having computed the principal variation path, it can be used as a solution and the other nodes s' can be searched with an alpha-beta window of $]-(e+1), -(e-1)[$ where e is the best value so far. If that search fails, the complete subtree is searched by using a full width alpha-beta window.

4.6. Synchronized state space search

A completely different approach to parallelizing the SSS* algorithm has bee taken by Diderich and Gengler [17]. Their ideas are based on a synchronization technique that has been successfully applied to the branch and bound algorithm [14]. The algorithm proposed has been called synchronized distributed state space search (SDSSS). It may be seen as an alternation of computation and synchronization phases, as shown by the following pseudo-code.

```
function SDSSS(T) return E is
begin
    ⟨Initialize the algorithm's data structures⟩
    loop
        ⟨Synchronization phase⟩
        ⟨Computation phase⟩
        exit loop when ⟨the solution has been found⟩
    end loop
end SDSSS
```

The algorithm has been designed for a distributed memory multiprocessor machine. Each processor manages its own local OPEN list of unvisited nodes.

The synchronization phase may be subdivided in three major parts. First, the processors exchange information about which nodes can be removed from the local OPEN lists. This corresponds to each processor sending the nodes for which the *purge* operation may be applied by all the other processors[11]. Next, all the processors agree on the globally lowest upper bound m^* for which nodes exist in some of the OPEN lists. Finally all the nodes having the same upper bound m^* are evenly distributed among all the processors. This operation concludes the synchronization phase.

The computation phase of the SDSSS algorithm may be described by the following pseudo-code.

⟨Computation phase⟩ ≡
  while ⟨there exists a node in the open list having an upper bound of m^*⟩ **loop**
    $(s, t, m^*) \leftarrow$ *remove*(OPEN)
    if $s = \text{root}(T) \wedge t = \text{SOLVED}$ **then**
      broadcast *the solution has been found*
      return m^*
    end if
    ⟨Apply the Γ operator to node s (see Sec. 3.3)⟩
  end loop

Experiments executing the SDSSS algorithm on an Intel iPSC/2 parallel machine have been conducted. For rumtrees of type $T(16, 5)$ a speedup of 11.4 has been measured for 32 processors. For trees with a smaller branching factor b, the results were less concluding.

4.7. A DISTRIBUTED GAME TREE SEARCH ALGORITHM

Feldman *et al.* [22] parallelized the alpha-beta algorithm for massively parallel distributed memory machines. Different subtrees are searched in parallel by different processors. The allocation of processors to trees is not done by some priority function but by imposing certain conditions on the nodes to be selectable. They introduce the concept of *younger brother waits*. This concept essentially says that in the case of a subtree rooted at s_1, where s_1 is the first son node of a node n, is not yet evaluated, then the other sons $s_2, \ldots, s_b$ of node n are not selectable. Younger brothers may only be considered after their elder brothers, which has as a consequence that the value of the elder brothers may be used to give a tight search window to the younger brothers.

This concept is nevertheless not sufficient to achieve the same good search window as the alpha-beta algorithm achieves. Indeed when node s_1 is computed, then the younger brothers may all be explored in parallel using the value of node s_1. Thus the node s_2 has the same search window as it would have in the sequential alpha-beta algorithm, but this is not true anymore for s_i, where $i \geq 3$. Indeed if nodes s_2 and s_3 are processed in parallel, they only know the value of node s_1, while in the sequential alpha-beta algorithm, the node s_3 would have known the value of both s_1 and s_2. This fact forces the parallel algorithm to provide an information dissemination protocol. In case the nodes s_2 and s_3 are evaluated on processors P and P', and processor P finishes its work before P', producing a better value than node s_1 did, then processor P will inform processor P' of this value, allowing it to

[11] This communication operation as well as all the other communication operations are optimized for the used topology of the architecture. Details on this topic may be found in [19].

continue with better information on the rest of its subtree or to terminate its work if the new value allows P' to conclude that its computation becomes useless.

The load distribution is realized by means of a dynamic load balancing scheme, where idle processors ask other processors for work. If a processor, call it the *master* processor, is asked for work by another processor, call it the *sub-contractor* processor, it hands out, if available, a node that verifies the *younger brother waits* condition together with the search window to consider. Both the master and the sub-contractor will then compute. At the moment the master processor has finished its local work, it may need the values of the subproblems that it sub-contracted. If theses values are available, the master can continue its computation. If on the contrary, some of the sub-contractors have not yet finished their work, the master may chose among two policies. The first one consists in doing some other work, waiting for the sub-contractors to be done. The second solution, actually the one chosen by Feldman et al. [23] and called *helpful master*, consists in having the master help the sub-contractors, that is, the master sub-contracts work for its own sub-contractors. The reason for this choice is as follows. If we suppose that the master was computing a useful tree T, then it is important that the value of T be known quickly. Thus any subtree of tree T that was sub-contracted should also be evaluated quickly. Hence, it is in the master's own interest to help its sub-contractors, rather than search other trees that, from its own point of view, are less useful than the sub-contractors' trees. Speedups as high as 100 have been obtained on a 256 processor machines (see [22]). In [25], Feldmann *et al.* have shown a speedup of 344 on a 1024 transputer network interconnected as a grid and a speedup of 142 on a 256 processor transputer de Bruijn interconnected network. These numbers were obtained by their program *Zugzwang*[12] for actual chess games. Their implementation did not use the concept of the helpful master.

4.8. A PARALLEL MINIMAX ALGORITHM WITH LINEAR SPEEDUP

In 1988, Althöfer [5] proved that it is possible, to develop a parallel minimax algorithm which achieves linear speedup in the average case. With the assumption that all minimax trees are binary win-loss trees, he exhibited such a parallel minimax algorithm. The algorithm proposed is essentially based on Thm. 2.1. In fact, depending on the value of ξ_2, his algorithm, while exploring the tree shown in Fig. 4.a), will put half of the processors on the subtree rooted at node s_1 and half of the processors on subtree rooted at node s_2 if $p > \xi_2$, and put half of the processors on subtree rooted at node s_1 and half of the processors on subtree rooted at node s_3 if $p < \xi_2$. This distribution scheme will almost certainly give the the correct minimax value. As the distribution principle yields linear speedups by construction, one may extend the algorithm to obtain an algorithm with linear speedup on average.

Althöfer's result is based on the fact that the set E is topologically closed and that the root value may be predicted accurately with high probability. This simply means that, depending on whether p is larger or smaller than ξ_2, one needs, with high probability, to explore only an α or a β tree.

In his paper [4] Althöfer also inspects the case when $p = \xi_2$. Without going into too much details, one may say that this construction is based on partitioning

[12] In 1992 the program *Zugzwang* has been second at the world computer chess championship.

processors over the tree in such a way that they simultaneously search a $T_\alpha \cup T_\beta$ tree, the number of processors being allocated in such a way that their probable execution times over the respectives tress being equal. This yields an asymptotically linear speedup algorithm. Once again the result is strongly connected to the asymptotic optimality of the alpha-beta minimax algorithm.

Böhm and Speckenmeyer [12] also suggested an algorithm which uses the same basic ideas as Althöfer in [5]. Their algorithm is more general in the sense that it needs only to know the distribution of the leaf values and is independent of the branching of the tree explored. No theoretical speedup results have been derived for the algorithm introduced.

In 1989, Karp and Zhang [33] proved that it is possible to obtain linear speedup on every instance of a random uniform minimax tree if the number of processors is close to the height of the tree.

4.9. Some other algorithms

Broder *et al.* [13] have studied the theoretical complexity of parallel minimax algorithms. They used a computation model in which the evaluation of a leaf node has unit cost and any other operation is for free. In such a model the evaluation of a best first uniform tree $T(b, h)$ has an optimal complexity of $\Theta(b^{h/2})$. In their paper they show that fast parallel evaluation of minimax trees is not always achievable. Let $\mathsf{Seq}(T(b, h))$ be the sequential running time of the alpha-beta minimax algorithm on tree $T(h, b)$ and let $\mathsf{Par}(T(b, h), \mathcal{A}, p)$ be the running time of the parallel algorithm $\mathcal{A}$ on a p processor machine.

Theorem 4.2 *For every parallel algorithm $\mathcal{A}$ that uses $p = b^{(2-\delta)\cdot h/2}$ processors, δ being a constant in $]0, 1[$, there exists and input tree $T(b, h)$ such that $\mathsf{Seq}(T(b, h)) \leq b^{(2-\delta)\cdot h/2}$ and*

$$\mathsf{Par}(T(b, h), \mathcal{A}, p) \geq \Omega\left(\left(\frac{\delta \log b}{h \log\log b}\right)^{\lfloor h/2 \rfloor}\right)$$

This means that for every parallel algorithm there exist some minimax trees for which a good speedup is not possible.

Schaefer [50] argued that an efficient parallel algorithm would be a composition of the PVSPLIT and the SCOUT algorithms, combined with a table manager and an efficient controller responsible for the computation resources.

5. Open Problems and Conclusion

While reading through the literature on minimax trees and associated algorithms one finds a lot of very interesting and astonishing results. But there is still some space for further improvements. For example, it would be interesting to have the results of Sec. 2 generalized for minimax trees of arbitrary shapes. One may also invent some new statistical models of minimax trees that are more close to the game trees explored by chess programs, for example. Having some theorems about the average case efficiency of more elaborated algorithms, like GSEARCH, aspiration search or any other minimax procedure would be very interesting. Furthermore, the problem of traversing minimax trees has never been approached through randomized algorithms, either of type Las Vegas or of type Monte Carlo. One may ask the question to what extend it would be possible to find efficient randomized minimax algorithms. Another approach would be to develop (polynomial time) approximation algorithms for searching minimax trees.

Not only in the sequential and theoretical area of minimax trees can we find inspirations for possible further work. The area of parallel minimax algorithms is also fruitful. As always, new parallel minimax algorithms of theoretical, as well as practical use, will help to better understand the area. But, to us, it seems very important to analyze the efficiency of old and new parallel minimax algorithms when applied to various models and types of minimax trees.

Finally it would be very useful if some kind of benchmark suite would be available to compare the efficiency of various minimax algorithms. Such a library would ideally be composed of tree generating functions for theoretical trees (best-first trees, worst-first trees, rumtrees, ...) as well as for minimax trees found in games like chess or checkers, to name just a few.

It is evident that this survey of open problems is neither exhaustive nor representatif. It give some of our ideas and thoughts on the subject.

In this paper we have surveyed the theoretical aspects of minimax trees and their associated sequential and parallel algorithm. Various structural as well as statistical aspects have been described. In the first part of this paper we concentrated on the pure theoretical aspects of minimax trees. The second part of this survey described and analyzed the most important sequential minimax algorithms whereas the third part represented a summary of the techniques commonly used when developing parallel minimax algorithms.

Acknowledgements

We thank A. Plaat and W. Pijls for their comments on an earlier draft of this paper.

References

1. Selim G. Akl, David T. Barnard, and Ralph J. Doran. Searching game trees in parallel. In *Proceedings of the 3rd biennial Conference of the Canadian Society for Computation Studies of Intelligence*, pages 224–231, November 1979.
2. Selim G. Akl, David T. Barnard, and Ralph J. Doran. Design, analysis, and implementation of a parallel tree search algorithm. *IEEE Transactions on Pattern Analysis and Machine Intelligence*, PAMI-4(2):192–203, March 1982.
3. Kenneth Almquist, Neil McKenzie, and Kenneth Sloan. An inquiry into parallel algorithms for searching game trees. Technical report 88-12-03, University of Washington, Department of Computer Science, Seattle, WA, December 1988.
4. Ingo Althöfer. On the complexity of searching game trees and other recursion trees. *Journal of Algorithms*, 9:538–567, 1988.
5. Ingo Althöfer. A parallel game tree search algorithm with a linear speedup. *Journal of Algorithms*. Accepted in 1992.
6. Ingo Althöfer. An incremental negamax algorithm. *Artificial Intelligence*, 43:57–65, 1990.
7. Bruce W. Ballard. The *-minimax search procedure for trees containing chance nodes. *Artificial Intelligence*, 21:327–350, 1983.
8. Gérard M. Baudet. *The Design and Analysis of Algorithms for Asynchronous Multiprocessors*. PhD thesis, Carnegie Mellon University, Pittsburgh, PA, 1978.

9. Gérard M. Baudet. On the branching factor of the alpha-beta pruning algorithm. *Artificial Intelligence*, 10:173–199, 1978.
10. Hans Jack Berliner and Carl Eberling. Pattern knowledge and search: The SUPREME architecture. *Artificial Intelligence*, 38(2):161–198, 1989.
11. Hans Jack Berliner, Gordon Goetsch, Murray S. Campbell, and Carl Eberling. Measuring the performance potential of chess programs. *Artificial Intelligence*, 43(1):7–21, April 1990.
12. Max Böhm and Ewald Speckenmeyer. A dynamic processor tree for solving game trees in parallel. In *Proceedings of the SOR '89*, 1989.
13. Andrei Z. Broder, Anna R. Karlin, Prabhakar Raghavan, and Eli Upfal. On the parallel complexity of evaluating game-trees. Technical report RR RJ 7729, IBM Research Division, October 1990.
14. Giovanni Coray and Marc Gengler. A parallel best-first branch-and-bound algorithm and its axiomatization. *Parallel Algorithms and Applications*, 2:61–80, 1994.
15. Van-Dat Cung and Catherine Roucairol. Parallel minimax tree searching. RR 1549, INRIA, November 1991. In French.
16. Nevin M. Darwish. A quantitative analysis of the alpha-beta pruning algorithm. *Artificial Intelligence*, 21:405–433, 1983.
17. Claude G. Diderich. Evaluation des performances de l'algorithme SSS* avec phases de synchronisation sur une machine parallèle à mémoires distribuées. Technical report LITH-99, Swiss Federal Institute of Technology, Computer Science Theory Laboratory, Lausanne, Switzerland, July 1992. In french.
18. Claude G. Diderich. A bibliography on minimax trees. Technical report 98, Swiss Federal Institute of Technology, Computer Science Theory Laboratory, Lausanne, Switzerland, May 1994. Previous versions of this report have been published in the "Bulletin of the EATCS", No. 49, February 1993 and in the "ACM SIGACT News", Vol. 24, No. 4, December 1993.
19. Claude G. Diderich and Marc Gengler. An efficient algorithm for solving the token distribution problem on k-ary d-cube networks. In *Proceedings of the International Symposium on Parallel Architectures, Algorithms, and Networks (ISPAN '94)*, December 1994.
20. Claude G. Diderich and Marc Gengler. Another view of minimax trees. Personal notes, 1994 April.
21. Edward A. Feigenbaum and Julian Feldman (Eds.). *Computers and Thought*. McGraw Hill, New York, NY, 1963.
22. Rainer Feldmann, Burkhard Monien, Peter Mysliwietz, and Oliver Vornberger. Distributed game tree search. *ICCA Journal*, 12(2):65–73, 1989.
23. Rainer Feldmann, Burkhard Monien, Peter Mysliwietz, and Oliver Vornberger. Distributed game tree search. In L. N. Kanal V. Kumar, P. S. Gopalakrishnan, editor, *Proceedings of Parallel Algorithms for Machine Intelligence and Vision*, pages 66–101. Springer-Verlag, 1990.
24. Rainer Feldmann, Peter Mysliwietz, and Burkhard Monien. A fully distributed chess program. In *Proceedings of Advances in Computer Chess 6*, pages 1–27. Ellis Horwood, 1990. Editor: D. Beal.
25. Rainer Feldmann, Peter Mysliwietz, and Burkhard Monien. Studying overheads in massively parallel min/max-tree evaluation (extended abstract). In *Proceedings of the ACM Annual Symposium on Parallel Algorithms and Architectures (SPAA'94)*, 1994.
26. E. W. Felten and S. W. Otto. Chess on a hypercube. In G. Fox, editor, *Proceedings of the the Third Conference on Hypercube Concurrent Computers and Applications*, volume II-Applications, pages 1329–1341, Passadena, CA, 1988.
27. Raphael A. Finkel and John P. Fishburn. Parallelism in alpha-beta search. *Artificial Intelligence*, 19:89–106, 1982.
28. S. H. Fuller, J. G. Gaschnig, and J. J. Gillogly. An analysis of the alpha-beta pruning algorithm. Technical report, Carnegie-Mellon University, Department of Computer Science, Pittsburgh, July 1973.
29. Rattikorn Hewett and Krishnamurthy Ganesan. Consistent linear speedup in parallel alpha-beta search. In *Proceedings of the Computing and Information Conference (ICCI'92)*, pages 237–240. IEEE Computer Society Press, May 1992.
30. Feng-Hsiung Hsu, T. S. Anantharaman, Murray S. Campbell, and A. Nowatzyk. *Computers, Chess, and Cognition*, chapter 5 – Deep Thought, pages 55–78. Springer-Verlag, 1990.
31. Toshihide Ibaraki. Generalization of alpha-beta and SSS* search procedures. *Artificial Intelligence*, 29:73–117, 1986.

32. Hermann Kaindl, Reza Shams, and Helmut Horacek. Minimax search algorithms with and without aspiration windows. *IEEE Transactions on Pattern Analysis and Machine Intelligence*, PAMI-13(12):1225–1235, December 1991.
33. Richard M. Karp and Yanjun Zhang. On parallel evaluation of game trees. In *Proceedings of the ACM Annual Symposium on Parallel Algorithms and Architectures (SPAA'89)*, pages 409–420, New York, NY, 1989. ACM Press.
34. Donald E. Knuth and Ronald W. Moore. An analysis of alpha-beta pruning. *Artificial Intelligence*, 6(4):293–326, 1975.
35. Richard E. Korf. Iterative deepening: An optimal admissible tree search. *Artificial Intelligence*, 27:97–109, 1985.
36. David Levy and Monty Newborn. *How Computers Play Chess*. Computer Science Press, Oxford, England, 1991.
37. T. Anthony Marsland and Murray S. Campbell. Parallel search of strongly ordered game trees. *ACM Computing Survey*, 14(4):533–551, December 1982.
38. T. Anthony Marsland and Fred Popowich. Parallel game-tree search. *IEEE Transactions on Pattern Analysis and Machine Intelligence*, PAMI-7(4):442–452, July 1985.
39. T. Anthony Marsland, Alexander Reinefeld, and Jonathan Schaeffer. Low overhead alternatives to SSS*. *Artificial Intelligence*, 31:185–199, 1987.
40. David Allen McAllester. Conspiracy numbers for min-max searching. *Artificial Intelligence*, 35:287–310, 1988.
41. Gerard P. Michon. *Recursive Random Games: A Probabilistic Model for Perfect Information Games*. PhD thesis, University of California at Los Angeles, Computer Science Department, Los Angeles, CA, 1983.
42. L. G. Mitten. Branch and bound methods: General formulation and properties. *Operations Research*, 18:24–34, 1970.
43. Judea Pearl. Asymptotical properties of minimax trees and game searching procedures. *Artificial Intelligence*, 14(2):113–138, 1980.
44. Judea Pearl. *Heuristics – Intelligent Search Strategies for Computer Problem Solving*. Addison-Wesley Publishing Co., Reading, MA, 1984.
45. Wim Pijls and Arie de Bruin. Another view of the SSS* algorithm. In *Proceedings of the International Symposium (SIGAL'90)*, Tokyo, Japan, August 1990.
46. Wim Pijls and Arie de Bruin. Searching informed game trees. In *Proceedings of Algorithms and Computation (ISAAC '92), LNCS 650*, pages 332–341, 1992.
47. Ronald L. Rivest. Game tree searching by min/max approximation. *Artificial Intelligence*, 34(1):77–96, 1987.
48. Igor Roizen and Judea Pearl. A minimax algorithm better than alpha-beta? Yes and No. *Artificial Intelligence*, 21:199–230, 1983.
49. Jonathan Schaeffer. The history heuristic. *ICCA Journal*, 6(3):16–19, 1983.
50. Jonathan Schaeffer. Distributed game-tree searching. *Journal of Parallel and Distributed Computing*, 6:90–114, 1989.
51. Jonathan Schaeffer. Conspiracy numbers. *Artificial Intelligence*, 43:67–84, 1990.
52. James H. Slagle and John K. Dixon. Experiments with some programs that search game trees. *Journal of the ACM*, 16(2):189–207, April 1969.
53. Igor R. Steinberg and Marvin Solomon. Searching game trees in parallel. In *Proceedings of the IEEE International Conference on Parallel Processing*, pages III–9 – III–17, 1990.
54. G. C. Stockman. *A Problem-Reduction Approach to the Linguistic Analysis of Waveforms*. PhD thesis, University of Maryland, May 1977. Published as computer science technical report TR-538.
55. G. C. Stockman. A minimax algorithm better than alpha-beta? *Artificial Intelligence*, 12(2):179–196, 1979.

AN ITERATIVE METHOD FOR THE MINIMAX PROBLEM *

LIQUN QI
School of Mathematics, University of New South Wales, Sydney 2052, Australia.

and

WENYU SUN
Department of Mathematics, Nanjing University, Nanjing 210008, China.

Abstract. In this paper an iterative method for the minimax problem is proposed. The idea is that we present a sequence of the extended linear-quadratic programming (ELQP) problems as subproblems of the original minimax problem and solve the ELQP problems iteratively. Since the ELQP problem can be solved directly or by using methods for linear variational inequality or linear complementarity problem, an iterative method for the minimax problem is obtained. The locally linear and superlinear convergence of the algorithm is established.

1. Introduction

In this paper we consider the following minimax problem

$$\min_{x\in X}\max_{y\in Y} L(x,y) \tag{1}$$

where $L(x, y)$ is a saddle function, i.e., L is a convex-concave function from $X \times Y$ to $[-\infty, +\infty]$, i.e., $L(\cdot, y)$ is a convex function on X for each $y \in Y$ and $L(x, \cdot)$ a concave function of $y \in Y$ for each $x \in X$, where X and Y are closed nonempty convex set in R^n and R^m, respectively,

$$X = \{x : c_i(x) \leq 0, \quad i = 1, \cdots, m_1\}, \tag{2}$$

$$Y = \{y : h_i(y) \leq 0, \quad i = 1, \cdots, m_2\}. \tag{3}$$

where $c_i : R^n \to R, h_i : R^m \to R$ are convex functions.

Many minimax problems often arise in engineering design (see Polak(1987) [14], Warren etc. (1967) [25]), computer-aided-design (see Polak etc. (1992) [15]) , circuit design (see Sussman-Fort (1989) [24]) and optimal control (see Rockafellar (1987)(1990) [19] [20]).

Typically, the standard convex programming can also be characterized in terms of minimax problem. For example, we consider

$$\begin{aligned} \min\ & f(x) \\ \text{s.t.}\ & c_i(x) \leq 0, i = 1, \cdots, l \end{aligned} \tag{4}$$

* This work was supported by the Australian Research Council.

D.-Z. Du and P. M. Pardalos (eds.), Minimax and Applications, 55–67.

where f and c_i, $(i = 1, \cdots, l)$ are convex functions. The Lagrangian function L of the above convex programming is given by

$$L(x, \lambda) = f(x) + \sum \lambda_i c_i(x). \tag{5}$$

Therefore $L(x, \lambda)$ in (5) is a convex-concave function on $\tilde{X} \times \tilde{Y}$, where

$$\tilde{X} = \{x : c_i(x) \leq 0, \quad i = 1, \cdots, l\}$$

$$\tilde{Y} = \{\lambda : \lambda_i \geq 0, \quad i = 1, \cdots, l\}.$$

A point (x^*, λ^*) is a Kuhn-Tucker vector of (4) if and only if (x^*, λ^*) is a saddle-point of Lagrangian function $L(x, \lambda)$ in (5). That is to find (x^*, λ^*) such that

$$L(x^*, \lambda) \leq L(x^*, \lambda^*) \leq L(x, \lambda^*), \quad \forall x, \lambda. \tag{6}$$

This implies (x^*, λ^*) is a Kuhn-Tucker vector of (4) if and only if

$$\begin{aligned} L(x^*, \lambda^*) &= \inf_x L(x, \lambda^*) = \sup_\lambda \inf_x L(x, \lambda) \\ &= \inf_x \sup_\lambda L(x, \lambda) = \sup_\lambda L(x^*, \lambda) = L(x^*, \lambda^*). \end{aligned} \tag{7}$$

There are several existing methods for the minimax problem (1). Oettli [11] introduced a method of feasible directions for the minimax problem. Murray and Overton [10] transformed the minimax problem into a nonlinear constrained optimization and used a projected Lagrangian method to solve the equivalent problem. Hald and Madsen [5] gave a combined LP and quasi-Newton methods. Polak et al. [15] presented a barrier function algorithm by means of defining a barrier function.

In this paper we transform the minimax problem into a sequence of Extended Linear-Quadratic Programming (ELQP) problems which are introduced by Rockafellar and Wets [21] in stochastic programming and which can be effectively solved by some existing algorithms. In addition, the extended linear-quadratic programming (ELQP) problem is equivalent to linear variational inequality problem. The problem of solving the linear variational inequality can be turned into a linear complementarity problem when X and Y are polyhedral convex sets. Therefore, we can use the existing effective algorithms for ELQP or linear complementarity or linear variational inequality to get solutions of ELQP problems, which converge to the solutions of the original minimax problem (1).

The organization of the remainder of this paper is as follows. In the next section we establish the ELQP problem as a subproblem of the minimax problem and show it is equivalent to the linear variational inequality problem. The locally linear and superlinear convergence of the sequence $\{x_k, y_k\}$ generated by means of iteratively solving the ELQP problem to the desired solution of (1) is established in section 3.

2. An ELQP Problem as a Subproblem

Consider the minimax problem (1). From Lemma 36.1 of [18], we have

$$\max_{y \in Y} \min_{x \in X} L(x, y) \leq \min_{x \in X} \max_{y \in Y} L(x, y). \tag{8}$$

Therefore, a point $(\bar{x}, \bar{y})$ is a saddle point of L if and only if the minimum of

$$\min_{x \in X}\{\max_{y \in Y} L(x, y)\} \tag{9}$$

is attained at $\bar{x}$, and the maximum of

$$\max_{y \in Y}\{\min_{x \in X} L(x, y)\} \tag{10}$$

is attained at $\bar{y}$, and these two extrema are equal, i.e.,

$$\max_{y \in Y}\{\min_{x \in X} L(x, y)\} = \min_{x \in X}\{\max_{y \in Y} L(x, y)\} \tag{11}$$

or

$$L(\bar{x}, \bar{y}) = \min_{x \in X} L(x, \bar{y}) = \max_{y \in Y} L(\bar{x}, y). \tag{12}$$

(9) and (10) also can be written as

$$(\mathcal{P}) \qquad \min_{x \in X} L(x, \bar{y}), \tag{13}$$

$$(\mathcal{D}) \qquad \max_{y \in Y} L(\bar{x}, y). \tag{14}$$

The saddle point condition

$$L(\bar{x}, y) \leq L(\bar{x}, \bar{y}) \leq L(x, \bar{y}), \qquad \forall x \in X, y \in Y \tag{15}$$

holds if and only if the convex function $L(\cdot, \bar{y})$ achieves its minimum at $\bar{x}$, i.e.,

$$0 \in \partial_x L(\bar{x}, \bar{y}) + N_X(\bar{x}), \tag{16}$$

and the concave function $L(\bar{x}, \cdot)$ achieves its maximum at $\bar{y}$, i.e.,

$$0 \in \partial_y L(\bar{x}, \bar{y}) + N_Y(\bar{y}), \tag{17}$$

where $\partial_x L(\bar{x}, \bar{y})$ and $\partial_y L(\bar{x}, \bar{y})$ denote the convex and concave subdifferentials of L at $(\bar{x}, \bar{y})$ about x and y respectively, and $N_X(\bar{x})$ and $N_Y(\bar{y})$ denote normal cones of X and Y at $\bar{x}$ and $\bar{y}$ respectively. Therefore the saddle point condition (15) is equivalent to

$$(0, 0) \in \partial L(\bar{x}, \bar{y}) + N_{X \times Y}(\bar{x}, \bar{y}), \tag{18}$$

where $\partial L(\bar{x}, \bar{y})$ denotes the saddle function subdifferential of L at $(\bar{x}, \bar{y})$ and $N_{X \times Y}(\bar{x}, \bar{y})$ denotes the normal cone of $X \times Y$ at $(\bar{x}, \bar{y})$. See [18]. If $X = R^n, Y = R^m$ and L is a continuously differentiable convex-concave function on $R^n \times R^m$, then (18) can be reduced to

$$\nabla L(\bar{x}, \bar{y}) = \begin{pmatrix} \nabla_x L(\bar{x}, \bar{y}) \\ \nabla_y L(\bar{x}, \bar{y}) \end{pmatrix} = 0 \tag{19}$$

which can be solved by Newton's method. Then the point $(\bar{x}, \bar{y})$ is the saddle point of the unconstrained minimax problem

$$\min_{x \in R^n} \max_{y \in R^m} L(x, y). \tag{20}$$

For the constrained minimax problem (1) with constraints (2) and (3), (13) and (14) are its equivalent problem pair. We introduce the following Lagrangian of (13) and (14)

$$\Phi(x,\mu) = L(x,\bar{y}) + \mu^T c(x),$$
$$\Phi^{(D)}(y,\lambda) = L(\bar{x},y) - \lambda^T h(y),$$

where μ and λ are multipliers. We also introduce a Lagrangian function

$$\hat{L}(x,\mu,y,\lambda) = L(x,y) + \mu^T c(x) - \lambda^T h(y),$$

which is associated to the minimax problem (1). So, the Kuhn-Tucker conditions associated to $(\mathcal{P})$ and $(\mathcal{D})$ are

$$\begin{cases} \nabla_x L(x,\bar{y}) + \nabla c(x)\mu = 0, \\ \mu \geq 0, \quad c(x) \leq 0, \\ \mu_i c_i(x) = 0, \quad i = 1,\cdots,m_1, \\ \nabla_y L(\bar{x},y) - \nabla h(y)\lambda = 0, \\ \lambda \geq 0, \quad h(y) \leq 0, \\ \lambda_i h_i(y) = 0, \quad i = 1,\cdots,m_2, \end{cases} \tag{21}$$

where $c : R^n \to R^{m_1}$ and $h : R^m \to R^{m_2}$ are the vector-valued functions with components $(c_i(x))_{i=1}^{m_1}$ and $(h_i(y))_{i=1}^{m_2}$, respectively. Therefore $\bar{z} = (\bar{x},\bar{\mu},\bar{y},\bar{\lambda})$ should satisfy the system of nonlinear equations

$$H(\bar{z}) = \begin{pmatrix} \nabla_x L(\bar{x},\bar{y}) + \nabla c(\bar{x})\bar{\mu} \\ \bar{\mu}_1 c_1(\bar{x}) \\ \vdots \\ \bar{\mu}_{m_1} c_{m_1}(\bar{x}) \\ \nabla_y L(\bar{x},\bar{y}) - \nabla h(\bar{y})\bar{\lambda} \\ \bar{\lambda}_1 h_1(\bar{y}) \\ \vdots \\ \bar{\lambda}_{m_2} h_{m_2}(\bar{y}) \end{pmatrix} = 0, \tag{22}$$

where $\bar{z} \in R^{n+m_1+m+m_2}$ and $H : R^{n+m_1+m+m_2} \to R^{n+m_1+m+m_2}$, which constitutes the equality part of the first order Kuhn-Tucker conditions. Then $\bar{z}$ can be solved by Newton's method from $H(z) = 0$.

In order to find a saddle point of the minimax problem (1), we construct a sequence of $(n+m)$-vector $\{(x_k, y_k)\}$ which are estimates of a saddle-point $(\bar{x},\bar{y})$ of the problem (1). This can be done by solving a sequence of extended linear-quadratic programming (ELQP) subproblems which can be effectively solved by the existing algorithms (for example, see [17] [19] [20] [27]).

Now we assume L is twice continuously differentiable, $c_i (i = 1,\cdots,m_1)$ and $h_i (i = 1,\cdots,m_2)$ are continuously differentiable. We consider an extended linear-quadratic programming subproblem

$$\min_{x\in X_k} \max_{y\in Y_k} L_k(x,y) \tag{23}$$

for (1) with

$$L_k(x,y) = \nabla_x L(x_k,y_k)^T(x-x_k) + \nabla_y L(x_k,y_k)^T(y-y_k) + \frac{1}{2}(x-x_k)^T P_k(x-x_k) - \frac{1}{2}(y-y_k)^T Q_k(y-y_k) - (y-y_k)^T R_k(x-x_k), \tag{24}$$

$$X_k = \{x| \quad c_i(x_k) + \nabla c_i(x_k)^T(x-x_k) \le 0, \quad i=1,\cdots,m_1\}, \tag{25}$$

$$Y_k = \{y| \quad h_i(y_k) + \nabla h_i(y_k)^T(y-y_k) \le 0, \quad i=1,\cdots,m_2\}, \tag{26}$$

where $P_k \in R^{n\times n}$, $Q_k \in R^{m\times m}$ and $R_k \in R^{m\times n}$ are approximations of the corresponding second order partial derivative matrices of the Lagrangian function $\hat{L}$ about x and y. We assume that P_k and Q_k are positive definite matrices.

Let (x_k^*, y_k^*) be a solution of the ELQP subproblem (23), where the context is clear, we briefly denote (x_k^*, y_k^*) by (x^*, y^*) for convenience. There is an associated pair of the primal-dual problems. The primal problem is

$$(\mathcal{P}_k') \quad \begin{array}{ll} \min_x & \nabla_x L(x_k,y_k)^T(x-x_k) + \frac{1}{2}(x-x_k)^T P_k(x-x_k) \\ & -(y^*-y_k)^T R_k(x-x_k) \\ \text{s.t.} & c_i(x_k) + \nabla c_i(x_k)^T(x-x_k) \le 0, \quad i=1,\cdots,m_1 \end{array} \tag{27}$$

where y^* is a vector such that $\max_y\{(y-y_k)^T[\nabla_y L(x_k,y_k) - R_k(x-x_k)] - \frac{1}{2}(y-y_k)^T Q_k(y-y_k)\}$ achives its maximum at y^*. The dual problem is

$$(\mathcal{D}_k') \quad \begin{array}{ll} \max_y & \nabla_y L(x_k,y_k)^T(y-y_k) - \frac{1}{2}(y-y_k)^T Q_k(y-y_k) \\ & -(y-y_k)^T R_k(x^*-x_k) \\ \text{s.t.} & h_i(y_k) + \nabla h_i(y_k)^T(y-y_k) \le 0, \quad i=1,\cdots,m_2 \end{array} \tag{28}$$

where x^* is a vector such that $\min_x\{(x-x_k)^T[\nabla_x L(x_k,y_k) - R_k^T(y-y_k)] + \frac{1}{2}(x-x_k)^T P_k(x-x_k)\}$ achieves its minimum at x^*.

In the following we give a simple but useful proposition.

Proposition 1 *Let $X = R^n, Y = R^m$. If P_k, Q_k and R_k are the corresponding second order partial derivative matrices of L at (x_k, y_k), then the point (x_k^*, y_k^*) is a saddle point of (23) if and only if (x_k^*, y_k^*) satisfies Newton's iteration of solving (19).*

Proof. From (24), if (x_k^*, y_k^*) is a saddle point of (23), then x_k^* is an optimal solution of the primal problem $(\mathcal{P}_k')$ and y_k^* is an optimal solution of the dual problem $(\mathcal{D}_k')$, and $\{(x_k^*, y_k^*)\}$ is a sequence satisfying Newton's iteration of solving (19). And vice versa. □

Remark 1. Proposition 1 indicates that, for an unconstrained minimax problem, the saddle point sequence of the ELQP subproblems is just a sequence of the Newton iteration for solving the saddle point condition (19). Now the problem we face is whether, for the constrained minimax problem (1), the saddle point sequence of ELQP subproblems (23) is a sequence of the Newton iteration for solving Kuhn-Tucker conditions (22). The positive answer for this problem will be given in the next section.

Remark 2. If let $\omega = (x, y)^T$,

$$\tilde{G}(\omega_k) = \begin{bmatrix} P_k & -R_k^T \\ -R_k & -Q_k \end{bmatrix} = \begin{bmatrix} \tilde{G}_{x,k} \\ \tilde{G}_{y,k} \end{bmatrix}, \quad \text{(say)} \tag{29}$$

which is as the second order partial derivative matrix of the Lagrangian function $\hat{L}$ about x and y at x_k and y_k or its approximation, then the function (24) in the ELQP problem becomes

$$L_k(\omega) = \nabla L_k(\omega_k)^T(\omega - \omega_k) + \frac{1}{2}(\omega - \omega_k)^T \tilde{G}(\omega_k)(\omega - \omega_k). \tag{30}$$

This implies that the rationale behind employing the ELQP subproblems in the minimax problem is that ELQP subproblem is just similar to the quadratic programming subproblems in SQP methods for solving constrained optimization. In terms of primal-dual problems, finding Kuhn-Tucker points of primal-dual problems of ELQP problem is just equivalent to finding Kuhn-Tucker points for quadratic programming subproblems of primal-dual problems of the original minimax problem.

By Theorem 2.3 of [20] we have the following fact. The saddle point optimality can be written equivalently as the following linear variational inequality

$$-\nabla_x L(x^*, y^*)(x - x^*) \leq 0, \quad \nabla_y L(x^*, y^*)(y - y^*) \leq 0. \tag{31}$$

Since the problem of solving the linear variational inequality is also equivalent to solving a linear complementarity problem, so we can use existing algorithms for solving the ELQP problem or the linear variational inequality or the linear complementarity problem to solve the minimax problem iteratively. Therefore our idea leads to an iterative method for solving the minimax problem (1).

3. Local and Superlinear Convergence

In this section we shall establish local and superlinear convergence of our iterative algorithm.

First, we state the algorithm as follows.

Algorithm 1 Step 1. *Start with an estimate (x_0, y_0) of a saddle point of Problem (1). Set $k = 0$.*

Step 2. *Having (x_k, y_k), find a saddle point (x_{k+1}, y_{k+1}) of the extended linear-quadratic programming subproblem (23). If there are more than one such saddle points , choose one which is the closest to (x_k, y_k).*

Step 3. *If (x_{k+1}, y_{k+1}) satisfies a prescribed convergence criterion, for example, $\frac{1}{2}\|H(z_{k+1})\|^2 \leq \epsilon$, stop. Otherwise set k:=k+1 and go to Step 2.*

The sequential ELQP method is essentially Newton's method for finding the saddle-point of minimax problem (1) by solving the nonlinear equations (22).

respectively. Note that $W_k = \nabla_z F(x_k, y_k; z)$ when $G_k = P_k, G_k^{(D)} = -Q_k$. Also let

$$W_G = \begin{bmatrix} G_k & \nabla c_1(\bar{x}) & \cdots & \nabla c_{m_1}(\bar{x}) & 0 & \cdots & \cdots & 0 \\ \bar{\mu}_1 \nabla c_1(\bar{x})^T & c_1(\bar{x}) & & & \vdots & & & \vdots \\ \vdots & & \ddots & & \vdots & & & \vdots \\ \bar{\mu}_{m_1} \nabla c_{m_1}(\bar{x})^T & & & c_{m_1}(\bar{x}) & 0 & \cdots & \cdots & 0 \\ 0 & \cdots & \cdots & 0 & G_k^{(D)} & \nabla h_1(\bar{y}) & \cdots & \nabla h_{m_2}(\bar{y}) \\ \vdots & & & \vdots & \bar{\lambda}_1 \nabla h_1(\bar{y})^T & h_1(\bar{y}) & & \\ \vdots & & & \vdots & \vdots & & \ddots & \\ 0 & \cdots & \cdots & 0 & \bar{\lambda}_{m_2} \nabla h_{m_2}(\bar{y})^T & & & h_{m_2}(\bar{y}) \end{bmatrix}. \quad (37)$$

By the perturbation theorem [12] we know that when $\|z_k - \bar{z}\| \leq \delta$, $\nabla_z H(z_k)$ is invertible and $\|\nabla_z H(z_k)^{-1}\| \leq \beta$. Obviously, if $\|G_k - \nabla_x^2 \Phi(x_k, \mu_k)\| \leq \frac{1}{6\beta}$ and $\|G_k^{(D)} - \nabla_y^2 \Phi^{(D)}(y_k, \lambda_k)\| \leq \frac{1}{6\beta}$, then $\|W_k - \nabla_z H(z_k)\| \leq \frac{1}{3\beta}$, W_k is also invertible and $\|W_k^{-1}\| \leq \frac{3}{2}\beta$.

In the following, we establish the local and superlinear convergence of Algorithm 1 for the minimax problem (1). In the following discussion, $G_k = P_k, G_k^{(D)} = -Q_k$.

Theorem 1 *Suppose that $(\bar{x}, \bar{y}) \in R^{n \times m}$ is a saddle point of (1) and satisfies the Jacobian uniqueness condition of $(\mathcal{P})$ and $(\mathcal{D})$. Let $L \in C^2(\bar{x}, \bar{y}), c_i, (i = 1, \cdots, m_1) \in C^2(\bar{x}), h_i, (i = 1, \cdots, m_2) \in C^2(\bar{y})$. Then If for any $r \in (0,1)$, there exist $\epsilon(r) > 0$ and $\delta(r) > 0$ such that if $\|z_k - \bar{z}\| \leq \epsilon(r), \|G_k - \nabla_x^2 \Phi(\bar{x}, \bar{\mu})\| \leq \delta(r)$, and $\|G_k^{(D)} - \nabla_x^2 \Phi^{(D)}(\bar{y}, \bar{\lambda})\| \leq \delta(r)$, then the saddle point sequence $\{(x_k^*, y_k^*)\}$ of the ELQP subproblems converges to $(\bar{x}, \bar{y})$ locally and linearly.*

Proof. Consider the function

$$T(x_k, y_k; z) = z - W_G^{-1} F(x_k, y_k; z). \quad (38)$$

For simplicity we write henceforth ϵ and δ for $\epsilon(r)$ and $\delta(r)$, respectively. Let $\|W_G^{-1}\| \leq \tau$. We can choose ϵ such that for all $z \in N(\bar{z}, \epsilon)$ and all $z_k \in N(\bar{z}, \epsilon)$ we have

$$\|W_G - W_k\| \leq \frac{1}{2\tau}.$$

Then

$$\begin{aligned} \|\nabla_z T(x_k, y_k; z)\| &= \|I - W_G^{-1} \nabla_z F(x_k, y_k; z)\| \\ &\leq \|W_G^{-1}\| \|W_G - W_k\| \\ &< \frac{1}{2}. \end{aligned} \quad (39)$$

This implies $T(x_k, y_k; z)$ is a contraction in $N(\bar{z}, \epsilon)$. Therefore we can choose x_k, y_k close enough to $\bar{x}, \bar{y}$, respectively, such that $\|F(x_k, y_k; \bar{z})\| \leq \frac{\epsilon}{2\tau}$. Then

$$\|T(x_k, y_k; \bar{z}) - \bar{z}\| \leq \frac{1}{2}\epsilon. \quad (40)$$

Thus there exists a unique fixed point $\hat{z}$ of $T(x_k, y_k; z)$ in $N(\bar{z}, \epsilon)$ and

$$\|\hat{z} - \bar{z}\| < 2\tau \|F(x_k, y_k; \bar{z})\|. \tag{41}$$

Now let z^* be a Kuhn-Tucker point of $(\mathcal{P}_k')$ and $(\mathcal{Q}_k')$. Then

$$F(x_k, y_k; z^*) = 0. \tag{42}$$

Since the zero of $F(x_k, y_k; \cdot)$ in $N(\bar{z}, \epsilon)$ is unique,

$$\hat{z} = z^*. \tag{43}$$

This means that the unique fixed point of $T(x_k, y_k; z)$ in $N(\bar{z}, \epsilon)$ is just the Kuhn-Tucker point of $(\mathcal{P}_k')$ and $(\mathcal{Q}_k')$.

Furthermore,

$$\begin{aligned}
\|F(x_k, y_k; \bar{z})\| \leq\ & \|\nabla_x L(x_k, y_k) + P_k(\bar{x} - x_k) - R_k^T(y^* - y_k) + \nabla c(x_k)\bar{\mu}\| \\
& + \sum_{i=1}^{m_1} \|\bar{\mu}_i(c_i(x_k) - c_i(\bar{x}) + \nabla c_i(x_k)^T(\bar{x} - x_k))\| \\
& + \|\nabla_y L(x_k, y_k) - Q_k(y - y_k) - R_k(x^* - x_k) + \nabla h(y_k)\bar{\lambda}\| \\
& + \sum_{i=1}^{m_2} \|\bar{\lambda}_i(h_i(y_k) - h_i(\bar{y}) + \nabla h_i(y_k)(\bar{y} - y_k))\| \\
\leq\ & \|\nabla_x L(x_k, y_k) + \nabla c(x_k)\bar{\mu} - \nabla_x^2 \Phi(\bar{x}, \bar{\mu})(x_k - \bar{x})\| \\
& + \|P_k - \nabla_x^2 \Phi(\bar{x}, \bar{\mu})\| \|x_k - \bar{x}\| \\
& + \|R_k^T(y^* - y_k)\| \\
& + \sum_{i=1}^{m_1} \|\bar{\mu}_i(c_i(x_k) - c_i(\bar{x}) + \nabla c_i(x_k)^T(\bar{x} - x_k))\| \qquad (44) \\
& + \|\nabla_y L(x_k, y_k) + \nabla h(y_k)\bar{\lambda} - \nabla_y^2 \Phi^{(D)}(y_k - \bar{y})\| \\
& + \|Q_k(\bar{y} - y_k) - \nabla_y^2 \Phi^{(D)}(\bar{y}, \bar{\lambda})\| \|y_k - \bar{y}\| \\
& + \|R_k(x^* - x_k)\| \\
& + \sum_{i=1}^{m_2} \|\bar{\lambda}_i(h_i(y_k) - h_i(\bar{y}) + \nabla h_i(y_k)(\bar{y} - y_k))\|.
\end{aligned}$$

We choose $\delta > 0$ and $\epsilon > 0$ such that when $x_k \in N(\bar{x}, \epsilon)$ and $y_k \in N(\bar{y}, \epsilon)$, we have

$$8\delta\tau \leq r,$$

$$\|\nabla_x L(x_k, y_k) + \nabla c(x_k)\bar{\mu} - \nabla_x^2 \Phi(\bar{x}, \bar{\mu})(x_k - \bar{x})\| \leq \delta \|x_k - \bar{x}\|,$$

$$\sum_{i=1}^{m_1} \|\bar{\mu}_i(c_i(x_k) - c_i(\bar{x}) + \nabla c_i(\bar{x} - x_k))\| \leq \delta \|x_k - \bar{x}\|,$$

$$\|\nabla_y L(x_k, y_k) + \nabla h(y_k)\bar{\lambda} - \nabla_y^2 \Phi^{(D)}(\bar{y}, \bar{\lambda})(y_k - \bar{y})\| \leq \delta \|y_k - \bar{y}\|$$

and

$$\sum_{i=1} m_2\|\bar{\lambda}_i(h_i(y_k) - h_i(\bar{y} + \nabla h_i(y_k)(\bar{y} - y_k))\|.$$

Hence we have

$$\|F(x_k, y_k; \bar{z})\| \leq 4\delta(\|x_k - \bar{x}\| + \|y_k - \bar{y}\|). \tag{45}$$

Then by (41) and (45), we get

$$\begin{aligned} \|z^* - \bar{z}\| &\leq 8\delta\tau(\|x_k - \bar{x}\| + \|y_k - \bar{y}\|) \\ &\leq r(\|x_k - \bar{x}\| + \|y_k - \bar{y}\|) \\ &\leq r\|z_k - \bar{z}\|. \end{aligned} \tag{46}$$

This shows that the sequence $\{z_k\}$ possesses locally linear convergence. □

We next proceed to establish the superlinear convergence theorem.

Theorem 2 *Suppose that $(\bar{x}, \bar{y}) \in R^{n\times m}$ is a saddle point of (1) satisfying the Jacobian uniqueness condition of $(\mathcal{P})$ and $(\mathcal{D})$. Suppose that the assumptions of Theorem 1 hold. Suppose $\{z_k\}$ is a sequence of points generated from the Algorithm 1 with respect to sequences of matrices $\{G_k\}$ and $\{G_k^{(D)}\}$. If furthermore,*

$$\lim_{k\to\infty} \frac{\|(G_k - \nabla_x^2\Phi(\bar{x}, \bar{\mu}))(x_{k+1} - x_k)\|}{\|z_{k+1} - z_k\|} = 0 \tag{47}$$

and

$$\lim_{k\to\infty} \frac{\|(G_k^{(D)} - \nabla_y^2\Phi^{(D)}(\bar{y}, \bar{\lambda}))(y_{k+1} - y_k)\|}{\|z_{k+1} - z_k\|} = 0, \tag{48}$$

then the convergence is Q-superlinear.

Proof. From (47) and (48), we have

$$\begin{aligned} &\|\nabla_x\Phi(x_{k+1}, \mu_{k+1})\| \\ = \;&\|\nabla_x\Phi(x_{k+1}, \mu_{k+1}) - \nabla_x\Phi(x_k, \mu_{k+1}) - G_k(x_{k+1} - x_k)\| \\ \leq \;&\|\nabla_x\Phi(x_{k+1}, \mu_{k+1}) - \nabla_x\Phi(x_k, \mu_{k+1}) - \nabla_x^2\Phi(\bar{z}, \bar{\mu})(x_{k+1} - x_k)\| \\ &+\|G_k - \nabla_x^2\Phi(\bar{x}, \bar{\mu})(x_{k+1} - x_k)\|. \end{aligned}$$

Therefore,

$$\lim_{k\to\infty} \frac{\|\nabla_x\Phi(x_{k+1}, \mu_{k+1})\|}{\|z_{k+1} - z_k\|} = 0. \tag{49}$$

Similarly,

$$\lim_{k\to\infty} \frac{\|\nabla_y\Phi^{(D)}(y_{k+1}, \lambda_{k+1})\|}{\|z_{k+1} - z_k\|} = 0. \tag{50}$$

Also, from the expression of $H(z)$, we have

$$\begin{aligned}
\|H(z_{k+1})\| &\le \|\nabla_x \Phi(x_{k+1}, \mu_{k+1})\| + \sum_{i=1}^{m_1} |\mu_i c_i(x_{k+1})| \\
&\quad + \|\nabla_y \Phi^{(D)}(y_{k+1}, \lambda_{k+1})\| + \sum_{i=1}^{m_2} |\lambda_i h_i(y_{k+1})| \\
&\le \|\nabla_x \Phi(x_{k+1}, \mu_{k+1})\| + \sum_{i=1}^{m_1} |\mu_i (c_i(x_{k+1}) - c_i(x_k) - \nabla c_i(x_k)(x_{k+1} - x_k))| \\
&\quad + \|\nabla_y \Phi^{(D)}(y_{k+1}, \lambda_{k+1})\| + \sum_{i=1}^{m_2} |\lambda_i (h_i(y_{k+1}) - h_i(y_k) - \nabla h_i(y_k)(y_{k+1} - \\
&\le \|\nabla_x \Phi(x_{k+1}, \mu_{k+1})\| + o(\|x_{k+1} - x_k\|) \\
&\quad + \|\nabla_y \Phi^{(D)}(y_{k+1}, \lambda_{k+1})\| + o(\|y_{k+1} - y_k\|).
\end{aligned}$$

Then from (49), (50) and (51) we obtain

$$\lim_{k\to\infty} \frac{\|H(z_{k+1})\|}{\|z_{k+1} - z_k\|} = 0. \tag{52}$$

Since $\nabla_z H(\bar{z})$ is nonsingular , then there exist $\delta_1, \delta_2 > 0$ such that for sufficiently large k, we have

$$\delta_1 \|z_{k+1} - \bar{z}\| \le \|H(z_{k+1})\| = \|H(z_{k+1}) - H(\bar{z})\| \le \delta_2 \|z_{k+1} - \bar{z}\|. \tag{53}$$

Using (52), we obtain

$$\lim_{k\to\infty} \frac{\|z_{k+1} - \bar{z}\|}{\|z_{k+1} - z_k\|} = 0, \tag{54}$$

which implies

$$\lim_{k\to\infty} \frac{\|z_{k+1} - \bar{z}\|}{\|z_{k+1} - \bar{z}\| + \|z_k - \bar{z}\|} = 0$$

and in turn implies

$$\lim_{k\to\infty} \frac{\|z_{k+1} - \bar{z}\|}{\|z_k - \bar{z}\|} = 0. \tag{55}$$

Therefore the convergence is Q-superlinear for the original minimax problem (1). □

References

1. F.H.Clarke, *Optimization and Nonsmooth Analysis*, (Wiley, New York, 1983).
2. V.F.Demyanov and V.N.Molozemov, *Introduction to Minimax*, (Wiley, New York, 1974).
3. A.V.Fiacco and G.P.McCormick, *Nonlinear Programming: Sequential Unconstrained Minimization Techniques*, (Wiley, New York, 1968).
4. U.M.Garcia-Palomares and O.L.Mangasarian, Superlinear convergent quasi-Newton algorithms for nonlinear constrained optimization problems, *Mathematical Programming* 11 (1976) 1-13.

5. J.Hald and K.Madsen, Combined LP and quasi-Newton methods for minimax optimization, *Mathematical Programming* 20 (1981) 49-62.
6. S.P.Han, Superlinearly convergent variable metric algorithms for general nonlinear programming problems, *Mathematical Programming* 11 (1976) 263-282.
7. S.P.Han, A globally convergent method for nonlinear programming, *Journal of Optimization Theory and Applications* 22 (1977) 297-309.
8. F.A.Lootsma, A survey of methods for solving constrained minimization problems via unconstrained minimization, in F.A.Lootsma, ed., *Numerical Methods for Nonlinear Optimization*, (Academic Press, New York, 1972) 313-347.
9. G.P.McCormick, Penalty function versus nonpenalty function methods of constrained nonlinear programming problems, *Mathematical Programming* 1 (1971) 213-238.
10. W.Murray and M.L.Overton, A projected Lagrangian algorithm for nonlinear minimax optimization, *SIAM Journal on Scientific and Statistic Computing* 1 (1980) 345-370.
11. W.Oettli, The method of feasible directions for continuous minimax problems, in A.Prekopa ed., *Survey of Mathematical Programming* Vol.1 (North-Holland, Amsterdam, 1979), 505-512.
12. J.M.Ortega and W.C.Rheinboldt, *Iterative Solution of Nonlinear Equations in Several Variables*, (Academic Press, New York, 1970).
13. J.S.Pang, Newton's method for B-differentiable equations, *Mathematics of Operations Research* 15 (1990) 311-341.
14. E.Polak, On the mathematical foundations of nondifferentiable optimization, *SIAM Review* 29 (1987) 21-89.
15. E.Polak, J.E.Higgins and D.Q.Mayne, A barrier function method for minimax problems, *Mathematical Programming* 54 (1992) 155-176.
16. L.Qi, Superlinearly convergent approximate Newton method for LC^1 optimization problems, *Mathematical Programming* 64 (1994), 277-294.
17. L.Qi and R.S.Womersley, An SQP algorithm for extended linear-quadratic problems in stochastic programming, Applied Mathematics Preprint, AM92/23, University of New South Wales.
18. R.T.Rockafellar, *Convex Analysis*, (Princeton, New Jersey, 1970).
19. R.T.Rockafellar, Linear-quadratic programming and optimal control, *SIAM Journal on Control and Optimization*, 25 (1987) 781-814.
20. R.T.Rockafellar, Computational schemes for large-scale problems in extended linear-quadratic programming, *Mathematical Programming* 48 (1990) 447-474.
21. R.T.Rockafellar and R.J.-B.Wets, A dual solution procedure for quadratic stochastic programs with simple recourse, in: A.Reinoza eds., *Numerical Methods, Lecture Notes in Mathematics 1005* (Springer-Verlag, Berlin, 1983) 252-265.
22. J.Stoer, Principles of sequential quadratic programming methods for solving nonlinear programs, in K.Schittkowski eds., *Computational Mathematical Programming*, (1985) 165-205.
23. W.Sun and Y.Yuan, *Optimization Theory and Methods*, (Science Press, Beijing, 1994).
24. S.E.Sussman-Fort, Approximate direct-search minimax circuit optimization, *International Journal for Numerical Methods in Engineering*, 28 (1989) 359-368.
25. A.D.Warren, L.S.Lasdon and D.F.Suchman, Optimization in engineering design, *Proc. IEEE* 55 (1967) 1885-1897.
26. R.S.Womersley, Local properties of algorithms for minimizing nonsmooth composite functions, *Mathematical Programming* 32 (1985) 68-89.
27. C.Zhu and R.T.Rockafellar, Primal-dual projected gradient algorithms for extended linear quadratic programming, *SIAM J.Optimization* 3 (1993) 751-783.

A DUAL AND INTERIOR POINT APPROACH TO SOLVE CONVEX MIN-MAX PROBLEMS

JOS F. STURM and SHUZHONG ZHANG
Econometric Institute
Erasmus University Rotterdam

Abstract. In this paper we propose an interior point method for solving the dual form of min-max type problems. The dual variables are updated by means of a scaling supergradient method. The boundary of the dual feasible region is avoided by the use of a logarithmic barrier function. A major difference with other interior point methods is the nonsmoothness of the objective function.

1. Introduction

Consider the following problem

$$\text{(P)} \qquad \min_{x\in\mathcal{X}} \max_{1\le i\le m} f_i(x)$$

where we assume that the functions $f_i(x)$, $1 \le i \le m$, are real valued convex functions defined on a convex and compact subset $\mathcal{X}$ of $\Re^n$.

Clearly, we have

$$\min_{x\in\mathcal{X}} \max_{1\le i\le m} f_i(x) = \min_{x\in\mathcal{X}} \max_{y\in S} y^T f(x)$$

where S is m-dimensional unit simplex given by

$$S := \{y \in \Re^m : \sum_{i=1}^{m} y_i = 1 \text{ and } y_i \ge 0,\ 1 \le i \le m\}$$

and the m-dimensional vector function $f(x)$ is given by

$$f(x) := (f_1(x), f_2(x), \cdots, f_m(x))^T.$$

Since the function $y^T f(x)$ is convex in x for fixed $y \in S$, and is concave in y for fixed $x \in \mathcal{X}$, it follows that (see e.g. Sion [6])

$$\min_{x\in\mathcal{X}} \max_{y\in S} y^T f(x) = \max_{y\in S} \min_{x\in\mathcal{X}} y^T f(x). \qquad (1)$$

From now on we shall concentrate on the dual problem of (P) given by

$$\text{(D)} \qquad \max_{y\in S} h(y),$$

where the dual objective function is defined as

$$h(y) := \min_{x\in\mathcal{X}} y^T f(x).$$

D.-Z. Du and P. M. Pardalos (eds.), Minimax and Applications, 69–78.

Note that the domain of h is S. Clearly, $h(y)$ is a concave function.

In two recent papers by Barros, Frenk, Schaible and Zhang [1, 2], fast algorithms for solving generalized fractional programming were constructed on the basis of a similar duality relation. The dual problem (D) can be derived using the Lagrangian function. For a thorough discussion on the Lagrange duality theory for convex programming, we refer to the book of Hiriart-Urruty and Lemaréchal [5].

Observe that Problem (D) has a very simple constraint set. However, the function $h(y)$ is in general non-differentiable. Throughout this paper we shall use an oracle to get an optimal solution $\bar{x}$ of the following problem:

$$\min_{x \in \mathcal{X}} y^T f(x) \tag{2}$$

where $y \in S$. Using this oracle we not only know the function value $h(y) = y^T f(\bar{x})$, but also an element belonging to the supergradient set. More precisely,

$$f(\bar{x}) \in \partial h(y)$$

where $\partial h(y)$ denotes the supergradient set of h at point y.

The basic underlying idea is that we first introduce a logarithmic barrier for Problem (D), and then apply a scaling and projection supergradient method maximizing the barrier function. Due to lack of differentiability in $h(y)$, the convergence analysis differs in flavor from usual path-following algorithms. The advantage of our approach is that we do not require any knowledge on the functions f_i, $i = 1, 2, \cdots, m$, and the structure of the constraint set $\mathcal{X}$. Remark that for the cases where m is relatively large compared to the dimension n, and the constraint set $\mathcal{X}$ is simple, solving (2) is much easier than solving the original problem.

The notation we use is as follows. The superscript of a vector is used to denote the iteration number, e.g. in the k-th iteration we have $y^{(k)}$; the subscript will denote the coordinate, e.g. the i-th coordinate of $y^{(k)}$ is $y_i^{(k)}$; capitalization of a vector will denote the diagonal matrix taking the elements from the vector in the diagonal, e.g. $Y^{(k)} = \operatorname{diag}(y_1^{(k)}, \cdots, y_m^{(k)})$. We denote the all-one vector by e, the Euclidean norm (the L_2 norm) simply by $\|\bullet\|$ and the L_∞ norm by $\|\bullet\|_\infty$.

We organize the presentation in the following way. In Section 2, we will introduce the search direction and present the new algorithm. The convergence analysis of the algorithm is carried out in Section 3 and some remarks concluding the discussion are made in Section 4.

2. The Scaling Supergradient Method

We introduce now the logarithmic barrier function

$$h_\mu(y) := h(y) + \mu \sum_{i=1}^{m} \log y_i .$$

Observe that $h_\mu(y)$ is a strictly concave function, for which the supergradient set is given by

$$\partial h_\mu(y) = \partial h(y) + \mu Y^{-1} e. \tag{3}$$

The concept of logarithmic barrier was introduced by Frisch [4] to steer the iterates away from the boundary. The optimizer of the barrier function will be a nearly optimal solution to (D) if the multiple μ of the barrier term is small, as it is shown in the following lemma.

Lemma 1 *If $\bar{y} \in S$ is such that $h_\mu(\bar{y}) = \max_{y \in S} h_\mu(y)$ then*

$$h(\bar{y}) \geq \max_{y \in S} h(y) - m\mu.$$

PROOF.
From the concavity of h_μ, it follows that

$$0 \in \partial h_\mu(\bar{y}),$$

i.e., there exists $\eta \in \partial h(\bar{y})$ such that

$$\eta + \mu \bar{Y}^{-1} e = 0.$$

By the concavity of h, we have for $y^* \in \arg\max_{y \in S} h(y)$ that

$$\begin{aligned} \max_{y \in S} h(y) &\leq h(\bar{y}) + \eta^T (y^* - \bar{y}) \\ &= h(\bar{y}) - \mu e^T \bar{Y}^{-1} (y^* - \bar{y}) \\ &= h(\bar{y}) + \mu(m - e^T \bar{Y}^{-1} y^*) \\ &\leq h(\bar{y}) + m\mu \end{aligned}$$

where we used $\bar{y}, y^* \in S$.

□

In this paper we shall maximize h_μ over S for a prefixed parameter $\mu > 0$. We shall fix $0 < \mu < \epsilon/m$ if an ϵ-optimal solution is desired.

Assume that the current iterate $y^{(k)} \in \overset{o}{S}$, where $\overset{o}{S}$ denotes the relative interior of S. Calling Oracle (2) we obtain

$$x^{(k)} \in \arg\min_{x \in \mathcal{X}} (y^{(k)})^T f(x).$$

Let $g^{(k)} := f(x^{(k)}) + \mu (Y^{(k)})^{-1} e$. Hence, by (3) we know that

$$g^{(k)} \in \partial h_\mu(y^{(k)}).$$

As a search direction we propose a scaled supergradient direction, which coincides with the supergradient direction of the function $h_\mu(Y^{(k)} z)$ on the domain $\{z : (y^{(k)})^T z = 1\}$. The scaling transformation $z = (Y^{(k)})^{-1} y$ is based on the idea of

Dikin's affine scaling algorithm [3] for linear programming. Remark that this scaling maps the current iterate $y^{(k)}$ into the all-one vector e.

To simplify notations, we write

$$P_v := I_m - \frac{1}{\|v\|^2} vv^T$$

to denote the orthogonal projection matrix onto the kernel of a given vector $v \in R^m$.

The scaled supergradient direction we propse is $Y^{(k)}d^{(k)}$, where

$$d^{(k)} := \frac{1}{\left\| P_{y^{(k)}} Y^{(k)} g^{(k)} \right\|} P_{y^{(k)}} Y^{(k)} g^{(k)}.$$

Remark that

$$Y^{(k)}d^{(k)} = \arg \max_{e^T w = 0} \{ (g^{(k)})^T w : \left\| (Y^{(k)})^{-1} w \right\| \leq 1 \}.$$

It is easily seen that $y^{(k)} + t_k Y^{(k)} d^{(k)} \in \overset{o}{S}$ if $| t_k | < 1$. In this paper, we require that

$$0 < t_k \leq \alpha < 1 \text{ for } k = 0, 1, \cdots$$

along with the classical conditions of the supergradient step length (cf. Shor [7]), viz.

$$\lim_{k \to \infty} t_k = 0$$

$$\sum_{k=0}^{\infty} t_k = \infty.$$

For simplicity we let $\alpha := \frac{1}{2}$. As an example, one may choose $t_k = \frac{1}{k+2}$ for $k = 0, 1, \cdots$.

Our scaling supergradient algorithm generates the following sequence of dual variables belonging to $\overset{o}{S}$,

$$y^{(0)} = \frac{1}{m} e$$

and

$$y^{(k+1)} := y^{(k)} + t_k Y^{(k)} d^{(k)} \text{ for } k = 0, 1, 2, \cdots.$$

In the next section, it will be shown that

$$\limsup_{k \to \infty} h_\mu(y^{(k)}) = \max_{y \in S} h_\mu(y).$$

3. Convergence Analysis

In the previous section, we have already seen that the sequence $\{y^{(k)}\}$ is contained in the relative interior of S. We shall now prove that our barrier method avoids the boundary so well that the sequence is actually contained in a closed subset of $\overset{o}{S}$.

By definition,

$$\left\| P_{y^{(k)}} Y^{(k)} g^{(k)} \right\| d^{(k)} = P_{y^{(k)}} Y^{(k)} f(x^{(k)}) + \mu P_{y^{(k)}} e.$$

Using $\min_{y \in S} \|y\|^2 = \frac{1}{m}$, it follows that

$$P_{y^{(k)}} e = e - \frac{y^{(k)}}{\left\| y^{(k)} \right\|^2} \geq e - m y^{(k)}.$$

Since $\mathcal{X}$ is convex and compact, all the convex functions f_i, $1 \leq i \leq m$, are uniformly bounded on $\mathcal{X}$. Letting $f_\infty := \max_{x \in \mathcal{X}} \|f(x)\|_\infty$, we have

$$\left\| P_{y^{(k)}} Y^{(k)} f(x) \right\| \leq f_\infty \left\| y^{(k)} \right\| \leq f_\infty$$

so that

$$\left\| P_{y^{(k)}} Y^{(k)} g^{(k)} \right\| d^{(k)} \geq \mu e - (f_\infty + m\mu) y^{(k)}.$$

This implies that

$$y_i^{(k+1)} \geq y_i^{(k)} \text{ for } i \text{ with } y_i^{(k)} \leq \frac{\mu}{f_\infty + m\mu}.$$

Since $0 < t_k \leq \frac{1}{2}$, we have

$$y^{(k+1)} \geq \frac{1}{2} y^{(k)}$$

for any k. Because $y^{(0)} = \frac{1}{m} e$, it follows that

$$\inf_k \min_{1 \leq i \leq m} y_i^{(k)} \geq c_1, \tag{4}$$

where $c_1 := \frac{1}{2} \frac{\mu}{f_\infty + m\mu}$.

Now we use (4) and the fact that all the limit points form a closed set contained in $\overset{o}{S}$ to conclude that there is one limit point, say $\bar{y}$, which attains the maximum function value in $h_\mu(y)$ among all the limit points. Let y^* be the maximum point of $h_\mu(y)$ in S. We shall now concentrate on proving $h_\mu(\bar{y}) = h_\mu(y^*)$.

The proof is done by contradiction. Suppose from now on that

$$h_\mu(\bar{y}) < h_\mu(y^*). \tag{5}$$

Let the upper level set of $h_\mu(y)$ at $\bar{y}$ be

$$L := \{ y \in S : \ h_\mu(y) \geq h_\mu(\bar{y}) \}.$$

By this construction, there will be no other limit point in $\overset{o}{L}$.

Clearly, $y^* \in \overset{o}{L}$. Moreover, there exists a positive number θ such that

$$B(y^*;\theta) \cap S \subseteq \overset{o}{L} \tag{6}$$

where $B(y^*;\theta)$ denotes a ball with center y^* and radius θ.

Now we turn to consider an iterative point $y^{(k)}$. Let the upper level set at $y^{(k)}$ be

$$L_k := \{y \in S : \; h_\mu(y) \geq h_\mu(y^{(k)})\}.$$

Due to the concavity of h_μ, the projected supergradient direction $P_e g^{(k)}$ provides a normal direction in S of a supporting hyperplane for L_k at $y^{(k)}$.

Let $y \in L_k$. The distance from y to the hyperplane is given by

$$\frac{(g^{(k)})^T(y - y^{(k)})}{\left\|P_e g^{(k)}\right\|}. \tag{7}$$

Let $\hat{y} \in \overset{o}{S}$. Define

$$\delta^{(k)} := (d^{(k)})^T (Y^{(k)})^{-1} (\hat{y} - y^{(k)}). \tag{8}$$

Lemma 2 *Let $r > 0$. If $B(\hat{y};r) \cap S \subseteq L_k$ then there exists some constant c_2 such that*

$$\delta^{(k)} \geq r c_2 (h_\mu(y^*) - h_\mu(y^{(k)})).$$

PROOF.
Consider the following supporting hyperplane of L_k,

$$\{z : \; (g^{(k)})^T P_e (z - y^{(k)}) = 0\}.$$

The distance from $\hat{y}$ towards this supporting hyperplane is

$$(g^{(k)})^T(\hat{y} - y^{(k)}) / \left\|P_e g^{(k)}\right\|.$$

As $B(\hat{y};r) \cap S \subseteq L_k$ this implies

$$(g^{(k)})^T(\hat{y} - y^{(k)}) \geq r \left\|P_e g^{(k)}\right\|.$$

Therefore,

$$\begin{aligned} \delta^{(k)} &= (d^{(k)})^T (Y^{(k)})^{-1} (\hat{y} - y^{(k)}) \\ &= (g^{(k)})^T(\hat{y} - y^{(k)}) / \left\|P_{y^{(k)}} Y^{(k)} g^{(k)}\right\| \\ &\geq r \frac{\left\|P_e g^{(k)}\right\|}{\left\|P_{y^{(k)}} Y^{(k)} g^{(k)}\right\|}. \end{aligned} \tag{9}$$

Using the Cauchy-Schwartz inequality and the supergradient inequality, we have

$$\left\|P_e g^{(k)}\right\| \geq (g^{(k)})^T(y^* - y^{(k)}) / \left\|y^* - y^{(k)}\right\| \geq \frac{h(y^*) - h(y^{(k)})}{\left\|y^* - y^{(k)}\right\|}. \tag{10}$$

As $y^{(k)}$ and y^* both belong to the unit simplex, it follows $\left\|y^* - y^{(k)}\right\| \leq \sqrt{2}$. Moreover, there holds

$$\left\|P_{y^{(k)}} Y^{(k)} g^{(k)}\right\| \leq \left\|Y^{(k)} g^{(k)}\right\| \leq \left\|Y^{(k)} f(x^{(k)})\right\| + \mu \left\|e\right\| \leq f_\infty + \sqrt{m}\mu. \tag{11}$$

From (9)-(11) it follows that

$$\delta^{(k)} \geq r c_2 (h_\mu(y^*) - h_\mu(y^{(k)}))$$

or $c_2 = \frac{1}{\sqrt{2}(f_\infty + \sqrt{m}\mu)}$. □

Define

$$\rho^{(k)} := \left\|\hat{Y}^{-1}(\hat{y} - y^{(k)})\right\|. \tag{12}$$

We have the following relation:

Lemma 3 *There holds*

$$(\rho^{(k+1)})^2 \leq (\rho^{(k)})^2 - 2t_k[\delta^{(k)} - (\rho^{(k)})^2 - (1 + \rho^{(k)})^2 t_k/2].$$

PROOF.
Since $y^{(k+1)} = y^{(k)} + t_k Y^{(k)} d^{(k)}$ we have

$$\begin{aligned}(\rho^{(k+1)})^2 &= \left\|\hat{Y}^{-1}(\hat{y} - y^{(k)} - t_k Y^{(k)} d^{(k)})\right\|^2 \\ &= (\rho^{(k)})^2 - 2t_k \delta^{(k)} + 2t_k (\hat{y} - y^{(k)})^T \hat{Y}^{-1}(I - \hat{Y}^{-1} Y^{(k)}) d^{(k)} \\ &\quad + t_k^2 \left\|\hat{Y}^{-1} Y^{(k)} d^{(k)}\right\|^2 \end{aligned} \tag{13}$$

Notice that

$$\left\|\hat{Y}^{-1} y^{(k)} - e\right\|_\infty \leq \left\|\hat{Y}^{-1} y^{(k)} - e\right\| = \rho^{(k)}. \tag{14}$$

Therefore, using $\left\|d^{(k)}\right\| = 1$ it follows

$$\left\|\hat{Y}^{-1} Y^{(k)} d^{(k)}\right\| \leq \left\|\hat{Y}^{-1} y^{(k)}\right\|_\infty \left\|d^{(k)}\right\| \leq 1 + \rho^{(k)}. \tag{15}$$

Similarly, using $\left\|d^{(k)}\right\| = 1$ and the Cauchy-Schwartz inequality, we have

$$\begin{aligned}|\,(\hat{y} - y^{(k)})^T \hat{Y}^{-1}(I - \hat{Y}^{-1} Y^{(k)}) d^{(k)}\,| &\leq \left\|(I - \hat{Y}^{-1} Y^{(k)}) \hat{Y}^{-1}(\hat{y} - y^{(k)})\right\| \\ &\leq \left\|(I - \hat{Y}^{-1} Y^{(k)}) e\right\|_\infty \left\|\hat{Y}^{-1}(\hat{y} - y^{(k)})\right\| \\ &= \left\|\hat{Y}^{-1} y^{(k)} - e\right\|_\infty \rho^{(k)} \\ &\leq (\rho^{(k)})^2,\end{aligned}$$

where the last inequality follows from (14).

Substituting the above inequality and the inequality (15) into (13) yields the desired result. □

Define

$$\hat{y}_\lambda := (1-\lambda)\bar{y} + \lambda y^*,$$

where $0 < \lambda < 1$. Let $\hat{y}$ be $\hat{y}_\lambda$. By (6) there exists $\tilde{h} > h_\mu(\bar{y})$ such that the ball $B(y^*;\theta) \cap S$ will be contained in the upper level set

$$\{y \in S : h_\mu(y) \geq \tilde{h}\}.$$

Using the concavity of h_μ, this implies

$$B(\hat{y}_\lambda; \lambda\theta) \cap S \subset \{y \in S : h_\mu(y) \geq (1-\lambda)h_\mu(\bar{y}) + \lambda\tilde{h}\}.$$

Since

$$\limsup_{k\to\infty} h_\mu(y^{(k)}) = h_\mu(\bar{y}) < \tilde{h} \tag{16}$$

we obtain from Lemma 2 and (16) that for given $0 < \lambda < 1$ there must exist k_1 such that for all $k \geq k_1$,

$$\delta^{(k)} \geq c_2\lambda\theta(h_\mu(y^*) - \tilde{h}). \tag{17}$$

On the other hand, by (12) we have

$$\begin{aligned}\rho^{(k)} &= \left\|\hat{Y}^{-1}(\hat{y} - y^{(k)})\right\| \\ &\leq \left\|\hat{Y}^{-1}(\hat{y} - \bar{y})\right\| + \left\|\hat{Y}^{-1}(\bar{y} - y^{(k)})\right\| \\ &= \lambda\sqrt{\sum_{i=1}^m \left(\frac{\bar{y}_i - y_i^*}{(1-\lambda)\bar{y}_i + \lambda y_i^*}\right)^2} + \left\|\hat{Y}^{-1}(\bar{y} - y^{(k)})\right\|. \end{aligned} \tag{18}$$

As $\bar{y} \neq y^*$ is a limit point, there is an unbounded set $K(\lambda)$ of integers such that

$$\left\|\hat{Y}^{-1}(\bar{y} - y^{(k)})\right\| \leq \lambda\sqrt{\sum_{i=1}^m \left(\frac{\bar{y}_i - y_i^*}{(1-\lambda)\bar{y}_i + \lambda y_i^*}\right)^2} \tag{19}$$

for all $k \in K(\lambda)$.

Based on (17) and (18), there exists a sufficiently small constant $\lambda_0 > 0$ such that when $\lambda = \lambda_0$ and $k \in K(\lambda_0)$, then

$$(\rho^{(k)})^2 < \min\{\frac{1}{3}c_2\lambda_0\theta(h_\mu(y^*) - \tilde{h}), 1\}. \tag{20}$$

Let k_1 be chosen according to (17) for $\lambda = \lambda_0$.

Because $\lim_{k\to\infty} t_k = 0$, there is $k_2 \in K(\lambda_0)$ with $k_2 \geq k_1$, such that for all $k \geq k_2$ we have

$$2t_k < \frac{1}{3}c_2\lambda_0\theta(h_\mu(y^*) - \tilde{h}).$$

In particular, for $k \geq k_2$ and if (20) holds, then we have

$$\frac{(1+\rho^{(k)})^2}{2}t_k < 2t_k < \frac{1}{3}c_2\lambda_0\theta(h_\mu(y^*) - \tilde{h}). \tag{21}$$

Using (17), (20), (21) and applying Lemma 3, it follows that

$$\begin{aligned}(\rho^{(k+1)})^2 &\leq (\rho^{(k)})^2 - 2t_k(1 - \frac{1}{3} - \frac{1}{3})c_2\lambda_0\theta(h_\mu(y^*) - \tilde{h}) \\ &= (\rho^{(k)})^2 - \frac{2}{3}t_k c_2\lambda_0\theta(h_\mu(y^*) - \tilde{h}) \end{aligned} \tag{22}$$

for $k = k_2$. This implies that $\rho^{(k_2+1)} < \rho^{(k_2)}$ and so (20) and (21) hold for $k := k_2+1$, and consequently (22) also holds for $k := k_2 + 1$. Recursively applying (22) yields a contradiction since $\sum_{j=k_2}^{\infty} t_j = +\infty$. This shows that inequality (5) cannot be true, which, in turn, proves the desired convergence result. To summarize, we present the following main theorem of this paper.

Theorem 1 *There holds*

$$\limsup_{k\to\infty} h_\mu(y^{(k)}) = \max_{y\in S} h_\mu(y).$$

4. Concluding Remarks

We have presented in this article an interior point method for solving a dual form of min-max type problems. An important question left is how to recover the primal solutions using approximately optimal dual variables and an approximately optimal objective value. We regard this as a topic for future research.

In a forthcoming paper, the authors will investigate a path-following scheme, extending the current results. Finally, we remark that our convergence proof fails for $\mu = 0$, in which case the method becomes comparable to the affine scaling algorithm for linear programming. It remains an open question whether the convergence still holds in that case.

References

1. A.I. Barros, J.B.G. Frenk, S. Schaible and S. Zhang, A new algorithm for generalized fractional programming, *Technical Report TI94-23*, Tinbergen Institute Rotterdam, 1994.
2. A.I. Barros, J.B.G. Frenk, S. Schaible and S. Zhang, How duality can be used to solve generalized fractional programming problems, 1994, submitted for publication.
3. I.I. Dikin, Iterative solutions of problems of linear and quadratic programming, *Soviet Mathematics Doklady* 8 (1967) 674-675.
4. K.R. Frisch, The logarithmic potential method for convex programming, Institute of Economics, University of Oslo, Oslo, Norway, 1955.

5. J.-B. Hiriart-Urruty and C. Lemaréchal, Convex analysis and minimization algorithms (vol. 1), Springer-Verlag, Berlin, 1993.
6. M. Sion, On general minimax theorems, *Pacific Journal of Mathematics* 8 (1958) 171-176.
7. N.Z. Shor, Minimization methods for non-differentiable functions, Springer-Verlag, Berlin, 1985.

DETERMINING THE PERFORMANCE RATIO OF ALGORITHM MULTIFIT FOR SCHEDULING

FENG CAO
Department of Computer Science, University of Minnesota
Minneapolis, MN 55455, USA.

Abstract. Scheduling n independent tasks nonpreemptively on m identical processors with the aim of minimizing the makespan is well-known to be NP-complete. Coffman, Garey and Johnson [1] described an algorithm–MULTIFIT and proved that it satisfies a bound of 1.22. Friesen [2] showed an example in which the upper bound is no less than $\frac{13}{11}$. Yue, Keller and Yu proved an upper bound of 1.2. Yue gave a proof for the upper bound of $\frac{13}{11}$, but the proof missed some cases. In this paper, a complete and simple proof is presented.

1. Introduction

Scheduling n independent tasks nonpreemptively on m identical processors with the aim of minimizing the makespan is well-known to be NP-complete. In 1978, Coffman, Garey and Johnson [1] describe an algorithm, MULTIFIT, abbreviated to MF. We follow the definitions and notations in this paper. Let Γ denote a given set of tasks, $\Gamma = \{T_1, T_2, ..., T_n\}$, each task having a length $l(T_i)$ and $l(\Gamma) = \sum_{i=1}^{n} l(T_i)$. Let $M_1, M_2, ..., M_m$ be m identical processors. A schedule is considered as a partition $P = \{P_1, P_2, ..., P_m\}$ of Γ and the makespan of the schedule is given by

$$f(P) = max_{1 \leq i \leq m} l(P_i)$$

Let τ_m^* be the minimal makespan time for given Γ and m. Let A be an algorithm and $F_A[\Gamma, m]$ be the makespan using A for given Γ and m. Define

$$R_m(A) = sup\{F_A[\Gamma, m]/\tau_m^* \mid \Gamma\}$$

MULTIFIT is based on the FFD algorithm on bin-packing problem. Given Γ and a bound C, let $OPT[\Gamma, C]$ denote the minimal number of bins that are needed and τ_m^* denote $min\{C \mid OPT[\Gamma, C] \leq m\}$. Define

$$r_m = inf\{r \mid FFD[\Gamma, r\tau_m^*] \leq m, \forall \Gamma\}$$

It was shown in [1] that

$$R_m(MF(k)) \leq r_m + (1/2)^k$$

For any pair positive integers $p > q$, a p/q minimal counterexample is that there exists Γ and a number m of bins satisfying the following conditions:
(i) $FFD[\Gamma, p] > m \geq OPT[\Gamma, q]$
(ii) for all lists Γ' satisfying (i), $|\Gamma'| \geq |\Gamma|$

D.-Z. Du and P. M. Pardalos (eds.), Minimax and Applications, 79–96.

(iii) $r_{m'} \leq p/q$, for m'=1, 2,..., $m-1$.

In this paper, we show that if $p = 120 - 3\delta$ and $q = 100$, there exists no p/q minimal counterexample with $\delta < \frac{20}{33}$, i.e. $r_m \leq \frac{13}{11}$.

We suppose (Γ, m) is a $(120-3\delta)/100$ minimal counterexample with $\delta < \frac{20}{33}$. $\Gamma = \{T_1, T_2, ..., T_n\}$ and $T_i \geq T_{i+1}$. We know $T_n = 20 + \Delta - 3\delta$ for some $\Delta > 0$. We discuss the cases of $\Delta > 5\delta$, $\frac{15}{4}\delta \leq \Delta \leq 5\delta$, $2.5\delta \leq \Delta < \frac{15}{4}\delta$ and $\Delta < 2.5\delta$. We provide a contradiction with $\delta < \frac{20}{33}$ for each case.

A bin B_i is called a $k-bin$ if there are k items packed into this bin before the first item in B_{i+1} is assigned. If there are more k items assigned to a $k-bin$, this bin is called a fallback $k-bin$; otherwise, a regular $k-bin$. Let $x_2 + \delta_2$ be the number of regular $2-bins$ in FFD. $\delta_2 = 1$ if the smaller item in the last regular $2-bin$ is less than $\frac{1}{2}(100-\Delta)$; otherwise $\delta_2 = 0$. Let y_2, t, s and v be the number of fallback $2-bins$, regular $3-bins$, $4-item$ bins and $5-item$ bins in FFD respectively. Let $s+w$ be the number of $4-item$ bins in the OPT packing.

We can assume that all items in the $regular-i$ bins except the last $(i-1)$ items are of the same size. Otherwise, we can change the size of the items in the $regular-i$ bins except the last $(i-1)$ items in Γ to that of the ith smallest item and obtain a new list Γ'. It is easy to see that the number of FFD bins remains the same for Γ'. Then the number of OPT bins for Γ' must remain the same. We consider Γ' instead of Γ.

Let l denote the size of the related item (sometimes we use the symbol of items as its size). We give each item the weight the same as [4], we will make some modification on some types and give weight of $Y_1 - type$ items later.

If $\Delta \geq 2.5\delta$,

Type	X_2	Y_2	X_3	Y_3	X_4
w	$\frac{1}{2}(100-\Delta)$	$max\{40-\frac{2}{5}\Delta, l-\Delta\}$	$\frac{1}{3}(100-\Delta)$	$l-\Delta$	$\frac{1}{4}(100-\Delta)$

Type	X_5	Z
w	$20-\frac{1}{5}\Delta$	1

If $\Delta < 2.5\delta$,

Type	X_2	Y_2	X_3	Y_3	X_4
w	$\frac{1}{2}(100-\Delta)$	$l-\Delta$	$\frac{1}{3}(100-\Delta)$	$l-\Delta$	$\frac{1}{4}(100-\Delta)$

Type	X_5	X_6	Z
w	$20-\frac{1}{5}\Delta$	$20-3\delta$	1

We introduced two lemmas in [4] in the following.

Lemma 1 *If $\Delta > 5\delta$, $t = y_2 + v + 1$ and $t \geq 2$.*

Lemma 2 *If $\Delta > \frac{5}{2}\delta$ and X_1 and X_2 are the items in the last regular $2-bin$ with $X_2 < \frac{1}{2}(100-\Delta)$, $X_i \in B_i^*$ for $i = 1,2$. Then $w(B_1^* \cup B_2^*) \leq 200 - \frac{6}{5}\Delta + 3\delta$ or $w(B_1^* \cup B_2^*) \leq 200 - \Delta$*

2. $\Delta > 5\delta$

Let S_3 be the number of the OPT bins which contains four items and at least one of them is from a regular $3 - bin$ except the last one. We prove an inequality about S_3.

Lemma 3 $S_3 \geq y_2 + 3v + 2$

Proof: 1°. For an OPT bin B^* which contains three items, there is at most one regular $3 - bin$ item. Otherwise,

$$l(B^*) > 40 + \Delta - 6\delta + \frac{2}{3}(100 - \Delta) > 100$$

A regular $3 - bin$ item can not be in the same OPT bin with a regular $2 - bin$ item. Otherwise,

$$\frac{1}{2}(100 - \Delta) + \frac{1}{3}(100 - \Delta) + 200 + \Delta - 3\delta > 100$$

A regular $3 - bin$ item must be in a OPT bin containing 4 items or a OPT bin which contains three items and one of them is from $Y_2 - bin$ or from the last regular $2 - bin$.

Since $t \geq 2$,

$$3(t-1) + 3 - \nu - (2y_2 + \delta_2) = y_2 + 3v + (3 - \nu - \delta_2) \geq 1$$

we have $S_3 \geq 1$.

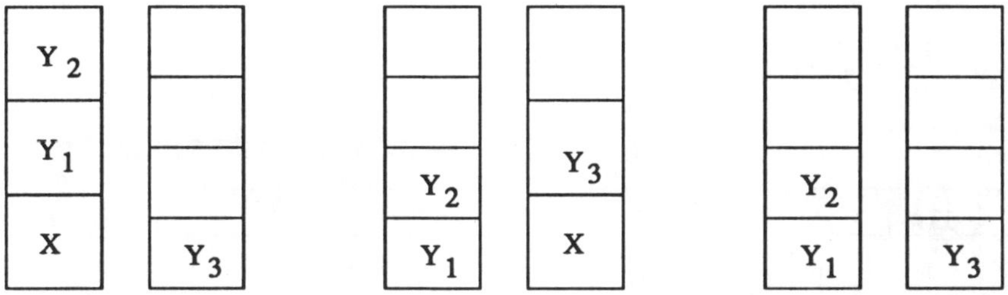

(a) (b) (c)

2°. Suppose Y_1, Y_2 and Y_3 are from the regular $3 - bins$. Claim that they must fall in the different OPT bins.

For (a),

$$l(B_1^* \cup B_2^*) > 100 - \Delta + 40 + \Delta - 6\delta + 3(20 + \Delta - 3\delta) > 200$$

For (b),

$$l(B_1^* \cup B_2^*) > 100 - \Delta + 40 + \Delta - 6\delta + 3(20 + \Delta - 3\delta) > 200$$

For (c),

$$l(B_1^* \cup B_2^*) > 100 - \Delta + 5(20 + \Delta - 3\delta) > 200$$

3°. Suppose X_1 and X_2 are from the same regular $2 - bin$ items, $X_1 \in B_1^*$ and $X_2 \in B_2^*$. Claim that there is at most one regular $3-bin$ item in $B_1^* \cup B_2^*$. Otherwise, $B_1^* \cup B_2^*$ contains two regular $3 - bin$ items and B_3^* is another OPT bin containing a regular $3 - bin$ item,

$$l(B_1^* \cup B_2^* \cup B_3^*) > 100 - \Delta + 100 - \Delta + 5(20 + \Delta - 3\delta) > 300$$

From 2° and 3°,

$$S_3 \geq y_2 + 3v + 2 \qquad \square$$

Theorem 1 $S_3 \geq 5$

Proof: It is trivial if $t \geq 4$ or $\nu \geq 1$ by Lemma 3.

i. $t = 2, y_2 = 1, \nu = 0$

Suppose B is a $Y_2 - type$ bin, $B = \{Y_1, Y_2, f\}$, $Y_i \in B_i^*$. We claim that there is at most one regular $3 - bin$ item in $B_1^* \cup B_2^*$.

$$l(B_1^* \cup B_2^* \cup B_3^*) > 100 - \Delta + 120 - 3\delta + 4(20 + \Delta - 3\delta) > 300$$

Suppose B_4^* contain the smallest item from regular $2-bins$. We claim $B_1^* \cup B_2^* \cup B_3^*$ contain at most one item from regular $3 - bins$. Otherwise,

$$l(B_i^* \cup B_3^* \cup B_4^*) > 100 - \Delta + 120 - 3\delta + 4(20 + \Delta - 3\delta) > 300 \quad i = 1, 2$$

Thus, we have $S_3 \geq 6 - 1 = 5$

ii. $t = 3, y_2 = 2, \nu = 0$

Similar to i, we have $S_3 \geq 5$. $\square$

It is easy to see that if an OPT bin contains 4 items and one of them is bigger than $\frac{1}{3}(100-\Delta)$, there are at least two items less than $\frac{1}{4}(100-\Delta)$. $\frac{1}{4}(100-\Delta) > 20+\Delta-3\delta$, i.e., $\Delta < \frac{12}{5}\delta + 4$.

$$w(\Gamma) + C \geq m(100 - \Delta)$$

$$w(\Gamma) + C \leq m(100 - \Delta) + 9\Delta - [(\frac{6}{5}\Delta - 3\delta)(10 - k) + k\Delta - 4\Delta] - (\frac{6}{5}\Delta - 3\delta)(10 - k)$$

$k = 0$,

$$w(T_n) \leq 33\delta - \frac{1}{5}\Delta < 20 - \frac{1}{5}\Delta$$

$k = 3$,

$$w(T_n) \leq 21\delta + \frac{2}{5}\Delta < 20 - \frac{1}{5}\Delta$$

This provides a contradiction.

3. $\frac{15}{4}\delta \leq \Delta \leq 5\delta$

Let β be the number of the items which are in the last regular $4-bin$ and are less than $\frac{1}{4}(100-\Delta)$. If $\beta = 0$, then $20 - \frac{1}{5}\Delta \leq 5\Delta$, this contradicts with $\delta < \frac{20}{33}$. Let $\overline{x}_i$ be the items in the last regular $4-bin$ and $\overline{x}_i \geq \overline{x}_{i+1}$.

1. $\beta = 1$
Define $w(\overline{x}_4) = \overline{x}_4$, $w(\overline{x}_i) = \frac{1}{4}(100-\Delta)$, $i = 1,2,3$. We analyze the possibility of OPT bins containing $\overline{x}_4$. It is easy to see that we should consider the following cases only:
(a) $\overline{x}_4 x_2 x_4 : \overline{x}_4 + \frac{1}{4}(100-\Delta)$
(b) $\overline{x}_4 x_3 x_3 : \overline{x}_4 + \frac{2}{3}(100-\Delta)$
(c) $\overline{x}_4 x_3 x_4 x_5 : \overline{x}_4 + \frac{47}{60}(100-\Delta)$
(d) $\overline{x}_4 x_3 x_5 x_5 : \overline{x}_4 + \frac{11}{15}(100-\Delta)$
(e) $\overline{x}_4 x_3 y_2$
  (e1) $w(y_2) = y_2 - \Delta$
  (e2) $w(y_2) = 40 - \frac{2}{5}\Delta$
(f) The OPT bin containing $\overline{x}_4$ contains another Z type item.
(g) The OPT bin containing $\overline{x}_4$ contains Y_3 type item.

For (a), (b), (c), (d) and (e2), we have

$$w(\Gamma) + C \geq m(100-\Delta)$$

$$w(\Gamma)+C \leq m(100-\Delta)+(9-4)\Delta+\frac{1}{4}(100-\Delta)-\overline{x}_4-\frac{13}{60}(100-\Delta)+\overline{x}_4-(\frac{6}{5}\Delta-3\delta)$$

i.e. $w(T_n) < 20 - \frac{1}{5}\Delta$
For (e1), (f) and (g),

$$w(\Gamma) + C \geq m(100-\Delta)$$

$$w(\Gamma) + C \leq m(100-\Delta) + (9-4)\Delta + \frac{1}{4}(100-\Delta) - \overline{x}_4$$

i.e. $w(T_n) < 20 - \frac{1}{5}\Delta$

2. $\beta = 2$
Now $\overline{x}_3$ and $\overline{x}_4$ are less than $\frac{1}{4}(100-\Delta)$, Define $w(\overline{x}_3) = \overline{x}_3$ and $w(\overline{x}_4) = \overline{x}_4$.
i. $\overline{x}_3$ and $\overline{x}_4$ are in the same OPT bin B^*.
It is easy to see that we should consider the following cases only:
(a) $\overline{x}_4\overline{x}_3 x_2$
(b) $\overline{x}_4\overline{x}_3 y_2$
(c) $\overline{x}_4\overline{x}_3 y_2 x_4$
(d) $\overline{x}_4\overline{x}_3 y_2 x_5$
(e) B^* contains one item of $Y_3 - type$.
(f) B^* contains another item of $Z - type$.
Define $w(\overline{x}_1) = \overline{x}_1$ and $w(\overline{x}_2) = \overline{x}_2$. We have

$$w(\Gamma) + C \geq m(100-\Delta)$$

$$w(\Gamma) + C \le m(100 - \Delta) + 9\Delta - 2\Delta + (\frac{6}{5}\Delta - 3\delta)$$

i.e.

$$w(T_n) < 20 - \frac{1}{5}\Delta$$

ii. $\overline{x}_3$ and $\overline{x}_4$ are in the different OPT bins.

Similarly, we only need to consider $\overline{x}_i x_3 x_4 x_5$. If $B_1^* = \{\overline{x}_4, x_3, x_4, x_5\}$ and $B_2^* = \{\overline{x}_4, x_3, x_4, x_5\}$,

$$l(B_1^* \cup B_2^*) > 100 - \Delta + \frac{2}{3}(100 - \Delta) + 2(20 + \Delta - 3\delta) > 200$$

By the assumption, we can consider that x_3 is from the last regular $3 - bin$ if $\overline{x}_i x_3 x_4 x_5$ exists. Thus, we have

$$w(\Gamma) + C \ge m(100 - \Delta)$$

$$w(\Gamma) + C \le m(100 - \Delta) + 9\Delta - 2\Delta + (\frac{6}{5}\Delta - 3\delta)$$

i.e. $w(T_n) < 20 - \frac{1}{5}\Delta$

3. $\beta = 3$

Similar to ii, we provide a contradiction.

4. $\frac{5}{2}\delta \le \Delta < \frac{15}{4}\delta$

If no $Y_1 - type$ bin occur, we have

$$w(\Gamma) + C \ge m(100 - \Delta) + w(T_n)$$

$$w(\Gamma) + C \le m(100 - \Delta) + 9\Delta - (\frac{6}{5}\Delta - 3\delta)$$

i.e. $w(T_n) < 20 - \frac{1}{5}\Delta$

We will discuss the different cases if $Y_1 - type$ bins occur.

1. $B_1 = \{x_1, f_1, f_2\}$ is $Y_1 - type$ bin, $B^* = \{x_1, h_1, h_2\}$ is the OPT bin containing x_1. Define:

$w(x_1) = 60 - \frac{3}{2}\delta$

$w(f_i) = 20 - \frac{1}{5}\Delta$ (i=1,2)

We have that $w(B_1) = 100 - \frac{3}{2}\delta - \frac{2}{5}\Delta \ge 100 - \Delta$. If there is no $Z-$ type item in B^*, it is easy to see that B^* can be balanced to $100 - \Delta$ by B_1.

2. $B_1 = \{x_1, y_1\}$ is a $Y_1 - type$ bin, $B_1^* = \{x_1, f_1, f_2\}$ is the OPT bin containing x_1.

(a) $B_2^* = \{y_1, t_1, t_2, t_3\}$ is the OPT bin containing y_1. Let $A = \{t_1, t_2, t_3, f_1, f_2\}$. If there is no $Z - type$ item in A, define $w(x_1) = x_1, w(y_1) = y_1$. Since $w(f_i) = 20 - \frac{1}{5}\Delta$ for i=1, 2, and $w(t_i) = 20 - \frac{1}{5}\Delta$ for i=1, 2, 3.

$w(B_1^* \cup B_2^*) = w(B_1^*) + 100 - \Delta$, $B_1^* \cup B_2^*$ can be balanced to $2(100-\Delta)$ by B_1.

If there is a $Z - type$ item in A, $w(B_1^* \cup B_2^*) \le 200 - 4(\frac{6}{5}\Delta - 3\delta)$

If there are at least two Z type items in A, $w(B_1^* \cup B_2^*) \le 200 - 2(\frac{6}{5}\Delta - 3\delta)$.

(b) $B_2^* = \{y_1, t_1, t_2\}$ is the OPT bin containing y_1. Let $A = \{t_1, t_2, f_1, f_2\}$ and t_i is not from $Y_1 - type$ bin.

If there is no $Z - type$ item in A, We only need to discuss $x_5x_5y_3x_3$, $w(x_5x_5y_3x_3) \le 100 - \frac{12}{5}\Delta + 6\delta$

But the weight of a $Y_3 - type$ bin is $110 - \frac{16}{5} - \frac{9}{4}\delta$. $w(A)$ can be balanced to $100 - \Delta$ by the weight of $Y_3 - type$ bin. Thus $w(B_1^* \cup B_2^*)$ can be balanced to $2(100-\Delta)$ by $w(A)$.

If there is a $Z - type$ item in A, denoted by z,

(b1) z is f_1 or f_2

$$\overline{\delta} = 40 + \frac{3}{2}\delta - (40 - \frac{2}{5}\Delta) = \frac{3}{2}\delta + \frac{2}{5}\Delta$$

(b2) z is t_1 or t_2

$$\overline{\delta} = 60 - \Delta + 6\delta - (60 - \frac{3}{5}\Delta) = 6\delta - \frac{2}{5}\Delta$$

(c) There are at least two $Z - type$ items in A.

If no (b2) occurs, we have

$$w(\Gamma) + C \ge m(100-\Delta) + w(T_n)$$

$$w(\Gamma)+C \le m(100-\Delta)+9\Delta+k_2(\frac{3}{2}\delta-\frac{3}{5}\Delta)+k_3[\Delta-4(\frac{6}{5}\Delta-3\delta)]+2k_4(3\delta-\frac{6}{5}\Delta)+(3\delta-\frac{6}{5}\Delta)$$

i.e. $w(T_n) \le 9\Delta + k_3[\Delta - 4(\frac{3}{2}\delta - \frac{3}{5}\Delta)] + (3\delta - \frac{6}{5})$

$k_3 = 0$ or $k_3 = 3$, $w(T_n) < 20 - \frac{1}{5}\Delta$

If (b2) occurs, the last regular $2 - bin$ denoted by $\{x_1, y_1\}$ can be considered to have the same element in the $Y_1 - type$ bin. Thus

$$w(\Gamma) + C \ge m(100-\Delta) + w(T_n)$$

$$w(\Gamma)+C \le m(100-\Delta)+7\Delta+k_2(\frac{3}{2}\delta-\frac{3}{5}\Delta)+k_3[\Delta-4(\frac{6}{5}\Delta-3\delta)]+2k_4(3\delta-\frac{6}{5}\Delta)+k_1(6\delta-\frac{7}{5}\Delta)$$

It is easy to see that $k_1 + k_2 + k_3 \le 3$. We have $k_1 \le 3$.

If $w(t_1) = 20 - \frac{1}{5}\Delta$, we only need to consider the case $w(\{t_1, t_2, f_1, f_2\}) > 100, w(t_2) > 40 + \frac{3}{5}\Delta \ge 40 + \frac{3}{2}\delta \ge Y_1$

If $w(t_1) = \frac{100}{3} - \frac{1}{3}\Delta$, $w(t_2) > \frac{80}{3} + \frac{11}{15}\Delta$

If $w(t_1) = 25 - \frac{1}{4}\Delta$, $w(t_2) > 35 + \frac{113}{20}\Delta$

$$w(T_n) \le 7\Delta + k_3[\Delta - 4(\frac{6}{5}\Delta - 3\delta)] + k_1(6\delta - \frac{7}{5}\Delta)$$

For $k_3 = 0, k_1 = 3; k_3 = 3, k_1 = 0$ or $k_3 = 0, k_1 = 3$; we have

$$w(T_n) \le 21\delta + \frac{2}{5}\Delta < 20 - \frac{1}{5}\Delta$$

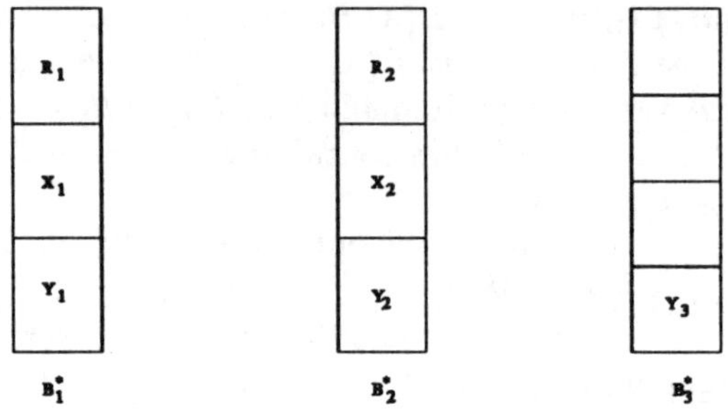

3. $B_2^* = \{y_1, y_1', t\}$ and $B' = \{x_1', y_1'\}$ is also of $Y_1 - type$. Let $B_3^* = \{x_1', f_3, f_4\}$ be the OPT bin containing x_1'. Denote A by $\{f_1, f_2, f_3, f_4, t\}$.

(a) there is no $Z - type$ item in A

$$w(A) = 5(20 - \frac{1}{5}\Delta) = 100 - \Delta$$

(b) there is only one $Z - type$ item in A, denoted by z.

(b1) If $z = f_i$ for some $1 \leq i \leq 4$.

$$\overline{\delta} = 40 + \frac{3}{2}\delta - (40 - \frac{2}{5}\Delta) = \frac{3}{2}\delta + \frac{2}{5}\Delta$$

(b2) $z = t$

$$w(B_1^* \cup B_2^* \cup B_3^*) \leq 300 - 4(\frac{6}{5}\Delta - 3\delta)$$

(c) there are two $Z - type$ items in A,

$$w(B_1^* \cup B_2^* \cup B_3^*) \leq 300 - 3(\frac{6}{5}\Delta - 3\delta)$$

(d) there are three $Z - type$ items in A,

$$w(B_1^* \cup B_2^* \cup B_3^*) \leq 300 - 2(\frac{6}{5}\Delta - 3\delta)$$

We can make an assumption: all the $Y_1 - type$ items are the same and the items in the last regular $2 - bin$ are also x_1 and y_1. If (b2), (c) or (d) occurs, we can assume that x_1 and y_1 are $Z - type$ items. We have

$$w(\Gamma) + C \geq m(100 - \Delta) + w(T_n)$$

$$\begin{aligned} w(\Gamma) + C \leq\ & m(100 - \Delta) + k_5[2\Delta - 4(\tfrac{6}{5}\Delta - 3\delta)] + k_6[\Delta - 3(\tfrac{6}{5}\Delta - 3\delta)] \\ & + 9\Delta + k_4[2(3\delta - \tfrac{6}{5}\Delta] + k_3[\Delta - 4(\tfrac{6}{5}\Delta - 3\delta)] \\ & + k_2(\tfrac{3}{2}\delta - \tfrac{3}{5}\Delta) + k_1(6\delta - \tfrac{7}{5}\Delta) \end{aligned}$$

It is easy to see that $k_2 + k_3 + k_4 + k_5 + k_6 \leq 3$.

By the assumption for the regular $3 - bin$ items and the regular $4 - bin$ items, $k_1 \leq 2$.

Consider the following cases:

$k_1 = 0, k_5 = 3, k_6 = 0, k_3 = 0$

$$w(T_n) \leq 7\Delta + 3[2\Delta - 4(\frac{6}{5}\Delta - 3\delta)] < 20 - \frac{1}{5}\Delta$$

$k_1 = 0, k_5 = 0, k_6 = 3, k_3 = 0$

$$w(T_n) \leq 7\Delta + 3[\Delta - 3(\frac{6}{5}\Delta - 3\delta)] < 20 - \frac{1}{5}\Delta$$

$k_1 = 0, k_5 = 0, k_6 = 0, k_3 = 3$

$$w(T_n) \leq 7\Delta + 3[\Delta - 4(\frac{6}{5}\Delta - 3\delta)] < 20 - \frac{1}{5}\Delta$$

$k_1 = 2, k_5 = 2, k_6 = 0, k_3 = 0$

$$w(T_n) \leq 7\Delta + 2[\Delta - 4(\frac{6}{5}\Delta - 3\delta)] + 2(6\delta - \frac{7}{5}\Delta) < 20 - \frac{1}{5}\Delta$$

$k_1 = 2, k_5 = 0, k_6 = 2, k_3 = 0$

$$w(T_n) \leq 7\Delta + 2[2\Delta - 4(\frac{6}{5}\Delta - 3\delta)] + 2(6\delta - \frac{7}{5}\Delta) < 20 - \frac{1}{5}\Delta$$

$k_1 = 2, k_5 = 0, k_6 = 2, k_3 = 2$

$$w(T_n) \leq 7\Delta + 2[2\Delta - 4(\frac{6}{5}\Delta - 3\delta)] + 2(6\delta - \frac{7}{5}\Delta) < 20 - \frac{1}{5}\Delta$$

We have that $\delta < \frac{20}{33}$ provides a contradiction for this part.

5. $\Delta < 2.5\delta$

Similar to Section 4, we assume that all the $Y_1 - type$ items and the items in the last regular $2 - bin$ are the same correspondingly. Denote x_1 and y_1 the bigger item and the smaller item respectively in a $Y_1 - type$ bin.

First, we consider the following cases:
If there exists a $6 - bin$, $6(20 - 3\delta) \geq 100 + 6\Delta$, extra weight of 7Δ can be obtained from this FFD bin.
If there exists a $Y_3 - type$ bin,

$$\frac{3}{4}(120 - 3\delta) - 3\Delta + 20 - 3\delta \geq 100 + \frac{3}{2}\Delta$$

Extra weight of $\frac{5}{2}\Delta$ can be obtained from this FFD bin.
If there exists a $Y_4 - type$ bin,

$$\frac{4}{5}(120 - 3\delta) - 4\Delta + 20 - 3\delta \geq 100 + 4\Delta$$

Extra weight of 5Δ can be obtained from this FFD bin.

We discuss the cases that there is no $Y_1 - type$ bin in FFD.

Let $\overline{x}_i$ $(1 \leq i \leq 5)$ be the $Z - type$ items in the last regular $5 - bin$ and $\overline{x}_i \geq \overline{x}_{i+1}$. Define $w(\overline{x}_i) = min(\overline{x}_i, \frac{1}{5}(100 - \Delta))$. If $w(\overline{x}_i) = \frac{1}{5}(100 - \Delta)$, we will not consider $\overline{x}_i$ as a $Z - type$ item again.

Suppose $\overline{x}$ be in the the last regular $5 - bin$ and $\overline{x} \leq \frac{1}{5}(100 - \Delta)$. Let B^* be the OPT bin containing $\overline{x}$.

1. There is no $Z - type$ item in B^*. Since $w(x_6) \leq x_6 - \Delta$ and $w(y_i) \leq y_i - \Delta$, we only need to consider the combinations of the regular bin items.

It is easy to see that $w(B^*) \leq 100 - \Delta$.

2. There is another $Z - type$ item in B^*, denoted by x'.

If $x' \leq \frac{1}{5}(100 - \Delta)$, since $w(x_6) \leq x_6 - \Delta$ and $w(y_i) \leq y_i - \Delta$, we only need to consider the combination of the regular bin items.

It is easy to see that $w(B^*) \leq 100 - \Delta$.

3. There is another two $Z - type$ items in B^*, denoted by x' and x''.

If $max(x', x'') < \frac{1}{5}(100 - \Delta)$, we have $w(B^*) \leq 100 - \Delta$ similarly.

4. There is another three $Z - type$ items in B^*, denoted by x' and x'' and x'''.

If $max(x', x'', x''') < \frac{1}{5}(100-\Delta)$, we only need to consider $B^* = \{\overline{x}, x', x'', x''', x_4\}$

If $w(B^*) \leq 100$, we have $x_4 \leq x_5 + \Delta$, i.e.

$$x_5 \geq \frac{1}{4}(100 - \Delta) - \Delta \geq 20 - \frac{1}{5}\Delta$$

From the above cases, we have

$$w(\Gamma) + C \geq m(100 - \Delta) + 4(\frac{6}{5}\Delta - 3\delta) + w(T_n)$$

$$w(\Gamma) + C \leq m(100 - \Delta) + 14\Delta - 4\Delta$$

we have $w(T_n) < 20 - 3\delta$.

We give the detail if $Y_1 - type$ bin occurs in FFD.

1. Let $B = \{x_1, s_1, s_2\}$ be a $Y_1 - type$ bin and $B^* = \{x_1, t_1, t_2\}$ be the OPT bin containing x_1. Suppose $s_1 \geq s_2$ and $t_1 \geq t_2$. Since B^* can not be dominated by B, we must have that $t_1 > s_1$ or $t_1 > s_2$ and $l(x_1 s_1 t_1) > 120 - 3\delta$.

$$l(B^* \cup B) > 120 - 3\delta + 60 - \frac{3}{2}\delta + 2(20 + \Delta - 3\delta)$$

If both t_1 and t_2 fall in the $Y_1 - type$ bin and $B_1 = \{x_1', t_1, p\}$ be the $Y_1 - type$ bin containing t_1. If $x_1' \geq x_1$, B^* is dominated by B_1. If $x_1' < x_1$,

$$l(B^* \cup B) > 120 - 3\delta + 60 - \frac{3}{2}\delta + 2(20 + \Delta - 3\delta)$$

Define $w(x_1) = x_1 - \Delta$ and $w(s_i) = s_i - \Delta$ for $i = 1, 2$.

$$w(B) \geq 120 + 2\Delta - \frac{21}{2}\delta - 3\Delta \geq 100 + 8\Delta$$

We obtain extra weight of 9Δ from $w(A)$.

2. Let $B = \{x_1, y_1\}$ be a $Y_1 - type$ bin and $B_1^* = \{x_1, s_1, s_2\}$ be the OPT bin containing x_1. Let $B_2^* = \{y_1, t_1, t_2, t_3\}$ be the OPT bin containing y_1.Define $A = \{t_1, t_2, t_3, s_1, s_2\}$, $w(x_1) = x_1$ and $w(y_1) = y_1$.

(i) There is no $Z - type$ item in A, Since $s_1 + s_2 < 40 + \frac{3}{2}\delta$, $s_i < \frac{1}{4}(100 - \Delta)$ for $i = 1, 2$.

(i,a) there is one item in A from $Y_i - type$ bins or $6 - bins$, $w(A)$ can be balanced to $100 - \Delta$.

(i,b) $t_j < \frac{1}{4}(100 - \Delta)$ for $j = 1, 2, 3$. Since $20 - 3\delta < 20 - \frac{1}{5}\Delta$, $w(A) < 100 - \Delta$.

(i,c) Suppose $t_1 \geq \frac{1}{4}(100 - \Delta)$, there are at least three items with weight less than $20 - 3\delta$ in A. We have $w(A) < 100 - \Delta$.

(ii) There is only one $Z - type$ item in A, denoted by z.
If $z < 20 - \frac{1}{5}\Delta$,

$$\frac{1}{4}(100 - \Delta) + 2(20 + \delta - 3\delta) + \frac{2}{5}(100 - \Delta) > 100 + \Delta$$

Similar to (i), $w(A)$ can be balanced to $100 - \Delta$. We assume that $z \geq 20 - \frac{1}{5}\Delta$ and all the other items be from the regular bins.

(ii,a) $max(t_1, t_2, t_3) < \frac{1}{4}(100 - \Delta)$
z must be from the last regular $4 - bin$ and $z < \frac{1}{4}(100 - \Delta)$.

(ii,b) $max(t_1, t_2, t_3) \geq \frac{1}{4}(100 - \Delta)$

$$\frac{1}{4}(100 - \Delta) + 2(20 + \delta - 3\delta) + \frac{2}{5}(100 - \Delta) > 100 + \Delta$$

There are at least three items less than $20 - \frac{1}{5}\Delta$. $w(A) < 100 - \Delta$ or $w(A)$ can be balanced to $100 - \Delta$.

(iii) There are two $Z - type$ items in A, denoted by z_1 and z_2.

(iii,a) $t_1 \geq \frac{1}{4}(100 - \Delta)$, we have $150 + 3\delta < 20$.

(iii,b) $max(t_1, t_2, t_3) < \frac{1}{4}(100 - \Delta)$ and $max(z_1, z_2) < 20 - \frac{1}{5}\Delta$
Similar to (i), $w(A)$ can be balanced to $100 - \Delta$.

(iii,c) $max(t_1, t_2, t_3) < \frac{1}{4}(100 - \Delta)$ and $z_1 \geq 20 - \frac{1}{5}\Delta > z_2$.
z_1 must be from a regular $4 - bin$, and $z_1 < \frac{1}{4}(100 - \Delta)$.

(iii,d) $max(t_1, t_2, t_3) < \frac{1}{4}(100 - \Delta)$ and $min(z_1, z_2) \geq 20 - \frac{1}{5}\Delta$
z_1 and z_2 must be from the last regular $4 - bin$, and $z_i < \frac{1}{4}(100 - \Delta)$ for $i = 1, 2$.

(iv) There are at least three $Z - type$ items in A,

(iv,a) $max(t_1, t_2, t_3) \geq \frac{1}{4}(100 - \Delta)$, we have $150 + 3\delta < 20$.

(iv,b) $max(t_1, t_2, t_3) < \frac{1}{4}(100 - \Delta)$ and the maximal item of all the $Z - type$ items is less than $20 - \frac{1}{5}\Delta$

Similar to (i), $w(A)$ can be balanced to $100 - \Delta$.

(iv,c) $max(t_1, t_2, t_3) < \frac{1}{4}(100 - \Delta)$ and the maximal item of all the $Z - type$ items is no less than $20 - \frac{1}{5}\Delta$. The maximal $Z - type$ item must be from the last regular $4 - bin$ or $5 - bin$.

3. B and B_1^* are the same as 2. Let $B_2^* = \{y_1, t_1, t_2\}$ be the OPT bin containing y_1 and t_i $(i = 1, 2)$ is not from $Y_1 - type$ bin. It is easy to see that $max(t_1, t_2) < y_1$.

Define $w(x_1) = x_1$, $w(y_1) = y_1$ and $A = \{t_1, t_2, s_1, s_2\}$.

(i) There is no $Z - type$ item in A,

(i,a) there is one item in A from $Y_i - type$ bins or $6 - bins$, $w(A)$ can be balanced to $100 - \Delta$.

(i,b) all items in A are from the regular bins : it is easy to see that $w(B_1^* \cup B_2^*)$ can be balanced to $2(100 - \Delta)$.

(ii) There is only one $Z - type$ item in A, denoted by z.

(ii,a) there is one item in A from $Y_i - type$ bins or $6 - bins$, $w(A)$ can be balanced to $100 - \Delta$.

(ii,b) if $z \le \frac{1}{5}(100 - \Delta)$, $w(A)$ can be balanced to $100 - \Delta$. So we assume that $z > \frac{1}{5}(100 - \Delta)$ and $w(A) > 100$ in the following.

(ii,c) $z \in \{s_1, s_2\}$.

$$\overline{\delta} = (40 + \frac{3}{2}\delta) - (40 - \frac{2}{5}\Delta)$$

(ii,d) $z \in \{t_1, t_2\}$.

(ii,d1) $w(t_1) = 20 - \frac{1}{5}\Delta$, $z \ge 100 - 3(20 - \frac{1}{5}\Delta)$

Since $y_1 \ge z$, $l(B_2^*) \ge 2(40 + \frac{3}{5}\Delta) + 20 - \frac{1}{5}\Delta > 100$

(ii,d2) $w(t_1) = \frac{1}{4}(100 - \Delta)$, $z \ge 100 - 2(20 - \frac{1}{5}\Delta) - \frac{1}{4}(100 - \Delta)$

z must be from the last regular $3 - bin$ and $w(x_3) < x_3 - \Delta$.

(ii,d3) $w(t_1) = \frac{1}{3}(100 - \Delta)$, $z \ge 100 - 2(20 - \frac{1}{5}\Delta) - \frac{1}{3}(100 - \Delta)$.

If z is from the last regular $4 - bin$, $w(x_4) < x_4 - \Delta$.

Claim that (d2) and (d3) can not be occured at the same time and (d3) occurs with the same type of regular items.

Otherwise, suppose that $B_5^* = \{y_1, x_3, z\}$ and $B_6^* = \{y_1, x_3, z'\}$, z is from the last regular $3 - bin$ and z' is no less than $\frac{80}{3} + \frac{11}{15}\Delta$.

$$l(B_5^* \cup B_6^*) \ge 2l(y_1) + 100 - \Delta + \frac{80}{3} + \frac{11}{15}\Delta > 200$$

We know that (ii,d2) occurs at most two times; (ii,d3) occurs one times for the regular $3 - bin$ item or three times for the regular $4 - bins$.

(iii) There are two $Z - type$ items in A, denoted by z_1 and z_2.

(iii,a) there is one item in A from $Y_i - type$ bins or $6 - bins$, $w(A)$ can be balanced to $100 - \Delta$.

(iii,b) $max(z_1, z_2) \le \frac{1}{5}(100 - \Delta)$, $w(A)$ can be balanced to $100 - \Delta$. We assume that $w(A) > 100$ in the following.

(iii,c) $z_1 \ge 20 - \frac{1}{5}\Delta > z_2$. z_2 must fall in B_1^*, z_1 is similar to the z in (ii,d).

(iv) There are more than two $Z-type$ items in A,

If the maximal $Z-type$ item is less than $20-\frac{1}{5}\Delta$, $w(A)$ can be balanced to $100-\Delta$. Otherwise, the discussion is similar to (ii,d).

4. $B=\{x_1,y_1\}$ and $B'=\{x_1,y_1\}$ are two Y_1-type bins. $B_1^*=\{x_1,s_1,s_2\}$, $B_2^*=\{x_1,s_3,s_4\}$, $B_3^*=\{y_1,y_1,t\}$ are OPT bins and $A=\{s_1,s_2,s_3,s_4,t\}$.

(i) There is no $Z-type$ item in A,

It is easy to see that the items in A can not be all from the regular bins. Otherwise it will be dominated.

(i,a) $t\geq\frac{1}{4}(100-\Delta)$,

$$\frac{1}{4}(100-\Delta)+20+3\delta-\Delta+3(20-\frac{1}{5}\Delta)>100+2\Delta$$

Therefore there are at least three items less than $20-\frac{1}{5}\Delta$ in A. Since $w(x_6)\leq x_6-\Delta$, $w(A)$ can be balanced to $100-\Delta$.

(i,b) $t<\frac{1}{4}(100-\Delta)$, there is an item y_4 from Y_4-type bin.

Assume the other items be from the regular $5-bins$. Otherwise, $w(A)$ can be balanced to $100-\Delta$.

Consider $y_4x_5x_5x_5x_5$. If $l(x_5x_5x_5x_5)>80-\frac{1}{5}\Delta$, $w(x_5x_5x_5x_5)<x_5x_5x_5x_5-\Delta$.

Suppose $x_5<20-\frac{1}{5}\Delta+\frac{1}{2}\delta$ and B'' be a Y_4-type bin,

$$w(B'')\geq 120-3\delta-(20+\frac{3}{10}\Delta)-4\Delta+20-3\delta\geq 100+6.5\Delta$$

Claim that (i,b) occurs at most three times. Otherwise, the total size is no less than $8l(x_1y_1)+120-3\delta+15(20-\frac{1}{5}\Delta)>1200$

$w(A)$ can be balanced to $100-\Delta$ by the extra weight of B''.

(ii) There is only one $Z-type$ item in A, denoted by z.

(ii,a) $t\geq\frac{1}{4}(100-\Delta)$, $z<20-\frac{1}{5}\Delta$.

Similar to (i,a), we know there are two items less than $20-\frac{1}{5}\Delta$, denoted by z_1 and z_2.

If z_i is from a regular $6-bin$ or a Y_4-type bin, $w(A)$ can be balanced to $100-\Delta$. Suppose that $x_5<20-\frac{1}{5}\Delta+\Delta$. If z_1 and z_2 are from Y_3-type bins, let B'' be a Y_3-type bin,

$$w(B'')\geq 120-3\delta-(20+\frac{4}{5}\Delta)-3\Delta+20-3\delta\geq 100+7\Delta$$

$w(A)$ can be balanced to $100-\Delta$ by the extra weight of B''.

(ii,b) $t\geq\frac{1}{4}(100-\Delta)$, $z\geq 20-\frac{1}{5}\Delta$

Similar to (i,a), $w(A)$ can be balanced to $100-\Delta$.

(ii,c) $t<\frac{1}{4}(100-\Delta)$, $z<20-\frac{1}{5}\Delta$

If there is one item from the regular $6-bins$ or the Y_4-type bins, $w(A)$ can be balanced to $100-\Delta$.

(ii,d) $t<\frac{1}{4}(100-\Delta)$, $z\geq 20-\frac{1}{5}\Delta$

z must be from the last regular $4-bin$. Suppose the other items are from the regular $5-bins$, (ii,d) can occur at most three times.

(iii) There are two $Z - type$ items in A, denoted by z_1 and z_2.

(iii,a) $t \geq \frac{1}{4}(100 - \Delta)$, $max(z_1, z_2) < 20 - \frac{1}{5}\Delta$. If the other two items are less than $20 - \frac{1}{5}\Delta$, $w(A)$ can be balanced to $100 - \Delta$. Otherwise, there is an item other than t is no less than $20 - \frac{1}{5}\Delta$.

$$\frac{1}{4}(100 - \Delta) + 20 - \frac{1}{5}\Delta + 3(20 + 3\delta - \Delta) < 100 + 2\Delta$$

we have $22\Delta + 3\delta < 20$.

(iii,b) $t \geq \frac{1}{4}(100 - \Delta)$, $max(z_1, z_2) > 20 - \frac{1}{5}\Delta$

The other two items must be less than $20 - \frac{1}{5}\Delta$. z_1 is from a regular $4 - bin$, if $Y_4 - type$ bin exists, $z_1 > 24 - \frac{3}{5}\delta$.

$$\frac{1}{4}(100 - \Delta) + 24 - \frac{3}{5}\delta + 3(20 + 3\delta - \delta) > 100 + 2\Delta$$

Then there is no $Y_4 - type$ bin, the other two items must be from $Y_3 - type$ bins and $6 - bins$. $w(A)$ can be balanced to $100 - \Delta$.

(iii,c) $t < \frac{1}{4}(100 - \Delta)$, $max(z_1, z_2) < 20 - \frac{1}{5}\Delta$

Similar to (ii,c), $w(A)$ can be balanced to $100 - \Delta$.

(iii,d) $t < \frac{1}{4}(100 - \Delta)$,$max(z_1, z_2) < 20 - \frac{1}{5}\Delta$.

we know that the bigger one of z_1 and z_2 must be from the last regular $4 - bin$. The smaller one is in A.

(iv) There are at least three $Z - type$ items in A,

(iv,a) $t \geq \frac{1}{4}(100 - \Delta)$, the maximal $Z - type$ item is less than $20 - \frac{1}{5}\Delta$.

(iv,a1) If there is an item other than t in A is bigger than $20 - \frac{1}{5}\Delta$, we have $22\Delta + 3\delta < 20$.

(iv,a2) if there is a non $Z - type$ item w and $w < 20 - \frac{1}{5}\Delta$. Similar to (ii,a), $w(A)$ can be balanced to $100 - \Delta$.

(iv,a3) All the $Z - type$ items are from the last regular $5 - bin$.

(iv,b) $t < \frac{1}{4}(100 - \Delta)$, the maximal $Z - type$ item is less than $20 - \frac{1}{5}\Delta$.

Similar to (iii,c), $w(A)$ can be balanced to $100 - \Delta$.

(iv,c) $t \geq \frac{1}{4}(100 - \Delta)$, the maximal $Z - type$ item is no less than $20 - \frac{1}{5}\Delta$.

(iv,c1) t is the maximal $Z - type$ item, the other $Z - type$ items is less than $20 - \frac{1}{5}\Delta$. The discussion is similar to (ii,a) and (iv,a).

(iv,c2) t is not of $Z - type$, the maximal $Z - type$ item must be from the last regular $4 - bin$ and is less than $\frac{1}{4}(100 - \Delta)$.

(iv,d) $t < \frac{1}{4}(100 - \Delta)$ and the maximal $Z - type$ item is no less than $20 - \frac{1}{5}\Delta$.

From the above analysis, we have:

Case 1. If there are four $Z - type$ items less than $20 - \frac{1}{5}\Delta$ and they fall into the same OPT bin with another item no less than $\frac{1}{4}(100 - \Delta)$, the weight of the OPT bin

$$4(20 + \Delta - 3\delta) + \frac{1}{4}(100 - \Delta) \leq 100$$

we have $30\Delta + 3\delta < 20$.

$$w(\Gamma) + C \geq m(100 - \Delta) + w(T_n)$$

$$w(\Gamma) + C \leq m(100 - \Delta) + 14\Delta + 6\Delta + 4\Delta$$

i.e. $w(T_n) + 3\delta \leq 24\Delta + 3\delta < 20$

Case 2. If [4,(iv,a)] occurs, $x_5 \geq 20 + \frac{1}{20}\Delta$ and all the $Z - type$ items less than $20 - \frac{1}{5}\Delta$ occur in [4,(iv,a)], Since $4x_5 \geq 4w(x_5) - \Delta$, $w(A) \leq 100$, i.e. $w(B_1^* \cup B_2^*)$ can be balanced to $200 - \Delta$ in [2,(ii,a)] and $w(A) \leq 100 + \Delta$, i.e. $w(B_1^* \cup B_2^* \cup B_3^*)$ can be balanced to $300 - \Delta$ in [4,(ii,d)].

If [4,(ii,d)] occur three times and [3,(ii,d2)] occurs, $x_4 \geq 100 - \Delta - 3(20 + \frac{9}{5}\Delta) \geq w(x_4) + 3\Delta$. The weight of [3,(ii,d2)] can be balanced to $100 - \Delta$ and $w(t) \leq t - \Delta$.

If [3,(ii,d3)] occurs and z is from the last regular $3 - bin$, $2x_3 \geq 100 - \Delta - z$, i.e. $w(x_3) \leq x_3 - \Delta$. If z is from the last regular $4 - bin$, $3(20 + \frac{9}{5}\Delta) + \frac{80}{3} + \frac{11}{15}\Delta + \frac{1}{2}\Delta$. The weight of [3,(ii,d3)] can be balanced to $100 - \Delta$.

We can assume the X_1 and Y_1 in [4,(ii,d)] are from the last regular $2 - bin$. We have

$$w(\Gamma) + C \geq m(100 - \Delta) + w(T_n) + 4(\frac{6}{5}\Delta - 3\delta)$$

$$w(\Gamma) + C \leq m(100 - \Delta) + 14\Delta - 2\Delta + 3\Delta - 2\Delta - 2\Delta$$

i.e. $w(T_n) + 3\delta \leq 33\delta < 20$

If [4,(ii,d)] occurs at least once, $x_4 \geq \frac{1}{3}[100 - \Delta - (20 + \frac{9}{5}\Delta)$. [3,(ii,d2)] occurs at most two times. We can assume that x_4 in one of [3,(ii,d2)] is from last regular $4 - bin$ and $4x_5 \geq 80 + \frac{1}{5}\Delta$. We do not need to pay attention to [3,(ii,d2)].

[3,(ii,d3)] occurs at most three times. If z is from the last regular $3 - bin$, the discussion is the same as before. If z is from the last regular $4 - bin$, $z \geq \frac{80}{3} + \frac{11}{15}\Delta$ and $z \geq \frac{1}{4}(100 - \Delta) + 2\Delta$.

Similarly, we have

$$w(\Gamma) + C \geq m(100 - \Delta) + w(T_n) + 4(\frac{6}{5}\Delta - 3\delta)$$

$w(\Gamma) + C \leq m(100 - \Delta) + 14\Delta - 2\Delta + 4\Delta - 4\Delta$

i.e. $w(T_n) + 3\delta \leq 33\delta < 20$

If [4,(ii,d)] doesn't occur, we have

$$w(\Gamma) + C \geq m(100 - \Delta) + w(T_n) + 4(\frac{6}{5}\Delta - 3\delta)$$

$$w(\Gamma) + C \leq m(100 - \Delta) + 14\Delta - 2\Delta - 2\Delta + 2\Delta$$

i.e. $w(T_n) + 3\delta \leq 33\delta < 20$

Case 3. Suppose $x_5 < 20 + \frac{1}{20}\Delta$, [4,(iv,a1)] or [4,(iii,a2)] occurs. Since $22\Delta + 3\delta < 20$, we have

$$w(\Gamma) + C \geq m(100 - \Delta) + w(T_n)$$

$$w(\Gamma) + C \le m(100 - \Delta) + 14\Delta + 6\Delta + 2\Delta$$

i.e. $w(T_n) + 3\delta \le 22\Delta + 3\delta < 20$

Case 4. [4,(ii,d)] and [4,(iii,d)] occur three times. We have

$$t_i \le 100 + 2\Delta - 3(20 - \frac{1}{5}\Delta) - (20 + \Delta - 3\delta)$$

$$x_4 > 100 - \Delta - (t_1 + t_2 + t_3) \ge 25 + \frac{1}{2}\Delta \ge w(x_4) + \frac{3}{4}\Delta$$

Subcase 4.1. If [3,(ii,d2)] occurs,
Since $z \ge 35 + \frac{13}{20}\Delta \ge \frac{1}{3}(100 - \Delta) + 2\Delta$, $w(x_3) \le x_3 - 2\Delta$.
[3,(ii,d2)] occurs at most twice. We can assume that x_4 in one [3,(ii,d2)] is from the last regular $4 - bin$. We get the extra weight $6\Delta + \frac{1}{4}\Delta$ from [4,(ii,d)], [4,(iii,d)] and [3,(ii,d2)].

If there exists x_3, we get the extra weight 3Δ for $w(x_3) \le x_3 - 2\Delta$. If there is no x_3, $x_4 \le w(x_4) + 2\Delta$. If x_4 and Y_1 are together, B^* can be balanced to $100 - \frac{7}{4}\Delta$. Since there are other two x_4's, we have

$$w(\Gamma) + C \ge m(100 - \Delta) + w(T_n) + 4(\frac{6}{5}\Delta - 3\delta)$$

$$w(\Gamma) + C \le m(100 - \Delta) + 14\Delta - 2\Delta + \frac{25}{4}\Delta - \frac{3}{2}\Delta - 5\Delta$$

i.e. $w(T_n) + 3\delta \le 33\delta < 20$

Subcase 4.2. If [3,(ii,d3)] occurs at most once
$z \ge \frac{80}{3} + \frac{11}{15}\Delta$. If z is from the last regular $3 - bin$, $y_1 \le x_3 + \Delta$ and

$$l(y_1 x_3 z) \ge \frac{80}{3} + \frac{11}{15}\Delta + 2y_1 - \Delta$$

i.e. $150\Delta + 3\delta < 20$. We can provide a contradiction with $\delta < \frac{20}{33}$.

If z is from the last regular $4 - bin$, $x_4 \ge z \ge \frac{80}{3} + \frac{11}{15}\Delta \ge w(x_4) + 2\Delta$. If there is a regular $4 - bin$ other than the last one, we get extra weight 4Δ. If there is no other regular $4 - bin$, consider x_3 in B^*: If there is an item from $6 - bin$ or $Y_i - type$ bin, $w(B^*) \le 100 - 2\Delta$. Similarly , we can get the extra weight of Δ from other two regular $3 - bin$ items. Assume that the x_3 in [3,(ii,d3)] is from the last regular $3 - bin$. we have

$$w(\Gamma) + C \ge m(100 - \Delta) + w(T_n) + 4(\frac{6}{5}\Delta - 3\delta)$$

$$w(\Gamma) + C \le m(100 - \Delta) + 14\Delta + 6\Delta - 5\Delta - 2\Delta - \Delta$$

i.e. $w(T_n) + 3\delta \le 33\delta < 20$

Subcase 4.3. If neither [3,(ii,d3)] nor [3,(ii,d2)] occurs,
Since $w(x_4) \le x_4 - \frac{3}{4}\Delta$, $x_4 \in B^*$.
If B^* contains x_3, we assume x_3 is from the last regular $3 - bin$ and x_4 is from the last regular $4 - bin$. We have

$$w(\Gamma) + C \ge m(100 - \Delta) + w(T_n) + 4(\frac{6}{5}\Delta - 3\delta)$$

$$w(\Gamma) + C \le m(100 - \Delta) + 14\Delta + 6\Delta - 5\Delta - 2\Delta - \Delta$$

i.e. $w(T_n) + 3\delta \le 33\delta < 20$
If B^* contains an item from $6 - bins$ or $Y_i - bins$, we can get extra weight of Δ. It is easy to see that $w(B^*) \le 100 - \Delta$. We also have

$$w(\Gamma) + C \ge m(100 - \Delta) + w(T_n) + 4(\frac{6}{5}\Delta - 3\delta)$$

$$w(\Gamma) + C \le m(100 - \Delta) + 14\Delta + 6\Delta - 5\Delta - 2\Delta - \Delta$$

i.e. $w(T_n) + 3\delta \le 33\delta < 20$

Case 5. [4,(ii,d)] and [4,(iii,d)] occur at most twice
Subcase 5.1. If [3,(ii,d2)] occurs,
If [4,(ii,d)] and [4,(iii,d)] occur at most once, or [4,(ii,d)] and [4,(iii,d)] occur twice and [2,(ii,a)] do not occur,

$$w(\Gamma) + C \ge m(100 - \Delta) + w(T_n) + 4(\frac{6}{5}\Delta - 3\delta)$$

$$w(\Gamma) + C \le m(100 - \Delta) + 14\Delta + 5\Delta - 5\Delta - 2\Delta$$

i.e. $w(T_n) + 3\delta \le 33\delta < 20$
If [4,(ii,d)] and [4,(iii,d)] occur twice and [2,(ii,a)] occurs
$x_4 \ge 100 - \Delta - 3(20 + \frac{14}{5}\Delta) - (20 + \frac{9}{5}\Delta) \ge 25 + \frac{3}{2}\Delta$
i.e. $w(x_4) \le x_4 - \frac{7}{4}\Delta$. [3,(ii,d2)] is balanced. We have

$$w(\Gamma) + C \ge m(100 - \Delta) + w(T_n) + 4(\frac{6}{5}\Delta - 3\delta)$$

$$w(\Gamma) + C \le m(100 - \Delta) + 14\Delta + 5\Delta - 5\Delta - 2\Delta$$

i.e. $w(T_n) + 3\delta \le 33\delta < 20$
Subcase 5.2. If [3,(ii,d3)] occurs, we only need to consider the case that z is from the last regular $4 - bin$.

If [4,(ii,d)] and [4,(iii,d)] occur at most once,

$$w(\Gamma) + C \geq m(100 - \Delta) + w(T_n) + 4(\frac{6}{5}\Delta - 3\delta)$$

$$w(\Gamma) + C \leq m(100 - \Delta) + 14\Delta + 5\Delta - 5\Delta - 3\Delta$$

i.e. $w(T_n) + 3\delta \leq 33\delta < 20$

If [4,(ii,d)] and [4,(iii,d)] occur twice, we assume that x_3 in [3,(ii,d3)] is from the last regular $3 - bin$. We have

$$w(\Gamma) + C \geq m(100 - \Delta) + w(T_n) + 4(\frac{6}{5}\Delta - 3\delta)$$

$$w(\Gamma) + C \leq m(100 - \Delta) + 14\Delta + 5\Delta - \Delta - 5\Delta - \Delta$$

i.e. $w(T_n) + 3\delta \leq 33\delta < 20$

From the above discussion, we have that $r_m \leq \frac{13}{11}$.

References

1. E.G. Coffman, M.R.Gray and D.J. Johnson, *An application of bin-packing to multiprocessor scheduling*, SIAM Journal of Computation, 7(1978)1-17
2. D.K. Friesen, *Tighter bounds for the MULTIFIT processor scheduling algorithm*, SIAM Journal of Computation, 13(1984)179-181
3. M. Yue, H Kellerer and Z.Yu, *A simple proof of the inequality* $R_M(MF(k)) \leq 1.2 + (1/2)^k$ *in multiprocessor scheduling*, Report NO. 124, Institut fur Mathematik, Technische Universitat Graz(1988), pp. 1-10.
4. M. Yue, *On the exact upper bound for the MULTIFIT processor scheduling algorithm*, Annals of Operation Research, 24(1990)233-250
5. E.G. Coffman, Jr., et al., *Approximation algorithms for bin-packing – an updated survey* , Algorithm Design and Computer System Design, (eds.) G. Ausiello et al., CISM Courses and Lectures 284(Springer, Vienna), pp. 49-106.

A STUDY OF ON-LINE SCHEDULING TWO-STAGE SHOPS*

BO CHEN
Warwick Business School, University of Warwick, Coventry, CV4 7AL, U.K.

and

GERHARD J. WOEGINGER
TU Graz, Institut für Mathematik, Kopernikusgasse 24, A-8010 Graz, Austria.

Abstract. We investigate the problem of on-line scheduling two-stage shops with the objective of minimizing the maximum completion time. We show that for two-stage open shops, no on-line algorithm can have a worst-case performance ratio less than $\frac{1}{2}(1+\sqrt{5}) \approx 1.618$ if preemption is not allowed. We provide an on-line heuristic algorithm with worst-case performance ratio 1.875. If preemption is allowed, however, we give an on-line algorithm with worst-case performance ratio 4/3, and show that it is a best possible on-line algorithm. On the other hand, for a two-stage flow shop or job shop, we prove that no on-line algorithm can outperform a simple one, which has a worst-case performance ratio of 2.

1. Introduction

We study the problem of scheduling a set $\mathcal{J} = \{J_1, J_2, \ldots, J_n\}$ of jobs in a two-stage shop, denoted by two machines A and B. Every job J_i consists of two *operations*, one of which should be processed on machine A for a_i time units and the other should be processed on machine B for b_i time units. At any time, each machine can handle at most one job and each job can be processed by at most one machine. In an *open shop*, the order in which the two operations of each job are executed is immaterial. In a *flow shop*, however, the operation for machine A must always be processed before the other operation for machine B. In a *job shop*, a job-dependent ordering for the two operations of each job is imposed, i.e., some jobs should be processed first on machine A than on B, while the other jobs should be first on B then on A. If *preemption* is allowed, then the processing of any operation may be interrupted and resumed at a later time. Our objective is to find a schedule which minimizes the *makespan*, i.e., the maximum job completion time. Following a standard classification scheme [5], the problems we are to investigate are respectively denoted by $O2||C_{\max}$, $O2|pmtn|C_{\max}$, $F2||C_{\max}$, and $J2||C_{\max}$.

The *off-line* versions of the four problems, in which all job information, such as the total number of jobs and their processing times, is fully known in advance, can all be solved to optimality in polynomial time. Gonzalez and Sahni [3] give a linear time algorithm for both $O||C_{\max}$ and $O|pmtn|C_{\max}$ problems. Johnson [7] shows how to solve the $F2||C_{\max}$ problem in $O(n \log n)$ time. Jackson [6] extends Johnson's result to the $J2||C_{\max}$ problem. If the number of machines is increased from two to

* This research has been supported by the Spezialforschungsbereich F 003 "Optimierung und Kontrolle", Projektbereich Diskrete Optimierung.

D.-Z. Du and P. M. Pardalos (eds.), Minimax and Applications, 97–107.

three, then all the problems immediately become (ordinarily/strongly) NP-complete if preemption is not allowed (see Gonzalez and Sahni [3], [4] and Garey, Johnson and Sethi [2]). In contrast to this, if preemption is allowed, then the open shop scheduling problem is polynomially solvable for any number of machines (Gonzalez and Sahni [3]).

In this paper, we are interested in the *on-line* versions of the problems, in which the jobs are not known *a priori*: job $J_i = (a_i, b_i)$ becomes known only when J_{i-1} has already been scheduled. As soon as job J_i appears, its two operations must *immediately* and *irrevocably* be assigned to their time slots, without any information on subsequent jobs. Because of this lack of information, on-line problems usually cannot be solved to optimality. The main resort for attacking on-line problems are therefore *heuristic algorithms* that yield near optimal solutions.

The usual measure for the quality of a heuristic algorithm H is its *worst-case (performance) ratio* R^H, which is the maximum ratio, for any possible job set $\mathcal{J}$, of the makespan of the schedule produced by H for $\mathcal{J}$ and the corresponding optimum makespan (for a precise definition, see Section 2). Hence, the worst-case ratio of a heuristic algorithm indicates a guaranteed performance of the algorithm. The results we will derive in this paper are the following.

(a) For any on-line algorithm H for $O2||C_{\max}$, we have $R^H \geq \frac{1}{2}(1+\sqrt{5}) \approx 1.618$. We design a heuristic algorithm IDLMIN with worst-case ratio 1.875. Hence, the gap between the lower and upper bound of the best worst-case ratio is reasonably small.
(b) For the on-line version of the $O2|pmtn|C_{\max}$ problem, we offer a complete solution: a heuristic algorithm ALG is suggested with a worst-case ratio of 4/3, and it is shown to be best possible.
(c) Finally, for the on-line version of $F2||C_{\max}$ and $J2||C_{\max}$, we prove that no on-line algorithm can outperform a simple one, which has a worst-case ratio of 2.

2. Definitions and Preliminaries

Denote each job J_i by an ordered pair (a_i, b_i) and represent the two operations of J_i by a_i (for machine A) and b_i (for machine B), if no confusion can arise. Let $\mathcal{J}_n = \{J_1, \ldots, J_n\}$ denote the job set of the first n jobs. Define $A_n = \sum_{i=1}^n a_i$ and $B_n = \sum_{i=1}^n b_i$. Denote by $\mathrm{OPT}(\mathcal{J}_n)$ or by OPT_n, in case the set $\mathcal{J}_n$ is clear from the context, the optimum makespan for $\mathcal{J}_n$. Trivial lower bounds on OPT_n in all four problems under investigation are

$$\mathrm{OPT}_n \geq \max\{A_n, B_n\} \tag{1}$$

and

$$\mathrm{OPT}_n \geq \max_{1\leq i\leq n} (a_i + b_i). \tag{2}$$

These trivial bounds are sufficiently strong to describe the off-line solution for the problems $O2||C_{\max}$ and $O2|pmtn|C_{\max}$.

Proposition 2.1 (Gonzalez and Sahni [3]) *For the off-line versions of the scheduling problems $O2||C_{\max}$ and $O2|pmtn|C_{\max}$, the optimum makespan fulfils*

$$\text{OPT}_n = \max\{A_n, B_n, \max_{1 \le i \le n} (a_i + b_i)\}.$$

The off-line solution for the flow shop or job shop problem is based on an appropriate sorting of the jobs, and there is no analogously simple representation for the optimum makespan (cf. Johnson [7] and Jackson [6]).

For the schedule of a job set $\mathcal{J}_n$ produced by a heuristic algorithm H, we denote the makespan by $H(\mathcal{J}_n)$ or by H_n. The quality of H is measured by its *worst-case (performance) ratio*

$$R^H = \sup\{H(\mathcal{J})/\text{OPT}(\mathcal{J}) : \mathcal{J} \text{ is a list of jobs}\}. \quad (3)$$

The following lemma shows that it is trivial to devise heuristic algorithms with a worst-case ratio of 2 for the problems under consideration.

Lemma 2.2 *There exists a simple on-line heuristic algorithm H with $R^H = 2$ for all the four problems $O2||C_{\max}$, $O2|pmtn|C_{\max}$, $F2||C_{\max}$, and $J2||C_{\max}$.*

Proof. Algorithm H proceeds as follows: As it receives a new job $J_n = (a_n, b_n)$, it reserves the time interval from $A_{n-1} + B_{n-1}$ to $A_n + B_n$ on both machines for this job. Clearly, J_n can be processed within this interval while obeying all restrictions on the processing order of its operations. Moreover, $H_n = A_n + B_n \le 2\text{OPT}_n$ by inequality (1). □

3. A Lower Bound for $O2||C_{\max}$

This section deals with the on-line version of the problem $O2||C_{\max}$. We derive a lower bound of 1.618 on the worst-case ratio of *any* heuristic algorithm and demonstrate that straightforward greedy heuristic algorithms cannot outperform the trivial algorithm described in Lemma 2.2.

Define $\phi = \frac{1}{2}(1 + \sqrt{5}) \approx 1.618$ to be the positive root of the equation $x^2 = x + 1$.

Lemma 3.1 *For any on-line heuristic algorithm H for the problem $O2||C_{\max}$, $R^H \ge \phi$ holds.*

Proof. Suppose to the contrary that there is an on-line algorithm H with $R^H = \phi - \varepsilon$ for some $\varepsilon > 0$. Consider how H schedules the job $J_1 = (1, \phi)$. There are two cases, depending on which machine processes the job first.

Case 1. Job J_1 is processed first on machine A. Then define another job $J_2 = (\phi, 0)$. This implies $\text{OPT}_2 = \phi + 1$ and $\phi + 1 \le H_2 \le (\phi - \varepsilon)\text{OPT}_2 < 2\phi + 1 - \varepsilon$, which implies that operation b_1 must be finished between time $\phi + 1$ and $2\phi + 1 - \varepsilon$. Hence, the idle time on B is less than $\phi + 1$ and the idle time on A is at most $\phi - \varepsilon$. Now $J_3 = (\phi, 1 + \phi)$ arrives. Since any of the two operations of J_3 cannot fit into the existing idle intervals on its machine, we conclude that $H_3 \geq H_2 + a_3 + b_3 \geq 3\phi + 2$. Considering that $\text{OPT}_3 = 2\phi + 1$, we have $H_3/\text{OPT}_3 \geq \phi$, a contradiction.

Case 2. Job J_1 is processed first on machine B. Then its starting time on machine A is $\phi + x$ for some $x \geq 0$. Let job $J_2 = (\phi + x + \delta, 0)$ arrive with $0 < \delta < \phi\varepsilon$. Since a_2 cannot fit into the existing idle interval, we get $H_2 \geq H_1 + a_2 = 2\phi + 2x + 1 + \delta$. However, $\text{OPT}_2 = \phi + x + 1 + \delta$, and again we have a contradiction: $H_2/\text{OPT}_2 > \phi - \varepsilon$. □

Next we examine a primitive greedy heuristic algorithm GREEDY that always tries to locally minimize the makespan: Whenever a new job $J_n = (a_n, b_n)$ arrives, GREEDY fits J_n into the current schedule in such a way that the incurred makespan is as small as possible. Although this description does not fully specify the behaviour of GREEDY, the example below demonstrates that *any* algorithm that behaves like that must fail.

Example 3.2 Choose an arbitrary integer k. Feed a job $J_1 = (0, 1)$ to GREEDY. Clearly, GREEDY schedules b_1 during the time interval $[0, 1]$. Then job $J_2 = (k, 1)$ arrives. GREEDY schedules a_2 during $[0, k]$ and b_2 during $[k, k + 1]$. Finally, job $J_3 = (1, k)$ arrives. Since the idle interval $[1, k]$ on machine B is too small for processing b_3, GREEDY has to schedule b_3 after time $k + 1$. The resulting makespan is $2k + 1$, whereas the optimum makespan is $k + 1$. As k tends to infinity, the ratio tends to 2.

In the next section, we will describe another type of greedy algorithm that does not try to minimize the makespan, but tries to minimize the idle time introduced between the operations.

4. An Algorithm for $O2||C_{\max}$

In this section we design an on-line heuristic algorithm IDLMIN for the $O2||C_{\max}$ problem, and prove that the (tight) worst-case ratio of IDLMIN equals $15/8 = 1.875$. The algorithm IDLMIN works as follows:

> In the schedule produced by IDLMIN, the job sequence on both machines is the same as that in which the jobs are released, i.e., operation a_i is always processed before operation a_{i+1}, and b_i is processed before b_{i+1}. Denote by X_i (respectively, Y_i) the completion time of operation a_i (respectively, operation b_i). Whenever a new job J_n is released, IDLMIN computes the numbers $a_n + 2X_{n-1}$ and $b_n + 2Y_{n-1}$. If $a_n + 2X_{n-1} \leq b_n + 2Y_{n-1}$, then operation a_n is processed before operation b_n, otherwise, after. Both operations are started as soon as possible. For example, if operation a_n is processed first, its processing starts at X_{n-1} and the processing of b_n starts at time $\max\{Y_{n-1}, X_{n-1} + a_n\}$.

The algorithm is designed to locally minimize the incurred idle time: Consider, e.g., the situation depicted in Figure 1. In the picture on the left-hand side, the incurred idle time is $d_\ell = X_{n-1} + a_n - Y_{n-1}$, and on the right-hand side, the idle time equals $d_r = Y_{n-1} + b_n - X_{n-1}$. Observe that IDLMIN schedules operation a_n

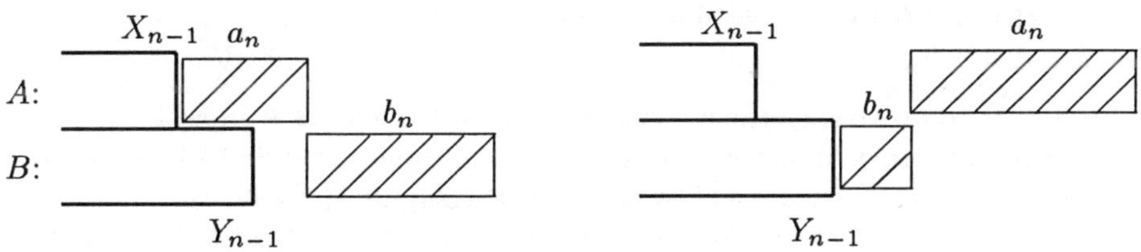

Fig. 1. Two examples for the behaviour of IDLMIN if $X_{n-1} < Y_{n-1}$.

before operation b_n if and only if $d_\ell \le d_r$. If $X_{n-1}+a_n \le Y_{n-1}$ or $Y_{n-1}+b_n \le X_{n-1}$, then IDLMIN processes J_n without introducing any idle time.

We denote by I_i the idle time right before job J_i. Note that if $I_i > 0$, then this idle time can be either on machine A or on machine B but only on one of them.

Lemma 4.1 *For any* $1 \le i \le n$, *we have* $I_i \le (a_i + b_i)/2$.

Proof. In case the scheduling of J_i does not introduce idle time, there is nothing to show. Hence assume $I_i > 0$. By the above discussion, for this case,

$$I_i = \min\{a_i + X_{i-1} - Y_{i-1}, b_i + Y_{i-1} - X_{i-1}\} \tag{4}$$

holds. The simple fact that, for any two numbers x and y, $\min\{x,y\} \le (x+y)/2$, yields that I_i is at most $(a_i + b_i)/2$. □

Lemma 4.2 *For any* $1 \le i \le n$, *we have* $X_i + Y_i \le \frac{3}{2}(A_i + B_i)$.

Proof. Since X_i (Y_i) equals the total processing time of the first i jobs and the total idle time between them on machine A (B), we have $X_i + Y_i = A_i + B_i + \sum_{k=1}^{i} I_k$, which, together with Lemma 4.1, completes the proof. □

Lemma 4.3 *For any* $1 \le i \le n$, *we have* $X_i + Y_i \le \min\{A_i + 3B_i, B_i + 3A_i\}$.

Proof. By symmetry we need only to prove $X_i + Y_i \le A_i + 3B_i$ for any i, which will be done by induction on i.

First consider the case $i = 1$. It is easy to see that $X_1 + Y_1 = a_1 + b_1 + \min\{a_1, b_1\} \le a_1 + 3b_1$. Now assume that the statement is true for $1 \le i \le k-1$. Let us consider the situation $i = k$. If $I_k = 0$ then by inductive assumption we simply have

$$X_k + Y_k = X_{k-1} + Y_{k-1} + a_k + b_k \le A_{k-1} + 3B_{k-1} + a_k + b_k \le A_k + 3B_k.$$

Therefore, we assume that $I_k > 0$. Then by (4), $I_k \le b_k + Y_{k-1} - X_{k-1}$, which leads to

$$X_k + Y_k = X_{k-1} + Y_{k-1} + I_k + a_k + b_k \le 2Y_{k-1} + a_k + 2b_k. \tag{5}$$

It is apparent that $X_{k-1} \ge A_{k-1}$ (all operations $a_1, a_2, \ldots, a_{k-1}$ are finished by X_{k-1}), which, together with Lemma 4.2, implies

$$Y_{k-1} \le \frac{1}{2}A_{k-1} + \frac{3}{2}B_{k-1}. \tag{6}$$

Combining (5) and (6) thus completes our proof of the lemma. □

Our next goal is to prove that the worst-case ratio of IDLMIN is at most $15/8 = 1.875$. Suppose to the contrary that that there exists a counter-example to this claim, i.e., a job sequence $\mathcal{J} = \{J_1, \ldots, J_n\}$ with $\text{IDLMIN}(\mathcal{J})/\text{OPT}(\mathcal{J}) > 15/8$. First, we will transform this counter-example step by step into another counter-example that *additionally* fulfils some properties.

Without loss of generality due to symmetry, we may assume that the schedule produced by IDLMIN terminates on machine B. Let $J_c = (a_c, b_c)$ denote the last job whose processing introduces idle time on machine B. (Trivially, such a job does exist: otherwise there is no idle time on machine B and the heuristic schedule is optimum). Additionally, we may assume that, in the counter-example, the following three properties are satisfied.

(P1) $B_n = \text{OPT}_n$.

Otherwise if $B_n < \text{OPT}_n$, append a job $(0, \text{OPT}_n - B_n)$ to the list $\mathcal{J}$, which does not increase the optimum makespan, but increases the makespan of the IDLMIN-schedule, and thus the ratio remains above 15/8.

(P2) $A_n = A_{c-1} + a_c$.

Otherwise, replace any job $J_i = (a_i, b_i)$ with $i \geq c+1$ by $(0, b_i)$. It is easy to see that this does not change the heuristic makespan, and does not increase the optimum makespan.

(P3) $a_c + b_c = \text{OPT}_n$.

Otherwise, let $d = \text{OPT}_n - a_c - b_c$. Replace J_c by the new job $(a_c + d, b_c + d)$ and leave all other jobs unchanged. This increases the optimum makespan OPT_n by d: A_n and B_n both are increased by d, the only job whose length increased is J_c and for J_c the total length now is $\text{OPT}_n + d$. On the other hand, the makespan produced by IDLMIN increases by $2d$: By definition of IDLMIN, the decision whether operation a_c or b_c is processed first only depends on the difference $a_c - b_c$ and on the value $2(X_{c-1} - Y_{c-1})$. Clearly, the latter value did not change and the difference between the lengths of the operations of J_c did not change. Hence, in the modified example, operation a_c still goes first and the heuristic makespan indeed increases by $2d$. We have

$$(\text{IDLMIN}_n + 2d)/(\text{OPT}_n + d) \geq \text{IDLMIN}_n/\text{OPT}_n \geq 15/8,$$

where we used the (trivial) inequality $\text{IDLMIN}_n/\text{OPT}_n \leq 2$, which follows, e.g., from $\text{IDLMIN}_n \leq A_n + B_n$. Observe that, in the modified example, properties (P1) and (P2) are still satisfied.

To summarize, we have

Lemma 4.4 *For heuristic algorithm* IDLMIN, *there exists a counter-example with worst-case ratio greater than* 15/8 *if and only if there exists such a counter-example that additionally fulfils properties (P1), (P2) and (P3).*

In the following, we thus assume that properties (P1), (P2) and (P3) are satisfied in the counter-example. Define $R = \sum_{i=c+1}^{n} b_i$ to be the overall length of all remaining operations on machine B after operation b_c. Then the makespan of the heuristic schedule equals $X_{c-1} + a_c + b_c + R$, and because of (P3),

$$X_{c-1} > \frac{7}{8}\text{OPT}_n - R. \tag{7}$$

Since operation a_c is processed before operation b_c, we have $a_c+2X_{c-1} \leq b_c+2Y_{c-1}$, which, together with (P3), yields

$$X_{c-1} - Y_{c-1} \leq b_c - \frac{1}{2}\text{OPT}_n. \tag{8}$$

Noticing that $B_n = B_{c-1} + b_c + R$, from Lemmas 4.2 and 4.3 and by (P1), (P2) and (P3), we derive that

$$X_{c-1} + Y_{c-1} \leq \frac{3}{2}A_n - \frac{3}{2}R \tag{9}$$

and

$$X_{c-1} + Y_{c-1} \leq 2\text{OPT}_n + A_n - 2b_c - 3R. \tag{10}$$

The sum of (8) and (9) minus twice of (7) yields

$$9\text{OPT}_n - 6A_n < 4b_c + 2R. \tag{11}$$

Similarly, the sum of (8) and (10) minus twice of (7) yields

$$4b_c + 4R < 4A_n - \text{OPT}_n. \tag{12}$$

By (11) and (12) we are lead to $9\text{OPT}_n - 6A_n < 4A_n - \text{OPT}_n$, which contradicts the simple fact (1).

Therefore, we have proved that there cannot exist any counter-example, which leads to the following theorem.

Theorem 4.5 *The heuristic algorithm* IDLMIN *for the* $O2||C_{\max}$ *problem has a worst-case performance ratio* 15/8.

Proof. By the above arguments, the worst-case ratio is at most 15/8 and it remains to show that this bound cannot be improved.

Consider the following example: First there is $J_1 = (0, 1)$. Then there come k jobs $J_i = (3, 1)$, $2 \leq i \leq k+1$. Check that IDLMIN constructs a schedule with $X_{k+1} = 3k$ and $Y_{k+1} = 3k + 1$. Then three more jobs $J_{k+2} = (k, k)$, $J_{k+3} = (2k + 2, 1)$ and $J_{k+4} = (2k, 6k + 5)$ arrive. For this job sequence, IDLMIN produces a makespan of $15k + 8$ whereas $\text{OPT}_{k+4} = 8k + 7$. If k tends to infinity, this ratio tends to 15/8. □

5. A Best Algorithm for $O2|pmtn|C_{\max}$

Now reconsider the on-line problem assuming that preemption is allowed. In this case we present a complete solution to $O2|pmtn|C_{\max}$.

For any schedule of the jobs $\{J_1, \ldots, J_n\}$, we denote by α_n (respectively, β_n) the overall size of all time intervals where only machine A (respectively, B) is busy, and denote by γ_n the overall size of all time intervals where both machines are busy. Observe that $A_n = \alpha_n + \gamma_n$ and $B_n = \beta_n + \gamma_n$ hold.

Lemma 5.1 *For any on-line heuristic algorithm H for the $O2|pmtn|C_{\max}$ problem, $R^H \geq 4/3$ holds.*

Proof. Suppose that $R^H < 4/3$ holds. In a first stage, we present two jobs $J_1 = J_2 = (1,1)$ to H. Since $\text{OPT}_2 = 2$, the algorithm produces a schedule with makespan less than $8/3$ and thus $\alpha_2 + \beta_2 + \gamma_2 < 8/3$, which, together with $\alpha_2 + \beta_2 + 2\gamma_2 = A_2 + B_2 = 4$, implies $\gamma_2 > 4/3$. Then in the second stage, a job $J_3 = (2,2)$ arrives. The makespan produced by H is at least $\gamma_2 + a_3 + b_3 > 16/3$ whereas $\text{OPT}_3 = 4$, a contradiction. □

In the remaining part of this section, we design and analyze a heuristic algorithm ALG whose worst-case ratio meets the 4/3 lower bound.

Algorithm ALG will never introduce simultaneous idle time on both machines, and hence its makespan equals

$$\text{ALG}_n = \alpha_n + \beta_n + \gamma_n = A_n + B_n - \gamma_n \tag{13}$$

for any n. Let us formally define ALG by induction. Suppose jobs $J_1, \ldots, J_{n-1}$ have been scheduled and a new job J_n is fed, where $n \geq 1$. Define

$$\begin{aligned} L_1^n &= \max\{\text{ALG}_{n-1}, \text{OPT}_n, \gamma_{n-1} + a_n + b_n\}, \\ L_2^n &= \frac{2}{3}(A_n + B_n), \\ U_1^n &= \text{ALG}_{n-1} + a_n + b_n, \\ U_2^n &= \frac{4}{3}\text{OPT}_n. \end{aligned}$$

We will show that

$$\max\{L_1^n, L_2^n\} \leq \min\{U_1^n, U_2^n\}. \tag{14}$$

Then, for any $L_1^n \leq d \leq U_1^n$, we can schedule job J_n, if necessary, with preemption, in such a way that, in the resulting schedule, there is no simultaneous idle time on both machines, and the makespan is d. In fact, let $p = U_1^n - d$, then $0 \leq p \leq a_n + b_n$. It is easy to check that it is always possible to insert p portion of the total processing $a_n + b_n$ of job J_n into the existing idle intervals. Hence, by concatenating the remaining portion of J_n to the schedule we get a schedule of makespan d. Now let ALG schedule the job J_n in such a way that the resulting schedule has a makespan of $\max\{L_1^n, L_2^n\}$, hence we always have $\text{ALG}_n \leq U_2^n$, as we desired.

Now let us go back to prove (14) by induction on n. If $n = 1$, since $L_1^1 = U_1^1 = a_1 + b_1$, $L_2^1 = \frac{1}{2}U_2^1 = \frac{2}{3}(a_1 + b_1)$, we are done. Suppose (14) is true for $n-1$, $n \geq 2$. Since $\text{ALG}_{n-1} \geq \frac{2}{3}(A_{n-1} + B_{n-1})$ and by (13), $\text{ALG}_{n-1} = A_{n-1} + B_{n-1} - \gamma_{n-1}$, we conclude that $\gamma_{n-1} \leq \frac{1}{3}(A_{n-1} + B_{n-1})$. Hence

$$\gamma_{n-1} + a_n + b_n \leq \frac{1}{3}(A_{n-1} + B_{n-1}) + a_n + b_n$$

$$= \frac{1}{3}(A_n + B_n) + \frac{2}{3}(a_n + b_n)$$
$$\leq \frac{4}{3}\text{OPT}_n = U_2^n,$$

which, together with $\text{ALG}_{n-1} = \max\{L_1^{n-1}, L_2^{n-1}\} \leq U_2^{n-1} \leq U_2^n$ and $\text{OPT}_n \leq U_2^n$, implies that $L_1^n \leq U_2^n$. Moreover,

$$L_2^n = \frac{2}{3}(A_n + B_n) \leq \frac{2}{3}(A_{n-1} + B_{n-1}) + a_n + b_n \leq \text{ALG}_{n-1} + a_n + b_n = U_1^n.$$

The proof of (14) is then completed by the self-evident facts that $L_1^n \leq U_1^n$ and $L_2^n \leq U_2^n$.

In conclusion, we have

Theorem 5.2 *On-line algorithm* ALG *is a best one for the* $O2|\text{pmtn}|C_{\max}$ *problem. Its worst-case ratio is* 4/3. □

6. On Flow and Job Shops

After considering the open shops, we turn our attention to the other shops, namely, flow shops and job shops. Our result in this section demonstrates a quite negative fact.

Theorem 6.1 *No on-line algorithm for the problem* $F2||C_{\max}$ *or the problem* $J2||C_{\max}$ *has a better worst-case ratio than the ratio of the simple algorithm described in Lemma 2.2, which is 2.*

Since a flow shop is a special job shop, to prove Theorem 6.1, it suffices to first suppose that there exists an on-line heuristic algorithm H for the $F2||C_{\max}$ problem with $R^H = 2 - \varepsilon$, for some real $\varepsilon > 0$, and then to demonstrate a contradiction.

Recall that every job must be processed first on machine A and then on machine B. Define a very small real number $\delta > 0$ such that $\varepsilon > 2\delta/(\delta + 1)$ holds. Choose a sufficiently large number n such that $2^{n+1}\varepsilon > (2 - \varepsilon)(n + 1)\delta$. Consider the job sequence $J_1, \ldots, J_{n+2}$ defined by $a_1 = 1$, $a_i = 2^{i-2}$ for $2 \leq i \leq n + 1$, $a_{n+2} = \delta$ and by $b_i = \delta$ for $1 \leq i \leq n + 1$ and $b_{n+2} = 2^n$. It is easy to see that for $1 \leq i \leq n + 1$, $A_i = 2^{i-1}$, $B_i = i\delta$ and $\text{OPT}_i = 2^{i-1} + \delta$ and that $\text{OPT}_{n+2} = 2^n + (n + 1)\delta$.

Lemma 6.2 *In the schedule produced by* H,
(i) *operation* a_i *is scheduled before operation* a_{i+1} *for every* $1 \leq i \leq n$,
(ii) *operation* b_i *is in the time interval* $[2^{i-1}, 2^i]$ *for every* $1 \leq i \leq n + 1$,
(iii) *operation* b_{n+2} *is started at time* $2^n + \delta$ *or later.*

Proof. (i) Suppose the statement is wrong and consider the smallest index i for which it fails. Then operations $a_1, \ldots, a_{i-1}$ and the operation a_{i+1} are all processed before operation a_i. Hence, the completion time of a_i is at least A_{i+1}. This yields that

$$H(\mathcal{J}_i) \geq A_{i+1} = 2^i > 2^i - \varepsilon + 2\delta - \delta\varepsilon \geq (2^{i-1} + \delta)(2 - \varepsilon) = (2 - \varepsilon)\text{OPT}(\mathcal{J}_i),$$

which is in contradiction to $R^H = 2 - \varepsilon$.

(ii) Because of (i), the processing of b_i cannot start before all operations a_j with $1 \leq j \leq i$ have been finished. Since H has worst-case ratio $2 - \varepsilon$, the processing of b_i cannot end later than $(2 - \varepsilon)\text{OPT}_i < 2^i$.

(iii) Statement (ii) implies that the processing of operations b_i with $1 \leq i \leq n+1$ leaves idle intervals only of lengths that are smaller than 2^n each. Hence, b_{n+2} must be processed after b_{n+1} has been completed. □

As a result of (iii) above, we see that the completion time of operation b_{n+2} is at least $2^{n+1} + \delta$. Hence, $H(\mathcal{J}_{n+2}) \geq 2^{n+2} + \delta$, which (by the choice of n) is larger than $(2 - \varepsilon)\text{OPT}_{n+2}$, a contradiction.

7. Discussion

We have studied the problems of on-line scheduling two-machine shops. For the preemptive open shop, the flow shop and the job shop, we have derived matching upper and lower bounds on the best possible worst-case performance ratios. For the non-preemptive open shop, we have shown that the best possible worst-case performance ratio is in the interval [1.618, 1.875]. Clearly, our results are just a first step towards a systematic investigation of on-line shop scheduling problems. Many questions remain open. We list some of them below.

(1) Tighten the bounds for the best worst-case ratio for the problem $O2||C_{\max}$. We feel that our lower bound 1.618 is closer to the truth than the derived upper bound. A first idea would be to improve our algorithm IDLMIN by reusing the introduced idle time intervals for processing later jobs.

(2) Consider more than two machines. We only investigated on-line problems for which the corresponding off-line versions are polynomially solvable. For problems with three machines, we do not even fully understand the combinatorial structures of their off-line solutions. Hence these problems are much harder to attack.

(3) Does *randomization* help in on-line shop scheduling problems? (cf. the paper [1].)

(4) How about other objectives? For example, instead of minimizing the maximum completion time, we could minimize the average completion time. What is the consequence of precedence constraints, release dates, due dates and other constraints for such on-line problems?

References

1. S. Ben-David, A. Borodin, R. Karp, G. Tardos and A. Widgerson, On the power of randomization in on-line algorithms, *Proc. 22nd Annual Symp. on Theory of Computing*, 1990, 379–386.
2. M. Garey, D.S. Johnson and R. Sethi, The complexity of flowshop and jobshop scheduling, *Math. Oper. Res.* **1**, 1976, 117–129.
3. T. Gonzalez and S. Sahni, Open shop scheduling to minimize finish time, *Journal ACM* **23**, 1976, 665–679.
4. T. Gonzalez and S. Sahni, Flowshop and jobshop schedules: complexity and approximation, *Oper. Res.* **26**, 1978, 36–52.

5. E.L. Lawler, J.K. Lenstra, A.H.G. Rinnooy Kan and D.B. Shmoys, Sequencing and scheduling: algorithms and complexity, *Handbooks in Operations Research and Management Science, Vol. 4: Logistics of Production and Inventory* (S.C. Graves, A.H.G. Rinnooy Kan and P. Zipkin, Eds.), North-Holland, 1993, 445–522.
6. J.R. Jackson, An extension of Johnson's results on job lot scheduling, *Naval Res. Logist. Quart.* **3**, 1956, 201–203.
7. S.M. Johnson, Optimal two- and three-stage production schedules with setup times included, *Naval Res. Logist. Quart.* **1**, 1954, 61–68.

MAXMIN FORMULATION OF THE APPORTIONMENTS OF SEATS TO A PARLIAMENT *

THORKELL HELGASON
Science Institute, Department of Mathematics, University of Iceland, Reykjavik, 3 Dunhagi, IS-107 Iceland

KURT JÖRNSTEN
Institute of Finance and Management Science, The Norwegian School of Economics and Buisiness Admistration, N-5035 Bergen-Sandviken, Norway

and

ATHANASIOS MIGDALAS
Division of Optimization, Department of Mathematics, Linköping Institute of Technology, S-581 83 Linköping, Sweden

Abstract.
We consider the problem of apportioning a fixed number of seats in a national parliament to the candidates of different parties that are elected in constituencies so as to meet party and constituency constraints. We propose a maxmin formulation of the problem. The purpose is to obtain a fair apportionment of the parliament seats, in the sense that the minimum number of voters behind any possible parliament majority is maximum. We illustrate the concepts with some examples, including an example from the last Icelandic election.

1. Introduction

In most European countries, and the Nordic states in particular, the elections to the national parliament are conducted on the basis of constituencies, i.e. disperse geographical and political regions, usually with some historical identity.

Each constituency is pre-assigned a certain *basic* number of eligible parliament seats, which, however, may increase during the actual allocation process, with some additional *epicratic* seats. Every political party participating in the elections nominates a list of candidates for each constituency.

The process of apportioning the parliament seats typically involves three steps. In the first step, the basic seats for each constituency are allocated to the regional lists of the political parties in proportion to the votes that they locally attracted.

In the second step, the epicratic seats are allocated to the parties on the basis of the national outcome. The third and final step apportions the epicratic seats of each party to the regional lists of that party observing, however, some constraints on the total number of seats assigned to each constituency.

Typically, the first step is carried out independently for each constituency and on the basis of the regional votes, as far as there is no national threshold value that must be overcome by the political parties in order for them to be represented in the

* Travel support obtained from the Nordic Council for Advanced Studies (NorFA) through the Nordic Mathematical Programming Network is gratefully acknowledged.

D.-Z. Du and P. M. Pardalos (eds.), Minimax and Applications, 109–118.

parliament. The apportionment is performed according to some standard greedy-type methods, e.g. the rules of d'Hondt and Sainte-Lague, or the largest remainder method [8]. These methods are known in the U.S.A. as the methods of Jefferson, Webster and Hamilton respectively [1].

The second step, where epicratic seats are assigned to the parties on the basis of the national outcome, is also curried out with some of the standard methods. Typically, the goal is to even out any *non-proportionality* (between the total number of basic seats allocated to a party and its national number of votes) that may have occurred during the first step.

The final step is the most complicated. It involves the allocation of the epicratic seats (assigned to a party during the second step) to the different constituencies. The allocation must respect the total number of seats and the seats already allocated to each party during the earlier steps of the process.

In this paper we study the third step of the process described above. What we basically propose is an apportionment in which the process described above is replaced by a two stage procedure. In the first stage, all basic seats are assigned to constituencies a priori, and the allocation of seats to parties, based on the national outcome, is performed according to any of the standard apportionment methods. In the second stage, a *two-dimensional apportionment* is performed based on a fairness measure.

Methods and formulations for two-dimensional apportionment are discussed in [1]-[2] and in [6]-[8]. In these papers, two-dimensional apportionment methods of the *divisor-type* are presented. Here we consider another objective for the apportionment process, namely that of fairness in the sense that the apportioned parliament seats lead to a parliament with the property that the minimum number of votes behind every possible absolute majority is as large as possible. We present a few examples which demonstrate that fairness in this sense cannot necessarily be achieved by the divisor methods. The properties of a fair apportionment is a subject for further investigation, since to our knowledge very little is known for both the *one-dimensional* and the two-dimensional case under such a fairness criterion.

This paper is organized as follows. In section 2 we introduce necessary notation and terminology and formulate two maxmin mathematical models for the two-dimensional apportionment. In section 3 we demonstrate the concepts on some illustrative examples, including an example from the last Icelandic elections. Closing, we draw some conclusions and propose interesting research topics in section 4.

2. Concepts and models

Let $C = \{1, \ldots, m\}$ be the index set of constituencies and $P = \{1, \ldots, n\}$ the index set of parties that participate in the elections. Denote by $x_{ij} \geq 0$ the number of seats that party j is allocated in constituency i after the entire process (steps one to three) has been executed.

Definition 1 *The nonnegative* $m \times n$ *integer matrix* $X = [x_{ij}]$ *is called an* apportionment *if*

$$\sum_{i \in C} \sum_{j \in P} x_{ij} = s, \tag{1}$$

where s *is the total number of seats in the parliament.*

Let the integer $m_{ij} \geq 0$ denote the number of seats allocated to party j in constituency i during the first step of the apportionment process. Then, by construction, the following holds true.

Proposition 1 m_{ij} *is a* lower bound *on* x_{ij}, *that is*

$$m_{ij} \leq x_{ij} \tag{2}$$

Proposition 2 *The row and column sums of the apportionment are bounded from above and below, that is, there exist nonnegative integers* $\{r_i\}$, $\{\rho_i\}$, $\{c_j\}$ *and* $\{\sigma_j\}$ *such that*

$$r_i \leq \sum_{j \in P} x_{ij} \leq \rho_i, \forall i \in C, \tag{3}$$

and

$$c_j \leq \sum_{i \in C} x_{ij} \leq \sigma_j, \forall j \in P. \tag{4}$$

Proof. It is a direct consequence of proposition 1 and the fact that the total number of parliament seats is fixed.

In general, the exact values of the row and column bounding constants are specified by the election law of the country. Typically, r_i is greater than or equal to the number of seats assigned to the ith constituency during the first stage of the process (in the Icelandic case, for instance, the low prescribes that $\rho_i = r_i + 1$), and $c_j = \sigma_j$ is set equal to the number of seats allocated to the jth party in the first and second stage of the process. Clearly, if the election law enforces $c_j = \sigma_j$, as for instance in Iceland, then proposition 2 implies that

$$\sum_{i \in C} x_{ij} = c_j, \forall j \in P, \tag{5}$$

where $\sum_{j \in P} c_j = s$. That is, re-allocation of seats allocated to a party during the first stage of the process is not permitted.

The column and row bounds implied by the election law must of course be consistent, i.e. satisfy the following conditions:

$$\sum_{i \in C} \sum_{j \in P} m_{ij} \leq s, \tag{6}$$

$$\sum_{j \in P} m_{ij} \leq \rho_i, \forall i \in C, \tag{7}$$

$$\sum_{i \in C} m_{ij} \leq \sigma_j, \forall j \in P, \tag{8}$$

$$\sum_{j \in P} c_j \leq s \leq \sum_{j \in P} \sigma_j, \tag{9}$$

and

$$\sum_{i \in C} r_i \leq \sum_{i \in C} \rho_i. \tag{10}$$

The number of epicratic seats equals to the slack in (7). In the case of Iceland, the election law enforces $\sum_{i \in C} r_i = s - 1$ and, therefore,

$$\sum_{j \in P} x_{ij} = r_i \tag{11}$$

for all constituencies but one for which

$$\sum_{j \in P} x_{ij} = r_i + 1. \tag{12}$$

However, this does not mean that the number of epicratic seats in Iceland is only 1. Indeed, 13 out of 63 seats in *Althing*, the Icelandic parliament, are of this type. Obviously in this case the upper bound $\{\rho_i\}$ are insignificant. In view of (11)-(12) we will subsequently, without loss of the generality, restrict our attention to the case where equalities hold in (3)-(4). The general case is of similar nature.

Proposition 3 *For given consistent parameters s, $\{r_i\}$, $\{\rho_i\}$, $\{c_j\}$ and $\{\sigma_j\}$, the polytope, T, defined by (3)-(4) and non-negativity conditions on $\{x_{ij}\}$, is nonempty and integral.*

Proof. Non-emptiness is a consequence of consistency. Integrality follows immediately from the total unimodularity of the constraint matrix, which is the classical transportation matrix.

An immediate consequence of the proposition is that any extreme point of the polytope T represents an apportionment (Definition 1).

Definition 2 *The set of all feasible integral points of the polytope T is denoted $\Pi(s, \{r_i\}, \{\rho_i\}, \{c_j\}, \{\sigma_i\})$, or for short Π, and is called a* parliament.

Clearly, a parliament consists of all feasible apportionments.

Let v_{ij} denote the number of votes that party j received in constituency i and let $V = [v_{ij}]$ be the corresponding $m \times n$ matrix.

Definition 3 *The pair $E = (\Pi, V)$ is called an* election.

The goal of the apportionment process discussed here is to find an apportionment X which, for a given election (Π, V), is in Π and, moreover, it is proportional to V in some sense.

We proceed now in formulating two apportionment models for the two-dimensional case in which the goal is to maximize the minimum number of votes behind any possible absolute majority in the parliament. This is a reasonable measure of fairness and it is different from the normal proportionality measures.

Let z_{ij} denote the number of elected parliament members from constituency i and party j that will be part of an absolute majority. For x_{ij} as before, the apportionment problem can be stated in terms of the following maxmin mathematical programming model, where we define $\infty \cdot 0 \equiv 0$:

$$\max_{x} \ \min_{z|x} \sum_{i \in C} \sum_{j \in P} \left(\frac{v_{ij}}{x_{ij}} \right) z_{ij} \tag{13}$$

s.t.

$$\sum_{j \in P} x_{ij} = r_i, \ \forall i \in C \tag{14}$$

$$\sum_{i \in C} x_{ij} = c_j, \ \forall j \in P \tag{15}$$

$$z_{ij} \leq x_{ij}, \ \forall i \in C, \ \forall j \in P \tag{16}$$

$$\sum_{i \in C} \sum_{j \in P} z_{ij} = \lfloor \frac{s}{2} \rfloor + 1 \tag{17}$$

$$x_{ij}, z_{ij} \geq 0 \text{ and integers}, \forall i \in C, \ \forall j \in P \tag{18}$$

An alternative formulation of the apportionment problem as a non-linear bilevel programming problem can be posed if we restrict our attention to apportionments X derived by divisor type methods. In [8], it is shown that a two-dimensional apportionment, based on divisor methods, can be found by solving the following linear multi-assignment problem:

$$\max \sum_{i \in C} \sum_{j \in P} \sum_{k \in K_{ij}} (\ln \frac{v_{ij}}{d_k}) x_{ij}^k \tag{19}$$

s.t.

$$\sum_{i \in P} \sum_{k \in K_{ij}} x_{ij}^k = c_j, \forall j \in P \tag{20}$$

$$\sum_{j \in P} \sum_{k \in K_{ij}} x_{ij}^k = r_i, \forall i \in C \tag{21}$$

$$0 \leq x_{ij}^k \leq 1, \forall k \in K_{ij}, \forall i \in C, \forall j \in P. \tag{22}$$

Here, d_k denotes the kth divisor. For the two-dimensional version of d'Hondt's apportionment method, for instance, the divisors $(d_1, d_2, \ldots)$ are chosen as $(1, 2, 3, \ldots)$, whereas in the apportionment method of Saint-Lague, the divisors are selected as $(1, 3, 5, \ldots)$. x_{ij} has been disaggregated into x_{ij}^k-variables, each representing the kth seat in constituency i for party j, where $|K_{ij}| = \min\{r_i, c_j\}$.

Proposition 4 *The solution to the linear multi-assignment problem (19)-(22) yields a two-dimensional apportionment in the sense of [3].*

The proof of this proposition is given in [8], where derivation and motivation of the objective function (19) is also presented.

In this case, requiring fairness in the apportionment of the seats leads to the following bilevel formulation:

$$\max_{x,z} \sum_{i\in C}\sum_{j\in P}\sum_{k\in K_{ij}} (\ln \frac{v_{ij}}{d_k}) x_{ij}^k \tag{23}$$

s.t. (19)-(22) and where Z solves the second level problem

$$\min_{z|x} \sum_{i\in C}\sum_{j\in P} (\frac{v_{ij}}{\sum_{k\in K_{ij}} x_{ij}^k}) z_{ij} \tag{24}$$

s.t.

$$z_{ij} \leq \sum_{k\in K_{ij}} x_{ij}^k, \forall i \in C, \forall j \in P \tag{25}$$

$$\sum_{i\in C}\sum_{j\in P} z_{ij} = \lfloor \frac{s}{2} \rfloor + 1 \tag{26}$$

$$z_{ij} \geq 0 \text{ and integer}, \forall i \in C, \forall j \in P \tag{27}$$

$$x_{ij}^k \in \{0,1\}, \forall k \in K_{ij}, \forall i \in C, \forall j \in P \tag{28}$$

In both models, for a fixed apportionment $X = B$, the problem of finding the absolute majority which is backed up by the minimum number of voters is expressed by the model below, where we assume that $\infty \cdot 0 \equiv 0$:

$$\min \sum_{i\in C}\sum_{j\in P} (\frac{v_{ij}}{b_{ij}}) z_{ij} \tag{29}$$

$$\text{s.t.} \sum_{i\in C}\sum_{j\in P} z_{ij} = \lfloor \frac{s}{2} \rfloor + 1 \tag{30}$$

$$0 \leq z_{ij} \leq b_{ij}, \text{and integer}, \forall i \in C, \forall j \in P \tag{31}$$

It is easily seen that the following result holds.

Proposition 5 *The integer program (29) - (31) is greedily solvable in polynomial time.*

On the other hand, for fixed $Z = P$ the first problem reduces to maximizing a convex function subject to linear constraints for a bipartite flow with lower bounds. This is a Global Optimization problem [9], and although of special structure it does remain hard in general [5]. Thus, although the integer program (29) - (31) is trivial, there is no obvious way of using this property in the design of an efficient projection algorithm.

In the case of the bi-level program, the complexity results are even more disappointing as even the linear case of it is, in general, an NP-hard problem [11]. Hence, whether proposition 5 can be used in the design of an optimal algorithm for the overall problem is at this stage an open question. Moreover, we demonstrate subsequently that the restriction to divisor type methods is indeed a restriction of the model.

TABLE I
Two-dimensional apportionment problem

	Party A	Party B	Party C	r_i
Const. I	110	210	230	4
Const. II	400	150	50	5
Const. III	280	220	180	6
c_j	6	5	4	s = 15

3. Illustrative examples

As a motivation to the maxmin model we demonstrate in this section, with the means of a few examples, that standard divisor methods of apportionment do not necessarily meet our objective of fairness.

The first example is on one-dimensional apportionments. Let $V = [500, 300, 100]$ be a vector of votes to three different parties in some constituency where 6 seats are to be apportioned. According to d'Hondt's rule, the apportionment is $X = [4, 2, 0]$. Hence, the minimum number of votes behind an absolute majority is 500. However, the apportionment of V according to the method of the largest reminder is $X = [3, 2, 1]$, which is also the apportionment according to the rule of Saint-Lague. Here the minimum number of votes behind an absolute majority is 566. Hence, d'Hondt's apportionment does not maximizes the objective function of the maxmin program.

In the second example, we consider the two-dimensional apportionment problem in table I. For this example, d'Hondt's rule gives the apportionment $X = \begin{pmatrix} 0 & 2 & 2 \\ 4 & 1 & 0 \\ 2 & 2 & 2 \end{pmatrix}$. The minimum number of seats for an absolute majority is equal to 790.

On the other hand, Saint-Lague's apportionment is $X = \begin{pmatrix} 1 & 1 & 2 \\ 3 & 2 & 0 \\ 2 & 2 & 2 \end{pmatrix}$ giving a minimum number 775 for an absolute majority. Hence, it contradicts the conclusion of the one-dimensional example.

The distribution of votes among parties and constituencies in the Icelandic election from 1991 [6] is given in table II.

The apportionment according to d'Hondt's rule is given in table III. This apportionment coincides with the one obtained under the current Icelandic election law, which prescribes the three step procedure mentioned in the introduction of the paper [8]. The table gives the current composition of the Icelandic parliament.

For this apportionment the minimum number of voters behind any absolute majority is 44018. However, this number does not satisfy our fairness criterion. An improved, in the maxmin sense, apportionment is shown in table IV. This is obtained by pivoting on the matrix (Table III) similarly to transport simplex method. For instance, increasing the number of seats in constituency **VL** for party **V** from 0 to 1 and decreasing the party's seats in constituency **RV** from 3 to 2, while increasing the number of seats for party **D** in **RV** from 9 to 10 and decreasing it in **VL**

TABLE II
Vote distribution in the Icelandic elections of 1991

	Party A	Party B	Party D	Party G	Party V	r_i
RV	9165	6299	28731	8259	7444	18
RN	9025	5386	15851	4458	2698	11
VL	1233	2485	2525	1513	591	5
VF	893	1582	1966	619	443	6
NV	739	2045	1783	1220	327	5
NE	1517	5383	3710	2794	750	7
AL	803	3225	1683	1519	348	5
SL	1079	3456	4577	2323	467	6
National result	24454	29861	60826	22705	13068	
c_j	10	13	26	9	5	$s = 63$

TABLE III
Composition of the Icelandic parliament

	Party A	Party B	Party D	Party G	Party V
RV	3	1	9	2	3
RN	3	1	5	1	1
VL	1	1	2	1	
VF	1	1	2	1	1
NV		2	2	1	
NE	1	3	2	1	
AL	1	2	1	1	
SL		2	3	1	

from 2 to 1, we obtain an absolute majority with 44632 voters.

The behavior of the standard apportionment methods illustrated in the previous examples appears also in non-European cases. For instance, in [1, Tables on pp. 167-168] several examples of Congressional elections are given and the one-dimensional apportionments according to Adams, Dean, Hill, Webster, Jefferson and Hamilton are presented. We experimented with the data for the 1900 Congressional allocation and calculated the minimum number of votes behind every possible majority. It turned out that Hill's method had the best result with 36251516 votes behind the minimum majority. However, it is trivial to construct a different apportionment in which the minimum number of votes behind a minimum majority is larger. For instance, shifting a seat from Nevada to Alabama gives an increase of 142199 voters behind the minimum majority. Note that these cases are dealing with states and not parties.

TABLE IV
A better, in the maxmin sense, composition of the Icelandic parliament

	Party A	Party B	Party D	Party G	Party V
RV	3	1	10	2	2
RN	3	1	5	1	1
VL	1	1	1	1	1
VF	1	1	2	1	1
NV		2	2	1	
NE	1	3	2	1	
AL	1	2	1	1	
SL		2	3	1	

4. Discussion

As shown in the previous section, apportionments based on divisor methods need not satisfy the maxmin fairness criterion. Neither seems to be any divisor method that dominates in this respect any other method in this category. Moreover, real life examples verify these conclusions. Therefore it seems appropriate to consider further studies of the proposed maxmin model. From the mathematical point of view, future research must involve the study of the model in order to classify it with respect to its complexity and develop efficient special purpose algorithms which obey the political restrictions. For instance, the pivoting strategy applied to the Icelandic case may be a good basis for the development of a heuristic procedure. Notably, the pivoting procedure does not alter the overall number of seats apportioned to a party, although it alters the constituencies where the seats are gained. In this sense it respects the resistance of political parties to loosing apportioned seats to their opponents.

Strategically, such an analysis is important as in some countries, Greece among them, the election law (i.e. the apportionment procedure) is not part of the country's constitution, and the party in power can change it before an election in an attempt to get some advantage. For instance in Greece, the socialist party gained absolute parliament majority in the last election, while the conservative party failed to do so in the election before, although they got the same percentage of votes. Clearly, the result of the last election was a consequence of an inappropriate strategy, seen with conservative eyes, of the conservative government, while the result of the elections before was a consequence of a good strategy, seen with socialist eyes, of the socialist government, which although it lost the governmental power could prohibit absolute majority of the conservatives in the parliament.

Other issues to be investigated are the properties of the suggested apportionment process in terms of monotonicity and consistency as well as whether they respect the quotas [3]. Furthermore, the suggested apportionment procedure needs to be checked in terms of bias towards small/large constituencies and parties. A major concern is whether the classical paradoxes (e.g. the *Alabama Paradox* and the *New State Paradox* [1]) occur.

References

1. M.L. Balinski and H.P. Young (1982) Fair Representation Meeting the Ideal of One Man, One Vote. Yale University Press, New Haven.
2. M.L. Balinski and G. Demange (1989) Algorithms for Proportional Matrices in Reals and Integers, Mathematical Programming 45, 193-210.
3. M.L. Balinski and G. Demange (1989) An Axiomatic Approach to Proportionality between Matrices, Mathematics of Operations Research 14, 700-719.
4. V. Elberling (1922) Om Samtidigt Valg af Flere Representater for Samme Kreds (in Danish). Published as Appendix K in the Report on the Danish Electoral Low Commission of 1921.
5. G.M. Guiswite and P.M. Pardalos (1993) Complexity Issues in Nonconvex Network Flow Problems, in "Complexity in Numerical Optimization", World Scientific, pp. 163-179
6. T. Helgason (1991) Apportionment of Seats in the Icelandic Parliament, Working paper Mar-1991, University of Iceland, Reykjavik, Iceland.
7. T. Helgason and K. Jörnsten (1991) Om Matrix Apportionments, Working paper Oct-1991, University of Iceland, Reykjavik, Iceland.
8. T. Helgason and K. Jörnsten (1994) Entropy of Proportional Matrix Apportionment, in "Proceedings from the Nordic Mathematical Programming Meeting in Linköping", K. Holmberg (ed.), Linköping University, Linköping, Sweden.
9. R. Horst and H. Tuy (1990) Global Optimization - Deterministic Approaches, Springer-Verlag, Berlin.
10. A. Hylland (1978) Allotment Methods - Procedures for Proportional Distribution of Indivisible Entities. Unpublished manuscript. John F. Kennedy School of Government, Harvard University. Reprinted as Working paper 1990/11 from the Norwegian School of Management, Oslo, Norway.
11. H. Tuy, A. Migdalas and P. Värbrand (1993) A Global Optimization Approach for the Linear Two-level Program, Journal of Global Optimization 13, 1-23

ON SHORTEST K-EDGE CONNECTED STEINER NETWORKS WITH RECTILINEAR DISTANCE

D. FRANK HSU
Department of Computer and Information Science, Fordham University, Bronx, New York 10458-5198, USA.

XIAO-DONG HU *
Institute of Applied Mathematics, and the Asian-Pacific Operations Research Center, Chinese Academy of Sciences, Beijing 100080, China.

and

YOJI KAJITANI
Graduate School of Information Science, Japan Advanced Institute of Science and Technology, Ishikawa, 923-12 Japan.

Abstract. In this paper we consider the problem of constructing a shortest k-edge connected Steiner network in the plane with rectilinear distance for $k \geq 2$. Given a set of P of points, let $l_k(P)$ denote the length of a shortest k-edge connected Steiner network on P divided by the length of a shortest k-edge connected spanning network on P. We prove that $\frac{3}{4} \leq \inf\{l_2(P)|P\}\frac{6}{7}$, $\frac{2}{3} \leq \inf\{l_3(P)|P\} \leq \frac{7}{8}$. We also show that if all points in P are on the sides of the rectilinear convex hull of P, then $l_k(P) = 1$, for k is even, and $l_k(P) > \frac{3k+1}{3k+3}$, for k is odd.

1. Introduction

Given a set of points P, let $N(V, E)$ be a network consisting of a set of vertices $V \subseteq P$ and a set of edges $E \subseteq V \times V$. $N(V, E)$ is called a Steiner network on P and in particular it is called a spanning network on P if $V = P$. The length of $N(V, E)$ is defined to be the total length of all edges in $N(V, E)$ and denoted by $l(N(V, E))$. A shortest k-connected Steiner network on P is a k-connected Steiner network on P with shortest length. Let $l_k(P)$ denote the length of a shortest k-edge connected Steiner network on P divided by the length of a shortest k-edge connected spanning network on P. For decades lots of research work has been done on shortest k-connected Steiner networks for $k = 1$, which is known as the Steiner Minimum Tree Problem. This includes the four important cases of euclidean, rectilinear, graphic and phylogenetic distances and some of their generalizations. In particular, the Gilbert-Pollak conjecture, which is $\inf\{l_1(P)|P\} = \sqrt{3}/2$ with euclidean distance, has been shown to be true by D.-Z. Du and F. K. Hwang [3]. In the case with rectilinear distance, it was shown by F. K. Hwang [11] that $\inf\{l_1(P)|P\} = 2/3$. See the survey paper on Steiner tree problem by F. K. Hwang and D. S. Richards [12]. Recently extensive study has also been made on shortest k-connected Steiner networks for $k \geq 2$, which is a special case of the so called generalized Steiner problem or survivable network design problem. In particular, C. L. Monma et al. [14] proved

* This author's work was support in part by the National Natural Science Foundation of China.

D.-Z. Du and P. M. Pardalos (eds.), Minimax and Applications, 119–127.

that $l_2(P) \geq 3/4$, for any P, where the length is defined by a nonnegative, symmetric function on $P \times P$ satisfying the triangle inequality. The readers are referred to the survey by M. Grötschel et al. [7]. More recently, the first two authors of this paper started an extensive study of the Steiner Network Problem. More specifically, they [9,10] have obtained bounds for the Steiner ratios $l_2(P)$ and $l_3(P)$ with the euclidean distance:

$$\frac{\sqrt{3}}{2} \leq \inf\{l_2(P)|P\} \leq 2\frac{\sqrt{3}+2}{\sqrt{3}+6} \text{ and } \frac{\sqrt{3}}{2} \leq \inf\{l_3(P)|P\} \leq \frac{3+\sqrt{3}}{5}.$$

In this paper we focus on shortest k-edge connected Steiner networks associated with the rectilinear distance (also known as Manhattan distance). The problem is defined as follows : given two points on the plane $p_i = (x_i, y_i), i = 1, 2$, where (x_i, y_i) is Cartesian coordinates of points $p_i, i = 1, 2$, the rectilinear distance between p_1 and p_2 is $|x_1 - x_2| + |y_1 - y_2|$. Applications of constructing shortest Steiner networks with rectilinear distance occur in printed circuit technology, or more generally in a variety of VLSI design problems. The rectilinear case is in some way similar to the euclidean case studied in [9,10]. However they are quite different in many other ways. In the euclidean distance case, the triangle inequality always holds strictly unless the three points studied are on one straight line. This property does not hold in general for the rectilinear distance. Consequently some new length reduction argument has to be introduced. On the other hand, since lines are restricted to run in just horizontal and vertical directions, some problems which remain open in [9,10] with respect to euclidean distance have been solved in rectilinear version. In this paper, we prove that $\frac{3}{4} \leq \inf\{l_2(P)|P\} \leq \frac{6}{7}$ and $\frac{2}{3} \leq \inf\{l_3(P)|P\} \leq \frac{7}{8}$. We also show that if all points in P are on the sides of rectilinear convex hull of P, then $l_k(P) = 1$, for k is even, and $l_k(P) \geq \frac{3k+1}{3k+3}$, for k is odd.

Finally, we remark that the Steiner ratio problem can be viewed as a max-min problem in the following way. Let $ST(P)$ be the set of k-edge connected Steiner networks on P and $SP(P)$ be the set of k-edge connected spanning networks on P. Then

$$\frac{1}{\inf_P l_k(P)} = \sup_P \max_{S \in ST(P)} \min_{T \in SP(P)} \frac{\ell(T)}{\ell(S)}$$

where $\ell(T)$ denotes the total length of network T.

2. Technical Preliminaries

For the convenience of readers, we state four general propositions about minimum-weight k-connected spanning networks with a nonnegative, symmetric weight function satisfying the triangle inequality. These were proved previously in [1,4,14] and will be used in our discussion in the next sections.

Lemma 1 [4] *For any set of points P, the minimum weight of a two-edge connected spanning network on P is equal to the minimum weight of a two-vertex connected spanning network on P.*

It is not difficult to see that there is no similar result for the case of k-connected spanning network when $k \geq 3$.

Lemma 2 [1] *For any set of points P, and $k \geq 2$, there exists a minimum-weight k-edge connected Steiner network satisfying the following conditions:*

(1) Every vertex has degree k or $k+1$;

(2) Removing any $1, 2, \cdots$, or k edges does not leave the resulting connected components all k-edge connected.

According to Lemma 2, we assume in the study of shortest k-edge connected Steiner network that

(♣) All k-edge connected Steiner networks to be considered satisfy condition (1).

From assumption (♣) we can easily deduce that

(†) Every two-edge connected Steiner network is also a two-vertex connected Steiner network.

(‡) There is no multiple edge in any shortest two-connected Steiner network.

Due to property (†), we will use the term 'two-connected' without specifying 'two-edge' or 'two-vertex'.

Lemma 3 [14] *For any set V of vertices, the weight of an optimal traveling salesman cycle of V (visiting all the vertices in V) is no greater than $\frac{4}{3}$ times the weight of an optimal two connected spanning network on V. Furthermore, this bound can be approached arbitrarily closely by a class of graphs with their canonical distance function.*

In fact, it is easy to find an example to show that the $\frac{4}{3}$-bound is also tight in the above sense with the rectilinear distance.

Lemma 4 [14] *For any set of vertices P, $l_2(P) \geq \frac{3}{4}$.*

Given a set of points P in the plane, let $N_k(V, E)$ denote a k-edge connected Steiner network on P. Points in P are called regular points and points in $V \setminus P$ are called Steiner points. In the following discussion, points in V are denoted by circles with black dots inside when either we do not know whether they belong to P, or what we claim is independent of that. In addition, for the simplicity of notation we will often use N in place of $N_k(V, E)$, unless we need to specify V, E and k, In order to simplify our arguments, we will only consider a special class of k-connected Steiner networks $N_k(V, E)$ on P which satisfy the following condition

(♠) There does not exist another k-connected Steiner network $N'_k(V', E')$ on P such that $l(N'_k(V', E')) \leq l(N_k(V, E))$ and $V' \subset V$.

Now we introduce some definitions with respect to rectilinear geometry by referring to parallel terminologies used in euclidean geometry. This will enable us to present our results strictly as well as concisely.

Definition 1 Given two points p and q in the plane, *the rectilinear edge* between p and q is defined as a set of all finite sequences of horizontal (vertical) line segments alternating with vertical (horizontal) lines such that

(1) p and q are connected by them;

(2) The length of all line segments in each sequence is equal to the rectilinear distance between p and q.

The length of edge between p and q is defined as the rectilinear distance between p and q.

Remark 1 According to above definition, the edge between p and q consists of infinite elements, each of which has the same length. This is in accordance with the well-known euclidean geometrical property that the straight line segment between two points is the shortest line connecting them. However, we will denote the edge between p and q by pq or qp, which is used to represent one certain element rather than the whole set and also indicated its length in mathematical formulae.

Definition 2 Let S be a set of points in the plane. S is called a *convex set with respect to rectilinear distance*, if for any two points s_1 and s_2 in S, there exists an edge s_1s_2 such that every point on edge s_1s_2 is in S.

Remark 2 The concept of the convex set in Definition 2 matches with that in the euclidean plane. Still more, it can be verified that if a set is convex set in the euclidean plane, then it is also a convex set with respect to the rectilinear distance. However, the converse is not true.

Definition 3 Let S be a set of points in the plane. *The convex hull* of S *with respect to rectilinear distance* is a set of smallest rectilinear convex sets containing S, where smallest set means a set with minimum size.

Remark 3 By contrast with corresponding concept in the euclidean plane, the convex hull of a set is a collection of convex sets, which may include infinite numbr of convex sets. We can prove that the total length of sides in each convex set is equal. Hence we denote the convex hull of S by $c(S)$, which will be used to indicate a single convex set in the convex hull. We further denote the sides of $c(S)$ by $\overline{c}(S)$, which can be considered as a cycle in a way similar to that we use to define $\mathcal{C}(N)$ in the next section.

From now on, edge, length, convex set and convex hull are all supposed to be understood in the rectilinear sense as we have defined above.

Before we proceed to the next section we recollect some basic properties of Steiner minimal trees. All leaves in a Steiner minimal tree are regular points. A Steiner tree is called a *full Steiner tree* if every regular point is a leaf. If a regular point in a Steiner tree is not a leaf, then the Steiner tree can be split at this regular point. In this way, any Steiner tree T can be decomposed into an edge-disjoint union of smaller full Steiner trees, which are called *full components* of T. *The size of a full component* is defined to be the number of regular points in the full component.

Lemma 5 [8] *All Steiner points in a Steiner minimal tree have degree either three or four.*

Lemma 6 [11] *For any set of points P, $l_1(P) \geq \frac{2}{3}$.*

3. Main Results

Most of our arguments presented in this paper will use the following two simple operations. And proofs are provided without details.

Definition 4 Let N be an k-edge connected Steiner network on P. Let s and t be two adjacent Steiner points of degree three, where s and t are also adjacent to two other points s_i and t_i, $i = 1, 2$, respectively. Then the operation of deleting s and t together with all edges incident to them and adding two edges s_1s_2 and t_1t_2 is called *Steiner lifting* of N at st. It is further called *admissible* if the resulting network remains k-edge connected.

Lemma 7 *If N is a shortest k-edge connected Steiner network on P, then N has no admissible Steiner lifting.*

Proof. It follows directly from the definitions.

Definition 5 Let N be an k-edge connected Steiner network on P. Let p be the position where edges ab and cd cross. Then the operation of replacing ab and cd with ac and bd is called *crossing lifting* of N at p. It is further called *admissible* if the resulting network remains k-edge connected.

Remark 4 The resultant network obtained by crossing lifting has fewer crossing than the original network. Moreover, its length is shorter than or equal to that of the original network.

The general description of an k-edge connected Steiner network is that a cycle encloses some connected subnetworks. More precisely, given a set of points P on the plane, let $N = N_k(V, E)$ be an k-edge connected Steiner network on P. As far as the length is concerned, we may assume that N is in $c(P)$. Now imagine that we stand at a regular point on $\overline{c}(P)$ facing $c(P)$. Start proceeding along line segments in N in such a way that when reaching a point in N or an intersection of two edges of N we take the line segment at our left side. At the end we will surely return to our starting point. Let $\mathcal{C}(N)$ denote the route we go through, which obviously is a cycle. Let $\mathcal{V}(N)$ represent all the positions on the route where edges meet or cross. In addition, let $\mathcal{V}^\star(N)$ denote all the places in $N \setminus \mathcal{C}(N)$ where edges meet or intersect. Now define a new network $\overline{N}(\overline{V}, \overline{E})$, where $\overline{V} = \mathcal{V}(N) \cup \mathcal{V}^\star(N)$, and $\overline{E}$ includes every edge in E unless it crosses other edges. In this case $\overline{E}$ includes it as several edges such that any two of them incident to exactly one crossing. It is easy to see that $\mathcal{C}(N) = \mathcal{C}(\overline{N}(\overline{V}, \overline{E}))$, and $l(N) = l(\overline{N}(\overline{V}, \overline{E}))$. We note that $\overline{N}(\overline{V}, \overline{E})$ may not satisfy condition (♣).

Lemma 8 *For any P, there exists a shortest two-connected spanning network on P without crossing.*

Proof. By contradiction argument. Suppose that there exists a set P_0 such that every shortest two-connected spanning network on P_0 has some crossings. Let $N(P_0, E)$ be the one with minimal number of crossings. Then we are able to produce a contradiction by finding an admissible crossing lifting of $N(P_0, E)$.

Remark 5 As a contrast, for some P, every shortest three-edge connected Steiner network $N(V, E)$ on P has some crossings. This means that V is a proper subset of $\mathcal{V}(N(V, E)) \cup \mathcal{V}^\star(N(V, E))$. In addition, $N(V, E) \setminus \mathcal{C}(N(V, E))$ may not be connected.

Lemma 9 *Let $N(V, E)$ be a three-edge connected Steiner network on P. Then $\overline{N}(\overline{V}, \overline{E}) \setminus \mathcal{C}(N(V, E))$ is a connected spanning network on $\mathcal{V}(N(V, E)) \cup \mathcal{V}^\star(N(V, E))$, and a connected Steiner network on $\mathcal{V}(N(V, E))$.*

Proof. If $\overline{N}(\overline{V}, \overline{E}) \setminus \mathcal{C}(N(V, E))$ is not connected, then there exist two edges on $\mathcal{C}(N(V, E))$ such that removing them will cause $N(V, E)$ to be disconnected.

3.1. Two-Connected Steiner Networks

In this part, N stands for a shortest two-connected Steiner network on P satisfying assumption (♣) and condition (♠). N is called basic if there is no cycle in $N \setminus \mathcal{C}(N)$, and nonbasic otherwise.

Theorem 1 *If N is basic, then*

(1) There are no two Steiner points in $N \setminus \mathcal{C}(N)$ which are adjacent to each other.

(2) $N \setminus \mathcal{C}(N)$ is a union of 3-size Steiner minimal trees.

(3) There is no regular point in $N \setminus \mathcal{C}(N)$ which is adjacent to two Steiner points.

Proof. (1) Suppose that there exists an edge ts in $N \setminus \mathcal{C}(N)$, where both s and t are Steiner points. Note that s and t both have degree three due to Lemma 5 and assumption (♣). We can produce a contradiction by showing that N has an admissible Steiner lifting at ts, which is impossible due to Lemma 7.

(2) It immediately follows from (1) and the definition of 3-size Steiner tree.

(3) By contradiction argument again. Suppose that there is a regular point r which is adjacent to two Steiner points s and t. Without loss of generality, let $rs \geq rt$. We can deduce that $N \setminus \{as, bs, rs, rt, ct, dt\} \cup \{ab, rc, rd\}$ is also a two-connected network on P. But its length is either shorter than (when $rs > rt$) or equal to (when $rs = rt$) the length of N, which contradicts condition (♠).

Theorem 2 *If all points of P are on the sides of the convex hull of P, i.e., $P \subset \overline{c}(P)$. Then N and $\overline{c}(P)$ are shortest two-connected spanning networks on P.*

Proof. Let N_0 be a shortest two-connected Steiner network on P without crossing. We can prove that $\mathcal{C}(N_0) = \overline{c}(P)$, which implies that $\overline{c}(P)$ is a shortest two-connected spanning networks on P. Consequently, condition (♠) deduces the conclusion that N is a spanning network.

Theorem 3 *Suppose that all points in P except one are on the sides of the convex hull of P. Then N is a shortest two-connected spanning network on P.*

Proof. Applying a similar analysis of Theorem 2 leads to the conclusion that the only possible way of cutting length short is to add two Steiner points adjacent to the unique point which is not on the sides of the convex hull of P. But using the same proof of Theorem 1 (3) can exclude this possibility.

Corollary 1 *If $|P| \leq 5$, then N is a shortest two-connected spanning network on P.*

Proof. Note that every point in P is on $\overline{c}(P)$ if $|P| \leq 4$, and then the corollary follows immediately from Theorems 2 and 3.

In fact, there is a simple example consisting of six points which can show that Theorem 3 and Corollary 1 can not be improved.

Theorem 4 $\frac{3}{4} \le \inf\{l_2(P)|P\} \le \frac{6}{7}$.

Proof. The left hand side of the inequality follows from Lemma 4. The right hand side of the inequality follows from a special class of set P_n such that the length of a shortest two-connected Steiner network on P_n divided by that of a shortest two-connected spanning network on P_n achieves $\frac{6}{7}$, as n approaches infinity.

Theorem 5 *There exists a polynomial-time approximation algorithm for constructing a shortest two-connected Steiner and spanning network with guaranteed worst-case performance ratios* $\frac{3}{2}$ *and 2, respectively.*

Proof. Christofides' heuristic [2] for TSP can be adopted here as a polynomial-time algorithm for constructing a shortest two-connected Steiner and spanning network with guaranteed worst-case performance ratios $\frac{3}{2}$ and 2, respectively.

3.2. Three-Edge Connected Steiner Networks

In this part, $N = N(V, E)$ stands for a shortest three-edge connected Steiner network on P satisfying assumption (♣) and condition (♠), unless it is specified otherwise.

Theorem 6 *Each Steiner point of* $N(V, E)$ *has degree three.*

Proof. Suppose that there is a Steiner point s which has degree four. Then we can produce a three-edge connected Steiner network $N'(V', E')$ with $l(N'(V', E')) \le l(N(V, E))$ and $V' \subset V$, which contradicts condition (♠).

Lemma 10 *There is no cycle in* N *uniquely composed of Steiner points.*

Proof. Suppose that there is such a cycle denoted by C. We may assume that there are no two parallel lines incident to a line on C. Otherwise, we are able to move the line between two parallel lines until it reaches a point in V. Thus $|C| = 4$. Let $C = \{s_1 \to s_2 \to s_3 \to s_4 \to s_1\}$. We can prove that N has an admissible Steiner lifting at edge $s_i s_{i+1}$, for some i. This contradicts Lemma 7.

Now according to Theorem 6 and Lemma 10, we are able to decompose N into an edge-disjoint union of full Steiner trees by splitting it at every regular point. Those full Steiner trees are called *full Steiner components* of N.

Lemma 11 *Let* T *be a full Steiner component of* N. *Then* N *has no cut set of size three which includes two edges in* T *unless it contains three edges incident to a common Steiner point.*

Proof. By contradiction, suppose that there exists a cut set $\{uv, u'v', xy\}$, where u, v, u' and v' are four points of T, and there are no two edges incident to each other (otherwise we will find a smaller cut set of N). It can be proved that N has an admissible Steiner lifting.

Theorem 7 $\frac{2}{3} \le \inf\{l_3(P)|P\} \le \frac{7}{8}$.

Proof. First we will show that for any P, $\frac{2}{3} \leq l_3(P)$. Given P, let N be a shortest three-edge connected Steiner network on P. We construct a spanning network N' on P by substituting each full Steiner component in N with its corresponding minimal spanning tree. Because of Lemma 11, we know that this procedure will not spoil three-edge connectivity. Thus N' is a three-edge connected spanning network on P. And then from Lemma 6 we deduce that $l(N') \leq \frac{3}{2}l(N)$.

We have constructed a special class of sets P_k such that the length of a shortest three-edge connected Steiner network on P_k divided by that of a shortest three-edge connected spanning network on P_k achieves $\frac{7}{8}$, as k approaches infinity.

Theorem 8 *Suppose that all points in P are on the sides of the convex hull of P. Then*

(1) There exists a shortest three-edge connected spanning network $N^\star$ on P with $\mathcal{C}(N^\star) = \bar{c}(P)$.

(2) There exists a shortest three-edge connected Steiner network $N^\star$ on P such that $\mathcal{C}(N^\star) = \bar{c}(P)$ and $N^\star \setminus \mathcal{C}(N^\star)$ is a Steiner minimal tree of P.

Proof. As $P \subset \bar{c}(P)$, we are able to indicate each point of P by r_i, for $i = 0, 1, \cdots, |P|-1$. For each i, r_i is adjacent to r_{i+1} and r_{i-1}, where $i+1$ and $i-1$ take modulo of $|P|$.

(1) Let N be a shortest three-edge connected spanning network on P. If $\mathcal{C}(N) \neq \bar{c}(P)$, then we are able to find an edge on $\bar{c}(P)$, say r_0r_1 such that there is a path in $\mathcal{C}(N)$ connecting its two endpoints r_0 and r_1, because $\bar{c}(P)$ encloses $\mathcal{C}(N)$. Denote this path by $r_0q_1q_2\cdots q_kr_1$, where $k \geq 1$. Note that q_j is either a point of P or an intersection of two edges in N, for each j. If q_1 is a point of P, then it follows that Lemma 2 that N has a cut set of size two, which contradicts that N is three-edge connected. Therefore q_1 is an intersection of two edges, say r_0r_i and r_mr_n, where $0 < m < i < n$. According to Lemma 9, we can make an admissible crossing lifting at q_1. Note that by Remark 4, we are able to repeat this argument until we obtain a shortest three-edge connected spanning network $N^\star$ on P with $\mathcal{C}(N^\star)$ on P with $\mathcal{C}(N^\star) = \bar{c}(P)$.

(2) Let N be a shortest three-edge connected Steiner network on P with minimal number of crossing. If $\mathcal{C}(N) \neq \bar{c}(P)$, then as above, we can find a path $r_0q_1\cdots q_kr_1$ in $\mathcal{C}(N)$ connecting r_0 with $r_1, k \geq 1$, where $\{r_0, r_1\} \subset P$ and $\{q_1, \cdots, q_k\} \subseteq V \setminus P$. Now consider $\overline{N}$ (refer to the beginning of this section for the construction of $\overline{N}$), and denote points adjacent to q_i by s_i, for $i = 1, \cdots, n$. Note that $s_i \in P$ implies that N has a cut set of size two and $s_i \in V \setminus P$ leads to that $\overline{N}$ has an admissible Steiner lifting at q_is_i. In addition, if s_i is a crossing in N, then $\overline{N}$ has an admissible crossing lifting at s_i. Therefore $\mathcal{C}(\overline{N}) = \mathcal{C}(N) = \bar{c}(P)$, and $\overline{N} \setminus \mathcal{C}(\overline{N})$ is a Steiner minimal tree of P.

Corollary 2 *As in theorem 8, suppose that $P \subset \bar{c}(P)$. Then $l_3(P) > \frac{5}{6}$.*

Proof. Let SMT(P) and MST(P) denote a Steiner minimal tree of P and a minimal spanning tree of P, respectively. Let N' and N indicate a shortest three-edge connected spanning and Steiner network on P satisfying conditions (1) and (2) in

Theorem 8, respectively. then we have

$$l(N) = l(\overline{c}(P)) + l(\mathrm{SMT}(P)) \geq l(\overline{c}(P)) + \frac{2}{3} l(\mathrm{MST}(P)).$$

Since a Hamilton cycle of P and a minimal spanning tree of P can compose a three-edge connected spanning network on P, we have $l(N') \leq l(\overline{c}(P)) + l(\mathrm{MST}(P))$. In addition, it is obvious that $l(\mathrm{MST}(P)) \leq l(\overline{c}(P))$. Therefore

$$l_3(P) = \frac{l(N)}{l(N')} \geq \frac{l(\overline{c}(P)) + \frac{2}{3} l(\mathrm{MST}(P))}{l(\overline{c}(P)) + l(\mathrm{MST}(P))} > \frac{5}{6}.$$

3.3. *k*-Edge Connected Steiner Networks

Most of our arguments used in the above analysis for the cases of $k = 2$ and 3 are based on Steiner lifting and crossing lifting, both of which are local operations on one Steiner point and one crossing together with other four points, respectively. Unfortunately that can not be applied to the case when $k \geq 4$. However, Theorems 2, 8 and Corollary 2 can be easily extended to the general case for $k \geq 4$.

Theorem 9 *If all points of P are on the sides of the convex hull of P, i.e., $P \subset \overline{c}(P)$, then for any $k \geq 2, l_k(P) = 1$ for k is even and $l_k(P) > \frac{3k+1}{3k+3}$ for k is odd.*

References

1. D. Bienstock, E. F. Brickell and C. L. Monma, Properties of *k*-connected networks, *SIAM J. on Discrete Mathematics*, 3 (1990) 320-329.
2. N. Christofides, Worst-case analysis of a new heuristic for the traveling salesman problem, *Report 388, Graduate School of Industrial Administration, Carnegie-Mellon University* (Pittsuburgh, PA, USA, 1976).
3. D.-Z. Du and F. K. Hwang, An approach for proving lower bounds: solution of Gilbert-Pollak's conjecture on Steiner ratio, *Proc. 31-th FOCS* (1990) 76-85.
4. G. N. Fredrickson and J. Ja'Ja, On the relationship between the biconnectivity augmentation and traveling salesman problem, *Theoretical Computer Science* 13 (1982) 189-201.
5. R. L. Graham, An efficient algorithm for determining the convex hull of a finite planar set, *Infor. Proc. Lett.*, 1 (1972) 132-133.
6. R. L. Graham and F. K. Hwang, Remarks on Steiner minimal trees, *Bull. Inst. Math. Acad. Sinica*, 4 (1976) 177-182.
7. M. Grötschel, C. L. Monma and M. Store, Design of survivable networks, In *Handbook in Operations Research and Management Science*, Eds: M. Ball, T. Magnanti, C. Monma, and G. Nemhauser (1992).
8. M. Hanan, On Steiner's problem with rectilinear distance, *SIAM J. Appl. Math.*, 14 (1966) 255-265.
9. D. F. Hsu and X.-D. Hu, Shortest two-connected Steiner networks with euclidean distance, *Technical Report* (JAIST, 1994).
10. D. F. Hsu and X.-D. Hu, Shortest three-edge connected Steiner networks with euclidean distance, *Technical Report* (JAIST, 1994).
11. F. K. Hwang, On Steiner minimal trees with rectilinear distance, *SIAM J. Appl. Math.*, 30 (1976) 104-114.
12. F. K. Hwang and D. S. Richards, Steiner tree problems, *Networks*, 22 (1991) 55-89.
13. J. B. Kruskal, On the shortest spanning subtree of a graph and the traveling salesman problem, *Proc. AMS*, 7 (1956) 48-50.
14. C. L. Monma, B. S. Munson and W. R. Pulleyblank, Minimum-weight two-connected spanning networks, *Math. Prog.*, 46 (1990) 153-171.
15. C. Papadimitriou and K. Steiglitz, *Combinatorial Optimization: Algorithms and Complexity*, Prentice-Hall, Inc., Englewood Cliffs, New Jersey, USA, 1982.

MUTUALLY REPELLANT SAMPLING*

SHANG-HUA TENG
Department of Computer Science, University of Minnesota, Minneapolis, Minnesota 55455, USA. Email steng@cs.umn.edu

Abstract. This paper studies the following class of sampling problem: Let $(D, || \ ||)$ be a metric space whose distance function $|| \ ||$ satisfies the triangular equality. Let k be an integer. We would like to find a sample subset of k elements of D that are mutually far away from each other. We study three versions of the "farness" condition and give optimal or close to optimal polynomial-time approximation algorithms for constructing such good samples. Because all three definitions measure different aspects of how one sample element "repel" its close neighbors to be chosen in the sample, we call these sampling problems "mutually repellant sampling". In applications, the metric space naturally models graphs or geometric domains. The definitions of farness are closed related with the condition that the k-sample best measures the "shape" of a graph or a geometric domain. Our results have applications in several graph partitioning algorithms and optimization heuristics. Furthermore, our algorithms are on-line in the sense that they do not have to know k in advance.

1. Introduction

Interestingly, this research was first motivated by an AI approach to optimization problems. In the Spring of 1994, Deniz Yuret, then a Ph.D student of MIT AI lab stopped by my office and explained to me one of his optimization software for finding an optimal solution when there are many local optima. The software Deniz described, (not completely precisely), uses the heuristic that first starts with an initial point, then finds a local optimum by some iterative improving methods, and then restarts from a point in the problem domain that is farthest away from all the local optima that have been found. This iterative process repeats until it runs out of time or finds an optimal solution. During our discussion, we realized that the procedure of finding the next starting point in an Euclidean domain, described by a finite constructive solid geometry (CSG) expression, is to apply Voronoi diagrams to discretize the problem of finding a point that is farthest away from a set of points (See Section 3 for more details).

One question that has motivated this research is: what are good initial points if we have k processors that can perform k local improvements in parallel. A reasonable and perhaps the most intuitive method is to choose k points that are "mutually far" away from each other. Then what would be a good definition of "mutual farness"?

The second interesting question is: The local search process of each processor may take different time. If one processor finishes before all others, what would be its good choice for the next starting point? Suppose we use k processors to perform

* Supported in part by the Graduate School Grant-in-Aid of Research, Artistry, and Scholarship of the University of Minnesota. Part of the work was started and has been done while the author was at the Department of Mathematics and the Lab. for Computer Science, MIT, Cambridge, MA.

D.-Z. Du and P. M. Pardalos (eds.), Minimax and Applications, 129–140.

some unknown $n \gg k$ number of asynchronous local search operations, how can we maintain, in an on-line fashion, the property that all starting points are mutually faraway from each other?

The second motivation of this research is the problem of graph partitioning, one of the most fundamental problems in parallel scientific processing, VLSI layout, circuit testing, and database query optimization. The simple version of graph partitioning is to divide the node set of a graph into two subsets of roughly equal size so that the number of edges that bridge the two is smallest possible. For parallel processing with k processors, we need to divide the node set of a computational graph into k subsets of roughly same size so that the interaction among subsets is minimized (in order to reduce the communication overhead).

Most of the previous graph partitioning methods [1, 2, 4, 12, 17, 19, 22] use recursive bisection, that is to first divide the graph into two and then recursively divide the subgraphs. However, Simon and Teng [20] have recently shown that recursive bisection may lead to "bad k-way partition". More attentions are given to more global and direct multiway partitioning strategies, with the multilevel method and the clustering method as the most appealing approaches [1, 10, 13, 21].

The multilevel approach first coarsens a graph, then finds a "good" partition for the coarsened graph, and then projects the partition back to the original graph. The clustering method, on the other hand, starts from k nodes, called *seeds of the partition*, and then grows each seed into a subgraph in a way that tries to minimize the boundary and to balance the number of nodes in each subgraphs.

The basic idea of coarsening is to find a sample of the graph that is small enough and that closely measures the "shape" of the graph. The most commonly used heuristics to find the seeds (in a clustering method) is once again to choose k nodes that are mutually faraway from each other.

Given a graph or a geometric domain (in two or three dimensions), which k nodes or points best represent the "shape" of the graph or the geometric domain? If $k = 2$, then perhaps two most distant nodes or points are the most intuitive choices, because they measure the diameter of the graph or the domain. In applications of classical graph theory, maximum and maximal independent sets are often used as a sample of the graph [1, 5, 15, 21]. Implicitly, the condition of *mutual farness* is used: In an independent set, each sample node excludes its neighbors from the sample. For unweighted graph, an independent set is a subset of nodes whose mutual distance is at least 2. Therefore, our definition of mutually repellant samples can be viewed as a generalization of independent sets. Most graph coarsening algorithms use independent set to repeatedly reduce the graph size.

In this paper, we study three versions of the sampling problem. We give polynomial time approximation algorithms with guaranteed performance for each version of the sampling problem. We also prove the optimality of one of our approximation algorithms by showing that it achieves the best possible approximation ratio unless $P = NP$.

This is a shortened version, written to explore more applications of this max-min optimization problem, of a companion paper *Approximating Maximum Distant Sampling* (submitted to the Journal of Algorithms).

In Section 2, we discuss three (related) definitions of mutual farness. Because all these three definitions measure different aspects of how one sampling point "repel" its close neighbors to be chosen in the sample, we call these sampling problem "mutually repellant sampling". We will also outline our results in Section 2. Sections 3, 4, and 5 present three approximation algorithms and prove their approximation ratios. Section 6 gives a tight lower bound on the approximation ratio of one of sampling problems. Section 7 discusses some potential applications of our algorithms and gives some open questions.

2. Mutually Repellant Sampling

We give an abstract definition of the sampling problem. Let $\Gamma = (D, || \, ||)$ be a *metric space* where D is called the *domain* and $||x - y||$ is a positive function that measures the *distance* between two elements $x, y \in D$. The metric satisfies the *triangular inequality* if for all $x, y, z \in D$ we have $||x - y|| \leq ||x - z|| + ||z - y||$.

For example, each graph $G = (V, E)$ defines a metric space where $D = V$ and the distance between two nodes in V is the length of the shortest path between the two nodes. Each region D in Euclidean space defines a metric space where the distance between two points is the Euclidean distance between the two points. Each point set P in a normed linear space defines a metric space with $D = S$, and the distance between points is the normed linear distance (e.g., Euclidean distance) between them. In these three cases, the metric space satisfies the triangular inequality.

Given a metric space with domain D, for each positive integer k, a k-sample is a set $S = \{s_1, ..., s_k\}$ of k elements (elements may repeat) of D. We now define three distance measures for mutual farness.

- **Min distance**: The min distance of S is equal to

$$\{\min_{i<j} ||s_i - s_j||\},$$

  the distance between the closest pair of elements in S.
- **Min-Selection distance with parameter l**: Let $1 \leq l < k$ be an integer. The lth *distance* of s_i is the lth smallest distance to other elements in S. The *min-selection-distance* with parameter l of S is the smallest lth distance among all elements in S. Clearly, the min distance is the min-selection distance with parameter 1.
- **Min-Average distance:** The average distance of S is equal to

$$\text{avg}(S) = \frac{\sum_{i \neq j} ||s_i - s_j||}{k(k-1)}.$$

With the three definitions of mutual farness, we have three versions of the sampling problem. In each case, we would like to find a k-sample that maximizes the farness measure. We will refer these three sampling problem as *max-min distance sampling*, *max-min-selection distance sampling*, and *max-average distance sampling*, respectively.

We give three approximation algorithms, one for each version. For both min distance and min-average distance, our algorithms achieve approximation ratio 1/2. Our algorithm for the min-selection distance achieves approximation ratio 1/4. By say an algorithm achieves *approximation ratio* $0 < \alpha < 1$, we mean that the farness distance of the resulting sample is at least α times that of an optimal k-sample. All our algorithms are *on-line* in the sense that they do not have to know k in advance.

It is worthwhile to point out that even though all our approximation algorithms are greedy constructions, the proofs of approximation ratio are quite different from one farness measure to another. We also show that our approximation for max-min distance sampling problem is the best possible for graphs: if $NP \neq P$, then there is no polynomial time algorithm that approximates the max-min distance sampling problem by a ratio better than 1/2.

For Euclidean domains, we will focus on convex domains defined by a linear number of faces. Our results can be generalized to any domain that can be described by a finite constructive solid geometry expression.

3. Max-Min Distance Sampling

In this section, we study the sampling problem with respect to the min distance. We will give both sequential and parallel approximation algorithms. Our algorithms achieve an approximation ratio of 1/2, which will proven to be best possible in Section 6.

3.1. The Algorithm

We analyze the following greedy sampling algorithm. Recall that we assume that the distance function satisfies the triangular equality.

Algorithm *Greedy MAX-MIN Distance Sampling*
1. Let q_1 be an arbitrarily element of D

2. **for** $(i = 2 : k)$ **do**
   – let q_i be the element x in D that maximizes $\min_{j=1}^{i-1} ||x - q_j||$.
3. **return** $Q_k = \{q_1, ..., q_k\}$.

Clearly the algorithm runs in polynomial time for graphs. We will show in the next subsection that the step of finding a point that maximizes the minimum distance to a set of points in Euclidean space can be solved in polynomial time using Voronoi diagrams.

Theorem 1 *For all k, Greedy MAX-MIN Distance Sampling achieves an approximation ratio of* 1/2.

Proof: We apply induction on k. When $k = 2$, the optimal solution chooses two most distant elements, say y_1 and y_2. Let q_1 be the first element chosen by the greedy algorithm. By triangular equality, we have $||y_1 - q_1|| + ||y_2 - q_1|| \geq ||y_1 - y_2||$. Therefore, $\max(||y_1 - q_1||, ||y_2 - q_1||) \geq ||y_1 - y_2||/2$. From the greedy algorithm, we have $||q_2 - q_1|| \geq \max(||y_1 - q_1||, ||y_2 - q_1||) \geq ||y_1 - y_2||/2$.

We now inductively assume that the theorem holds for $k-1$. Let $P_k = \{p_1, ..., p_k\}$ be the optimal k-sample. Let $Q_k = \{q_1, ..., q_k\}$ be the k-sample generated by the greedy algorithm. For a set S of elements in D, let $\min(S)$ denote the distance between the closest pair of elements in S.

From the induction hypothesis, we have for all $1 \leq i \leq k$, $\min(Q_{k-1}) \geq \min(P_k - \{p_i\})/2$ and hence $\min(Q_{k-1}) \geq \min(P_k)/2$. This implies that the mutual distances among elements in Q_{k-1} are at least $\min(P_k)/2$. To complete the proof, we need only show that $\min_{j=1}^{k-1} ||q_k - q_j|| \geq \min(P_k)/2$, that is, the newly added element is not too close to other elements in Q_{k-1}.

Because q_k is chosen to be the one that is farthest from Q_{k-1}, it is sufficient to show that there exists an element x in D such that the minimum distance from x to Q_{k-1} is at least $\min(P_k)/2$. We will show that there is an element in P_k whose minimum distance to Q_{k-1} is at least $\min(P_k)/2$, i.e., to show that $\max_{i=1}^{k} \min_{j=1}^{k-1} ||p_i - q_j|| \geq \min(P_k)/2$.

We will prove the above statement by contradiction by assuming

$$\max_{i=1}^{k} \min_{j=1}^{k-1} ||p_i - q_j|| < \min(P_k)/2.$$

By the Pigenhole principle, there exists $1 \leq s \neq t \leq k$ and $1 \leq j \leq k-1$, such that both $||p_s - q_j||$ and $||p_t - q_j||$ are less than $\min(P_k)/2$. It implies, by the triangular equality, that $||p_s - p_t|| < \min(P_k)$, which is a contradiction. Therefore,

$$\min_{j=1}^{k-1} ||q_k - q_j|| \geq \max_{i=1}^{k} \min_{j=1}^{k-1} ||p_i - q_j|| \geq \min(P_k)/2.$$

□

Because the algorithm makes decision on each element independent of the knowledge of k, the algorithm is on-line with respect to k.

3.2. Voronoi Diagram for Euclidean Domains

The key computational step in the Greedy MAX-MIN Distance Sampling algorithm is to find an element in D that maximizes the distance to the element set Q_{k-1}. For a region in the Euclidean space, we can not search such a point enumeratively. We now show how to use Voronoi diagrams [6, 18] to find such a point efficiently. We reformulate the problem using geometric terms: Given a point set $Q = \{q_1, ..., q_{k-1}\}$ in D, find a point q_k in the domain D that maximizes the distance to Q.

For illustration, we assume D is a convex region given be a linear number of facets in $\mathbb{R}^d$. Let V_i be the Voronoi cell for q_i. Recall that the *Voronoi cell* [6, 18] of q_i contains all points whose distance to q_i is at most equal the distance to any other point in Q. The point q_i is called the *center* of V_i. A *Voronoi point* is a point that achieves an equal minimum distance to a maximal point subset of Q. Each Voronoi cell V_i is a convex polytope and with Voronoi points as its 0-dimensional faces. There are two types of Voronoi cells: finite cells and infinite cells. We only need to consider points in D. Therefore, the intersection of D with each Voronoi cell V_i is finite convex polytope. For simplicity, we again use V_i to refer this intersection.

By the convexity of V_i, the point in V_i that is farthest from q_i must be a corner vertex of V_i. Therefore, the point in D that maximizes the distance to Q must be either a Voronoi point or the intersection point of the Voronoi diagram of Q with the boundary of D. In $\mathbb{R}^d$, there are at most $O(k^{d/2})$ such points [6]. Therefore, we can use the Voronoi diagram of Q (and its intersection with D) to solve the problem in polynomial time.

3.3. Parallel NC Construction for Graph Domains

In this section, we give an parallel NC algorithm for the approximation of max-min distance sampling in a graph domain. Since each step of finding the next point can be found in $O(\log n)$ time after a preprocessing step that finds all pair-wise shortest paths. If k is polylogarithmic in n, then the algorithm given in Section 3.1 has a direct NC parallel implementation. We now give a parallel algorithm that runs in polylogarithmic time in n, using polynomial number of processors when k is large.

The basic idea is to reduce the problem to a maximal independent set problem. Let $G = (V, E)$ be a graph. For each positive integer $\Delta < n$, let G_Δ be the *distance* Δ *graph* of G, that is, two nodes u and v are connected by an edge in G_Δ if the distance between them is no more than Δ. The following lemma follows the definition of G_Δ and min distance of a k-sample.

Lemma 1 *If G_Δ has an independent set of size k, then G has a k-sample of min distance at least $\Delta + 1$.*

However, finding a maximum independent set is very expensive. Our NC algorithm makes use of the following lemma about G_Δ.

Lemma 2 *If the optimal k-sample of G has min distance equal to Δ, letting $\Delta_1 = \lceil \Delta/2 \rceil - 1$, then each maximal independent set of G_{Δ_1} has size at least k.*

Proof: The proof is essentially the same as that of Theorem 1. Let $Q = \{q_1, ..., q_l\}$ be the maximal independent set with the smallest size. If $l \geq k$, the the lemma is true. Otherwise, let $P = \{p_1, ..., p_k\}$ be an optimal k-sample. As in Theorem 1 we can show that there is a p_j whose distance to Q is at least $\Delta_1 + 1$, and hence $Q + \{p_j\}$ is an independent set, which is a contradiction to the assumption that Q is maximal. Therefore $l \geq k$. □

It follows from Luby [16] that a maximal independent set of graph can be found in NC. The idea then is to perform binary search to find Δ_1 in polylogarithmic time. Therefore, we can compute a 1/2 approximation of the max-min distance sampling in NC.

4. Max-Min-Selection Distance Sampling

Let S be a k-sample of a metric space $\{D, || \;||\}$. Let $1 \leq l \leq k$ be an integer. Recall that the lth distance of an element $s \in S$ is the lth smallest distance to other elements in S. The min-selection-distance with parameter l of S is the smallest lth distance among all elements in S. The problem in this section is to find a k-sample that maximizes the min-selection-distance.

We will show that the following greedy-based algorithm achieves an approximation ratio of 1/4. We conjecture that the approximation ratio of this algorithm is 1/2.

Algorithm *Greedy MAX-Min-Selection Distance Sampling*

1. Let q_1 be an arbitrarily element of D

2. **for** $(i = 2 : k)$ **do**
   - **if** $i < l$, **then** choose the next element q_i to be the one that maximizes the minimum distance to $Q_{i-1} = \{q_1, ..., q_{i-1}\}$
   - **else** let q_i be the element x in V that maximizes the lth distance of x in $\{q_1, ..., q_{k-1}, x\}$.
3. **return** $Q_k = \{q_1, ..., q_k\}$.

To show that the algorithm has an approximation ratio of 1/4, we first prove the following lemma. Let $P_k = \{p_1, ..., p_k\}$ be the optimal k-sample. Let $Q_k = \{q_1, ..., q_k\}$ be the k-sample generated by the greedy algorithm above.

Lemma 3 *Let Δ be the min-lth-selection distance of P_k. Then there exists i such that the lth selection distance of p_i in $Q_{k-1} + \{p_i\}$ is at least $\Delta/2$.*

Proof: To prove by contradiction, we assume that the lemma is false. For each i there are l elements in Q_{k-1} whose distance to p_i is less than $\Delta/2$. There are k elements in P_k but only $k-1$ in Q_{k-1}. Therefore, there must be a q_j in Q_{k-1} so that there are at least $l+1$ elements in P_k whose distance to q_j is less than $\Delta/2$ (Pigenhole Principle). Without loss of generality, assume $p_1, p_2, ..., p_{l+1}$ are less than $\Delta/2$ distance away from q_j. By the triangular equality, p_1 is less than Δ distance away from $p_2, ..., p_{l+1}$, which is a contradiction, and hence the lemma is true. □

We now show that the greedy algorithm is has a 1/4-approximation ratio.

Theorem 2 *For all k, Greedy MAX-MIN-Selection Distance Sampling achieves an approximation ratio of* 1/4.

Proof: First observe that the lth min-selection distance of P_{k-1} is no more than that of P_k, because deleting the element in P_k with the smallest lth-selection distance can only increase the lth min selection distance.

We now prove the theorem by induction on k. The basis case is that when $k = l+1$. Notice that the distance between q_1 and q_2 is at least 1/2 of the diameter of the domain. Therefore, by triangular equality, for any $3 \leq i \leq k$, $\max(||q_i - q_1||, ||q_i - q_2||)$ is at least 1/4 of the diameter of the domain. So the lth distance of any elements in Q_k is at least 1/4 of the diameter which is at least as larger as the lth min-selection distance in the optimal solution. So at base case, we have approximation equal to 1/4.

We now assume the theorem is true for $k-1$ for $(k \geq l+2)$. By Lemma 3, the lth min-selection distance of q_k is at least $\Delta/2$, where Δ is the min-lth-selection distance of P_k. We now prove by contradiction by assuming that the min-lth-selection distance of Q_k is less than $\Delta/4$. By the induction hypothesis, the min-lth-selection distance of Q_{k-1} is at least 1/4 of that of P_{k-1}, and hence is at least $\Delta/4$. Therefore, there must be an $i < k$ such that the lth min-selection distance of q_i is less than $\Delta/4$, implying

1. $||q_k - q_i|| < \Delta/4$.
2. There exists other $l-1$ elements in Q_{k-1} that have distances less than $\Delta/4$.

Conditions 1 and 2 implies that there are l elements in Q_{k-1} whose distances to q_k are less than $\Delta/2$, contradicting with what we have proved that the lth min-selection distance of q_k is at least $\Delta/2$. Therefore the lemma is true. □

Because the algorithm makes decision on each element independent of the knowledge of k, the algorithm is on-line with respect to k.

5. Max-Average Distance Sampling

In this section, we present a polynomial time approximation algorithm for maximizing the average distances among k sample elements and show that the algorithm has an approximation ratio of 1/2.

Let $(D, || \ ||)$ be a metric space and let $S = \{s_1, ..., s_k\}$ be a k-set of D. Recall that the average distance of S is equal to

$$\text{avg}(S) = \frac{\sum_{i \neq j} ||s_i - s_j||}{k(k-1)}.$$

Let the *total distance* of S be

$$\text{total}(S) = \sum_{i<j} ||s_i - s_j||.$$

Clearly $\text{avg}(S) = 2\text{total}(S)/(k(k-1))$ and thus maximizing the average distance is the same as maximizing the total distance.

We now show that the following greedy algorithm achieves an approximation ratio of 1/2 for the total distance and hence for the average distance.

Algorithm *Greedy MAX-Average Distance Sampling*

1. Let q_1 be an arbitrarily element of D
2. **for** $(i = 2 : k)$ **do**
   – let q_i be the element x in V that maximizes the following quantity
   $$\sum_{j=1}^{i-1} ||x - q_j||.$$
3. **return** $Q_k = \{q_1, ..., q_k\}$.

Theorem 3 *For all k, Greedy MAX-Average Distance Sampling achieves an approximation ratio of* 1/2.

Proof: We prove the theorem by induction on k. Clearly when $k = 2$, the theorem is true because the total (average) distance of the optimal solution is equal to the diameter of D and the greedy algorithm finds a pairs of elements whose distance is at least 1/2 of the diameter. Now assume the theorem is true for $k - 1$.

Let $P_k = \{p_1, ..., p_k\}$ be the optimal k-sample. Let $Q_k = \{q_1, ..., q_k\}$ be the k-sample generated by the greedy algorithm. We need to show total$(Q_k) \geq$ total$(P_k)/2$.

Note that the induction hypothesis implies

$$k\text{total}(Q_{k-1}) \geq (k-2)\text{total}(P_k)/2.$$

To see this, observe that for each $1 \leq i \leq k$, total$(Q_{k-1}) \geq$ total$(P_k - \{p_i\})/2$ (the induction hypothesis). The only distances of P_k missing in $P_k - \{p_i\}$ are those from p_i to $P_k - \{p_i\}$. Sum over all $1 \leq i \leq k$, we have

$$k\text{total}(Q_{k-1}) \geq (k-2)\text{total}(P_k)/2,$$

because each distance $||p_s - p_t||$ miss only twice, once for $i = s$ and once for $i = t$.

By definition, total$(Q_k) =$ total$(Q_{k-1}) + \sum_{j=1}^{k-1} ||q_k - q_j||$. To show total$(Q_k) \geq$ total$(P_k)/2$, it is then sufficient to show $\sum_{j=1}^{k-1} ||q_k - q_j|| \geq$ total$(P_k)/k$. Because Greedy MAX-Average Distance Sampling chooses q_k that maximizes $\sum_{j=1}^{k-1} ||q_k - q_j||$, all we need to show is that there is an element x in D such that $\sum_{j=1}^{k-1} ||x - q_j|| \geq$ total$(P_k)/k$. We will restrict attention only to P_k, i.e., we want to show that there exists $1 \leq t \leq k$, that $C_t = \sum_{j=1}^{k-1} ||p_t - q_j|| \geq$ total$(P_k)/k$.

By the triangular equality, for each pair $1 \leq i, j \leq k$, for each $1 \leq l \leq k - 1$,

$$||p_i - p_j|| \leq ||p_i - q_l|| + ||p_j - q_l||,$$

implying $||p_i - p_j|| \leq (C_i + C_j)/(k-1)$.

Therefore

$$\begin{aligned} \text{total}(P_k) &= \sum_{i<j} ||p_i - p_j|| \\ &\leq C_1 + C_2 + ... + C_k \\ &\leq k(\max_{t=1}^{k} C_t). \end{aligned}$$

Thus $(\max_{t=1}^{k} C_t) >$ total$(P_k)/k$. □

Because the algorithm makes decision on each element independent of the knowledge of k, the algorithm is on-line with respect to k.

6. Lower Bounds

In this section, we observe that if $P \neq NP$, then there is no polynomial time approximation algorithm for the max-min distance sampling problem of ratio better than 1/2, even when we restrict the input graph to be planar. As a corollary, the problem of the max-min distance sampling problem is NP-hard.

We make reduction from the MAXIMUM INDEPENDENT SET problem which is defined as: Does a graph $G = (V, E)$ have an independent set of size k? where a set $S \subseteq V$ is an *independent set* of G if no two vertices in S are joined by an edge in E. MAXIMUM INDEPENDENT SET is NP-complete (Karp [14]) and remains NP-complete for cubic planar graphs (Garey, Johnson, and Stockmeyer [8]).

Notice that the smallest pairwise distance in an independent set is at least 2, therefore, the max-min distance sampling problem has a solution of size k with smallest distance greater than 1 iff the graph has an independent set of size k. Thus, it is NP-hard.

Lemma 4 *The smallest pairwise distance in of any maximal independent set is either 2 or 3.*

Proof: Let S be a maximal independent set of G. If $u, v \in S$ are the pair of vertices in S that has the smallest distance, say l. If $l = 2$, then the lemma holds. We now assume $l > 2$. Let $uw_1, ..., w_{l-1}, v$ be the shortest path between u and v. Clearly, $w_1, ..., w_{l-1}$ do not belong to S, for otherwise u and v will not be the pair of vertices in S that has the smallest distance. Moreover, $w_1, ..., w_{l-1}$ do not join any other vertices in S because we assume $l > 2$. Therefore, if $l > 4$, we can put w_2 in the independent set. Thus, the smallest pairwise distance in of any maximal independent set is either 2 or 3. □

Theorem 4 *If $NP \neq P$, there is no polynomial time approximation algorithm for the max-min distance sampling problem of approximation ratio better than* $1/2$.

Proof: We first prove the theorem for a general graph $G = (V, E)$. Let G' be the graph obtained from G by adding an additional node to V and connecting it to all nodes in V. As long as G is not a complete graph, each maximum independent set of G is an maximum independent set of G' and vice versa. Moreover, all maximum independent sets have min distance equal to 2. Suppose a maximum independent set of G has size k. Then if we have a polynomial time approximation algorithm that achieves approximation ratio better than $1/2$, it should return a k-sample whose min distance is also 2. Therefore, If $NP \neq P$, then there is no polynomial time approximation algorithm for the max-min distance sampling problem of approximation ratio better than $1/2$.

To obtain a similar result for planar graphs, we need a planar transformation of the planar graph so that all maximum independent sets have min distance equal to 2. One method is to choose a finite degree node v_0 (each planar graph has a node of degree no more than 5). Let $v_1, ..., v_t$ ($t \leq 5$) be the neighbors of v_0. Attach with each node from $v_0, ..., v_t$ a chain of length 2 to obtain a modified graph G'. If G has a maximum independent set of size k, then the modified graph G' has a maximum independent set of size $k+t+1$. In a maximum independent set of G', one of $v_0, ..., v_t$ must be chosen as well as the tail of the length-2 chain attached to the node. Therefore each maximum independent set of G' has min distance equal to 2. It follows from the result that MAXIMUM INDEPENDENT SET is NP-complete for planar graph that if $NP \neq P$, there is no polynomial time approximation algorithm for the max-min distance sampling problem of approximation ratio better than $1/2$ for planar graphs. □

7. Applications and Open Questions

As described in the introduction, the result of this paper can be used as the first step in a clustering based graph partitioning algorithm, e.g., [10]. Most of the current clustering based graph partitioning algorithms use some heuristic approaches to pick the k seeds, in which some of the seeds may be too close to others. Our algorithm ensures the quality of the k-sample which is used as the seeds for clustering based partitioning. The on-line property of our algorithm will make the partitioning code more adaptive: to generate a $k+1$-way partition, we can reuse the seeds for the k-way partition. Incrementally, we only need to compute one more seed, and hence improve the efficiency of the partitioning algorithm. Our sampling algorithms can also be applied to multilevel methods [1, 10, 13, 21], in shortening the hierarchical coarsening phase. Instead of repeatedly coarsen a graph by taking maximal independent set, our algorithm gives a more direct way to sample a graph. We plan to experimentally investigate these applications of our sampling algorithms. Another project on our list is to incorporate our sampling procedures into the AI based optimization software to support parallel search.

Marshall Bern of Xerox PARC has pointed out that our max-min distance sampling procedure is the same to that used by Gonzalez [11] in his 2-approximation algorithm for the minmax k-clustering problem, which is to decompose a graph or a point set into k subgraphs or subsets of smallest possible diameters. Gonzalez's algorithm starts with k seeds that have the largest possible min distance, and partitions the graph by a simultaneous Breadth-First-Search (BFS) partitioning. He has shown that the 2-approximation ratio is the best possible unless $P = NP$. Feder and Greene [7, 3] extended Gonzalez's work to point sets in fixed dimensional Euclidean space and showed that it is NP-hard to achieve 1.969 approximation ratio for the minmax k-clustering problem on point sets. Bern conjectured that the minmax k-clustering problem and the max-min distance sampling problem are equivalent with respect to the approximation ratio.

One mutual farness condition is to measure the lth smallest distance of a k-sample. Notice that this measure is different from the min-selection distance of parameter l. The lth smallest distance is chosen from the $k(k-1)/2$ pairwise distances formed by the k-sample.

It remains to be see whether there is a polynomial time constant ratio approximation algorithm for this distance measure. Section 3 gave a 1/2 approximation algorithm for the case when $l = 1$. Using a similar proof technique to that of Section 4, we can give a 1/3 approximation algorithm for $l = 2$ and $l = 3$ (see the companion paper). We conjecture that a polynomial time 1/2-approximation algorithm exists for this measure.

We now list some open questions motivated by this research.

- It is known that MAXIMUM INDEPENDENT SET problem is solvable in polynomial time for bipartite graphs and perfect graphs. Is the max-min distance sampling problem solvable in polynomial time for these graphs.
- Develop a polynomial time 1/2-approximation algorithm for the max-min-selection distance sampling problem.
- Design a polynomial time 1/2-approximation algorithm for max lth smallest distance sampling problem.

– What is the relationship between max-min distance sampling and clustering?

Acknowledgments I would like to thank Deniz Yuret for motivating this research. I would like thank Marshall Bern, Edmond Chow, Ding-Zhu Du, Dan Li, Vivek Sarin, Peng-Jun Wan for helpful discussions at various stages during this work and for motivating related future research projects.

References

1. S. T. Barnard and H. D. Simon. A Fast Multilevel Implementation of Recursive Spectral Bisection for Partitioning Unstructured Problems. *Concurrency: Practice and Experience* 6(2), pp101–117, 1994.
2. M. J. Berger and S. Bokhari. A partitioning strategy for nonuniform problems on multiprocessors. *IEEE Trans. Comp.*, C-36:570-580, 1987.
3. M. Bern and D. Eppstein. Approximation algorithms for geometric problems. Xerox Palo Alto Research Center, to appear, 1995.
4. G. E. Blelloch, A. Feldmann, O. Ghattas, J. R. Gilbert, G. L. Miller, D. R. O'Hallaron, E. J. Schwabe, J. R. Shewchuk and S.-H. Teng. Automated parallel solution of unstructured PDE problems. *CACM*, to appear, 1993.
5. T. F. Chan and B. Smith. Domain decomposition and multigrid algorithms for elliptic problems on unstructured meshes. *Contemporary Mathematics*, 1–14, 1993.
6. H. Edelsbrunner. *Algorithms in Combinatorial Geometry*, volume 10 of *EATCS Monographs on Theoretical CS*. Springer-Verlag, 1987.
7. T. Feder and D. H. Greene, Optimal algorithms for approximate clustering. *ACM 20th STOC*, 434–444, 1988.
8. M. R. Garey and D. S. Johnson. *Computers and Intractability: A Guide to the Theory of NP–completeness.* Freeman, San Francisco, 1979.
9. J. R. Gilbert, G. L. Miller, and S.-H. Teng. Geometric mesh partitioning: implementation and experiments. ICPP, 1995, to appear.
10. T. Goehring and Y. Saad. Heuristic algorithms for automatic graph partitioning. University of Minnesota Supercomputer Institute, Minneapolis,MN 55415, UMSI 94-29, Feb. 1994.
11. T. Gonzalez. Clustering to minimize the maximum intercluster distance. *Theoretical Computer Science*, 38, 293–306, 1985.
12. B. Hendrickson and R. Leland. An improved spectral graph partitioning algorithm for mapping parallel computations. Technical Report, Sandia Lab, 1992.
13. B. Hendrickson and R. Leland. The Chaco user's guide, Version 1.0. Technical Report SAND93-2339, Sandia National Laboratories, Albuquerque, NM, 1993.
14. R. M. Karp. Reducibility among combinatorial problems, in *Complexity of Computer Computation*, (R. E. Miller and Thatcher, Eds.), pages 85–103. Plenum, New York, 1972.
15. D. G. Kirkpatrick. Optimal search in planar subdivisions. *SIAM J. Computing*, 12 (1), 28–35, 1983.
16. M. Luby. A simple parallel algorithm for the maximal independent set problem, *SIAM J. Comput.*, 1036–1053, November, 15 (4), 1986.
17. G.L. Miller, S.-H. Teng, W. Thurston, and S.A. Vavasis. Automatic Mesh Partitioning. To appear in *Proceedings of the 1992 Workshop on Sparse Matrix Computations: Graph Theory Issues and Algorithms*, Institute for Mathematics and its Applications, 1992.
18. F. P. Preparata and M. I. Shamos. *Computational Geometry An Introduction.* Texts and Monographs in Computer Science. Springer-Verlag, 1985.
19. H. D. Simon. Partitioning of unstructured problems for parallel processing. *Computing Systems in Engineering* 2:(2/3), pp135-148, 1991.
20. H. D. Simon and S.-H. Teng How good is recursive bisection. *SIAM, J. Scientific Computing*, accepted and to appear 1995 Also as NASA Ames Report RNR-93-013, August 1993.
21. S.-H. Teng. A geometric approach to parallel hierarchical and adaptive computing on unstructured meshes". In *Fifth SIAM Conference on Applied Linear Algebra*, pp 51–57, J.G. Lewis, ed., SIAM, Philadelphia, 1994.
22. R. D. Williams. Performance of dynamic load balancing algorithms for unstructured mesh calculations. Concurrency, 3 (1991) 457.

GEOMETRY AND LOCAL OPTIMALITY CONDITIONS FOR BILEVEL PROGRAMS WITH QUADRATIC STRICTLY CONVEX LOWER LEVELS*

LUIS N. VICENTE
Department of Computational and Applied Mathematics, Rice University, Houston, Texas, USA 77251

and

PAUL H. CALAMAI
Department of Systems Design Engineering, University of Waterloo, Waterloo, Ontario, Canada N2L 3G1

Abstract. This paper describes necessary and sufficient optimality conditions for bilevel programming problems with quadratic strictly convex lower levels. By examining the local geometry of these problems we establish that the set of feasible directions at a given point is composed of a finite union of convex cones. Based on this result, we show that the optimality conditions are simple generalizations of the first and second order optimality conditions for mathematical (one level) programming problems.

1. Introduction

A bilevel program is defined as the problem of minimizing a function f (the upper level function) in two different vectors of variables x and y subject to (upper level) constraints, where the vector y is an optimal solution of another constrained optimization problem (the lower level problem) parameterized by the vector x. References [2] and [17] survey the extensive research that has been done in bilevel programming.

It is interesting to note that any minimax problem can be stated as a bilevel programming problem since the minimax problem

$$\min_x \max_y f(x,y)$$

is equivalent to the following bilevel program

$$\begin{array}{c} \min_{x,y} \ f(x,y) \\ \text{subject to } (x,y) \in \{(x,y) : \ y \in \ \text{argmin} \ \{-f(x,y)\}\}. \end{array}$$

Bilevel programs with quadratic strictly convex lower levels have often been studied in the literature. Branch and bound solution strategies have been proposed by Bard [4] and Edmunds and Bard [11] for those problems with strictly convex and separable convex upper level objectives respectively, and by Al-Khayyal, Horst and

* Support for this work has been provided by INVOTAN, FLAD and CCLA (Portugal) and by the Natural Sciences and Engineering Research Council of Canada.

D.-Z. Du and P. M. Pardalos (eds.), Minimax and Applications, 141–151.

Pardalos [1] when upper level objectives are concave. Whereas Edmunds and Bard recommend a cutting plane algorithm for computing *global* solutions, algorithms for computing *local star* minima and *local* minima, when upper level functions are strictly convex or concave, have been proposed by Vicente, Savard and Júdice [18]. The generation of test problems with quadratic strictly convex upper levels can be accomplished using a technique described by Calamai and Vicente [8].

Different optimality conditions have been derived for bilevel programs. Bard [3] used the equivalence with a particular mathematical program having an infinite and parametric set of constraints in an attempt to establish such conditions. However a counterexample was discovered by Clarke and Westerberg [9]. Bi and Calamai [5] replaced the lower level problem with the corresponding Karush Kuhn Tucker conditions which they then incorporated into the upper level objective via an exact penalty to arrive at a mathematical program for which necessary and sufficient conditions were derived. A number of authors have employed nonsmooth analysis to develop necessary and sufficient conditions (see references in [17]). Gauvin and Savard [13] used the concept of the steepest descent direction to define necessary conditions for these problems. In what follows we derive both necessary and sufficient conditions based on analyzing the set of feasible directions defined by the special geometry of bilevel programs.

It is well known that the set of feasible directions at a point in a feasible polyhedral region is simply a convex cone. For the minimization of a function over a polyhedral set one can derive optimality conditions at a given feasible point by establishing criteria that guarantee the absence of both first order descent directions, and stationary directions of negative curvature, in the convex cone of feasible directions. We generalize this concept to bilevel programs by showing that the set of feasible directions at a given point is composed of a finite union of convex cones and by establishing first and second order optimality conditions by analyzing the feasible directions in each of these convex cones.

In order to exploit these ideas it is important to design algorithms that can compute the convex cones of feasible directions. However, since verifying whether a given point is a local minimum of a linear bilevel program is an NP-Hard problem [18] algorithms for computing these convex cones are typically inefficient.

In section 2 we formulate the problem and analyze its geometry by proving that the sets of feasible directions are finite unions of convex cones. Algorithms to compute the convex cones are discussed in section 3. In section 4 we derive optimality conditions for this problem and generalize the concept of a projected gradient. The paper concludes with section 5 where we report conclusions and present some directions for future work in this area.

2. Problem Statement and Geometry

Consider the following bilevel program

$$\begin{array}{ll} \min_{x,y} & f(x,y) \\ \text{subject to} & (x,y) \in \{(x,y) : \; y \in \operatorname{argmin} \{q_x(y) : \; Ax + By \le c\}\} \end{array} \tag{1}$$

where $x \in \Re^{nx}$, $y \in \Re^{ny}$, $f : \Re^{nx+ny} \to \Re$ is a mapping defined on a closed convex set, $c \in \Re^{nc}$, $A \in \Re^{nc \times nx}$, $B \in \Re^{nc \times ny}$ and

$$q_x(y) = \frac{1}{2} y^T Q y + y^T R x + r^T y$$

with $Q \in \Re^{ny \times ny}$, $R \in \Re^{ny \times nx}$, $r \in \Re^{ny}$ and $q : \Re^{ny} \to \Re$. The matrix Q is assumed to be symmetric positive definite and thus the lower level problem

$$\begin{array}{c} \min_y \; q_x(y) \\ \text{subject to } By \leq c - Ax \end{array}$$

is a quadratic strictly convex program in y. Variables x (resp. y) are called the upper level (resp. lower level) variables. In the same way the function $f(x, y)$ (resp. $q_x(y)$) is called the upper level (resp. lower level) objective function.

The relaxed feasible region of problem (1) defined by

$$\Omega = \{(x, y) : \; x \in \Re^{nx}, \; y \in \Re^{ny}, \; Ax + By \leq c\} \tag{2}$$

is assumed to be nonempty and the set

$$\Upsilon = \{(x, y) \in \Omega : \; y \in \text{ argmin } q_x(y)\}$$

is called the induced region of problem (1). At an induced region point $(x, y) \in \Upsilon$, a direction (d_x, d_y), $d_x \in \Re^{nx}$, $d_y \in \Re^{ny}$, is called an induced region direction if there exists $\bar{\alpha} > 0$ such that $(x + \alpha d_x, y + \alpha d_y) \in \Upsilon$ for all $\alpha \in (0, \bar{\alpha})$. Induced region directions are feasible directions for problem (1).

In what follows we adopt the convention that if i is any index and v is any vector then $(v)_i$ denotes the i-th component of v.

If we partition the lower level constraints into two index sets $\mathcal{A}$ and $\mathcal{A}^c$ then the face $F_{\mathcal{A}}$ of Ω corresponding to $\mathcal{A}$ is given by

$$F_{\mathcal{A}} = \{(x, y) \in \Omega : \; (Ax + By - c)_i = 0, \; i \in \mathcal{A}\}$$

and the relative interior of $F_{\mathcal{A}}$ is given by

$$ri(F_{\mathcal{A}}) = \{(x, y) \in F_{\mathcal{A}} : \; (Ax + By - c)_i < 0, \; i \in \mathcal{A}^c\}.$$

Since Ω is a nonempty polyhedral set, the relative interiors of nonempty faces of Ω form a partition of Ω (Rockafellar [19], Theorem 18.2). Thus every point $(x, y) \in \Omega$ is uniquely associated with that nonempty face $F_{\mathcal{A}}$ of Ω that satisfies $(x, y) \in ri(F_{\mathcal{A}})$.

If $F_{\mathcal{A}}$ is nonempty then a point $(x, y) \in ri(F_{\mathcal{A}})$ is in the induced region Υ if and only if there exists multipliers $\lambda \geq 0$ such that

$$Qy + Rx + r + B_{\mathcal{A}}^T \lambda = 0 \tag{3}$$

where $B_{\mathcal{A}}$ is the row partition of B corresponding to the indices in $\mathcal{A}$.

To further exploit this relationship we let $k = nx + ny$ and introduce a projection P, from the primal dual space $\Re^{k+\kappa}$ to the primal space $\Re^k$, where κ is some positive integer related to the rank of $\mathcal{A}$. This projection maps a given set $\Psi \subset \Re^{k+\kappa}$ onto the set

$$P(\Psi) = \{p \in \Re^k : (p, \lambda) \in \Psi \text{ for some } \lambda \in \Re^\kappa\}.$$

Properties of this projection operator that are useful for our analysis follow. The proof of the first property is trivial and is not included.

Property 1 *If $\Psi \subset \Re^{k+\kappa}$ is convex then $P(\Psi)$ is convex.*

Property 2 *If $\Psi \subset \Re^{k+\kappa}$ is polyhedral then $P(\Psi)$ is polyhedral.*

Proof. Since Ψ is convex, Ψ is polyhedral if and only if it is finitely generated (Rockafellar [19], Theorem 19.1). But this is true if and only if there exists a finite collection of points $(p_1, \lambda_1), \ldots, (p_n, \lambda_n)$ with $p_i \in \Re^k$ and $\lambda_i \in \Re^\kappa$, $i = 1, \ldots, n$, and a fixed integer m, $0 \le m \le n$, such that $(p, \lambda) \in \Psi$ implies

$$(p, \lambda) = \sum_{i=1}^{m} \sigma_i (p_i, \lambda_i) + \sum_{i=m+1}^{n} \sigma_i (p_i, \lambda_i)$$

where the scalars σ_i satisfy $\sum_{i=1}^m \sigma_i = 1$ and $\sigma_i \ge 0$ for $i = 1, \ldots, n$.

If a given vector p is in $P(\Psi)$ then (p, λ) is in Ψ for some $\lambda \in \Re^\kappa$. Hence $\Psi = P(\Psi) \oplus \Lambda$ for some set $\Lambda \subset \Re^\kappa$ and $P(\Psi)$ is finitely generated. □

We now establish the main result of this section.

Theorem 1 *The set $T(x^*, y^*)$ of induced region directions at $(x^*, y^*) \in \Upsilon$ is a finite union of convex cones*

$$T(x^*, y^*) = \bigcup_{l \in \mathcal{L}} T_l(x^*, y^*)$$

where $\mathcal{L}$ is some finite index set and $T_l(x^*, y^*)$, $l \in \mathcal{L}$, are convex cones of induced region directions at (x^*, y^*).

Proof. Let $|| \cdot ||$ be a norm defined on $\Re^k$. Define a δ-neighbourhood $\mathcal{N}^*$ of (x^*, y^*)

$$\mathcal{N}^* = \{(x, y) : x \in \Re^{nx}, \ y \in \Re^{ny}, \ ||(x - x^*, y - y^*)|| < \delta\}$$

where $\delta > 0$ is chosen so that all points in $\Omega_{\mathcal{N}^*} = \Omega \cap \mathcal{N}^*$ have as active constraints a subset of the constraints that are active at (x^*, y^*) and let the index set $\mathcal{A}$ represent the set of active constraints at (x^*, y^*).

Since Ω can be partitioned into its interior $int(\Omega)$ and the frontier $fr(\Omega)$ we consider the following distinct cases:

Case 1. $fr(\Omega) \cap \mathcal{N}^* = \emptyset$. In this case $(x^*, y^*) \in int(\Omega)$, $\mathcal{A} = \emptyset$, $Qy^* + Rx^* + r = 0$, $F_{\mathcal{A}} = \Omega$ and $(x, y) \in int(\Omega)$ for all $(x, y) \in \Omega_{\mathcal{N}^*}$. Therefore if we define $\mathcal{L} = \{1\}$ then

$$T_1(x^*, y^*) = \{(d_x, d_y) : d_x \in \Re^{nx}, \ d_y \in \Re^{ny}, \ Qd_x + Rd_y = 0\}$$

defines the convex cone of induced region directions at (x^*, y^*).

Case 2. $fr(\Omega) \cap \mathcal{N}^* \ne \emptyset$. In this case $(x^*, y^*) \in fr(\Omega)$ and $Qy^* + Ry^* + r + B_{\mathcal{A}}^T \lambda_{\mathcal{A}}^* = 0$ for some $\lambda_{\mathcal{A}}^* \ge 0$, where $B_{\mathcal{A}}$ is defined as in (3).

Let $\{\mathcal{A}_l\}_{l\in L}$ and $\{F_l\}_{l\in L}$ identify, respectively, all subsets of $\mathcal{A}$ and the corresponding faces of Ω (i.e., $F_l = \{(x,y) \in \Omega : (Ax + By - c)_i = 0, i \in \mathcal{A}_l\}$). In addition, for each $l \in L$ let $n_l = |\mathcal{A}_l|$ and let the index set $\mathcal{A}_l^c$ identify all constraints that define Ω other than those in $\mathcal{A}_l$.

For each $l \in L$ with $n_l > 0$ consider the face F_l and define Ψ_l to be the set of all points (x, y, λ), $x \in \Re^{nx}$, $y \in \Re^{ny}$, $\lambda \in \Re^{n_l}$, satisfying

$$Qy + Rx + r + B_{\mathcal{A}_l}^T \lambda = 0 \text{ and } \lambda \geq 0$$

where $B_{\mathcal{A}_l}$ is the row partition of B corresponding to the indices in $\mathcal{A}_l$. By property 2, $P(\Psi_l)$ is a polyhedral set. Thus, if $P(\Psi_l)$ is nonempty then there exists a positive integer m_l, two matrices $U_l \in \Re^{m_l \times nx}$ and $V_l \in \Re^{m_l \times ny}$ and a vector $w_l \in \Re^{m_l}$ such that $P(\Psi_l)$ can be expressed as $\{(x,y) : x \in \Re^{nx}, y \in \Re^{ny}, U_l x + V_l y - w_l \leq 0\}$. The induced region points that are contained in $ri(F_l)$ are those points that are also in the set $P(\Psi_l)$, i.e., the intersection of the induced region with the relative interior of the face F_l is given by

$$\begin{aligned} P(\Psi_l) \cap ri(F_l) = \{(x,y) \in \Omega : \ & (Ax + By - c)_i = 0,\ i \in \mathcal{A}_l, \\ & (Ax + By - c)_i < 0,\ i \in \mathcal{A}_l^c, \\ & U_l x + V_l y - w_l \leq 0\}. \end{aligned}$$

If $P(\Psi_l) \cap ri(F_l)$ is nonempty and $(x^*, y^*) \in P(\Psi_l)$ then the closure of the set of directions (d_x, d_y), $d_x \in \Re^{nx}$, $d_y \in \Re^{ny}$, satisfying $(x^* + \alpha d_x, y^* + \alpha d_y) \in P(\Psi_l) \cap ri(F_l)$ for all $\alpha \in (0, \alpha_l)$ for some $\alpha_l > 0$, is given by

$$\begin{aligned} T_l(x^*, y^*) = \{(d_x, d_y) : \ & (Ad_x + Bd_y)_i = 0,\ i \in \mathcal{A}_l, \\ & (Ad_x + Bd_y)_i \leq 0,\ i \in \mathcal{A} \backslash \mathcal{A}_l, \\ & (U_l d_x + V_l d_y)_j \leq 0,\ j \in \mathcal{J}_l\}. \end{aligned}$$

where $\mathcal{J}_l = \{j \in \{1, \ldots, m_l\} : (U_l x^* + V_l y^* - w_l)_j = 0\}$. This set is the convex cone of induced region directions at (x^*, y^*) on the face F_l.

For $l \in L$ with $n_l = 0$ we have $F_l = \Omega$ and $ri(F_l) = int(\Omega)$. In this situation if $P(\Psi_l) \cap ri(F_l)$ is nonempty and $(x^*, y^*) \in P(\Psi_l)$, where $P(\Psi_l) = \{(x,y) : x \in \Re^{nx}, y \in \Re^{ny}, Qy + Rx + r = 0\}$, then the closure of the set of directions (d_x, d_y), $d_x \in \Re^{nx}$, $d_y \in \Re^{ny}$, satisfying $(x^* + \alpha d_x, y^* + \alpha d_y) \in P(\Psi_l) \cap ri(F_l)$, for all $\alpha \in (0, \alpha_l)$ for some $\alpha_l > 0$, is given by

$$T_l(x^*, y^*) = \{(d_x, d_y) : Qd_x + Rd_y = 0,\ (Ad_x + Bd_y)_i \leq 0,\ i \in \mathcal{A}\}.$$

This set is the convex cone of induced region directions at (x^*, y^*) on the face Ω. Thus, in case 2, $\mathcal{L} = \{l \in L : P(\Psi_l) \cap ri(F_l) \neq \emptyset \text{ and } (x^*, y^*) \in P(\Psi_l)\}$. □

In order to illustrate theorem 2 consider the induced region defined by the set

$$\Upsilon = \{(x_1, x_2, y) \in \Omega : y \in \text{argmin } q_x(y)\}$$

where $\Omega = \{(x_1, x_2, y) \in \Re^3 : -x_2 + y \leq 0,\ x_2 + y \leq 4\}$ and $q_x(y) = y^2/2 - x_1 y$. At the induced region point $(x_1^*, x_2^*, y^*) = (2, 2, 2)$, both constraints defining Ω are active. Hence $\mathcal{A} = \{1, 2\}$ and, in the neighbourhood $\mathcal{N}^*$ of (x_1^*, x_2^*, y^*), we have to analyze four faces F_1, F_2, F_3 and F_4 corresponding respectively to the four subsets $\mathcal{A}_1 = \{1\}$, $\mathcal{A}_2 = \{2\}$, $\mathcal{A}_3 = \{1,2\}$ and $\mathcal{A}_4 = \emptyset$ of $\mathcal{A}$.

The intersection of the relative interiors of the four faces with the induced region is given by the following sets

$$\begin{aligned}
P(\Psi_1) \cap ri(F_1) &= \{(x_1, x_2, y) \in \Omega : -x_2 + y = 0,\ x_2 + y - 4 < 0, \\
&\qquad -x_1 + y \leq 0\} \\
P(\Psi_2) \cap ri(F_2) &= \{(x_1, x_2, y) \in \Omega : -x_2 + y < 0,\ x_2 + y - 4 = 0, \\
&\qquad -x_1 + y \leq 0\} \\
P(\Psi_3) \cap ri(F_3) &= \{(x_1, x_2, y) \in \Omega : -x_2 + y = 0,\ x_2 + y - 4 = 0, \\
&\qquad -x_1 + y \leq 0\} \\
P(\Psi_4) \cap ri(F_4) &= \{(x_1, x_2, y) \in \Omega : -x_2 + y < 0,\ x_2 + y - 4 < 0, \\
&\qquad -x_1 + y = 0\}
\end{aligned}$$

Since each of these sets in nonempty and $(x_1^*, x_2^*, y^*) \in P(\Psi_i)$ for all $i \in \{1,2,3,4\}$ we have at (x_1^*, x_2^*, y^*) four convex cones of induced region directions given by

$$\begin{aligned}
T_1(x_1^*, x_2^*, y^*) &= \{(d_{x_1}, d_{x_2}, d_y) \in \Re^3 : -d_{x_2} + d_y = 0, d_{x_2} + d_y \leq 0, -d_{x_1} + d_y \leq 0\} \\
T_2(x_1^*, x_2^*, y^*) &= \{(d_{x_1}, d_{x_2}, d_y) \in \Re^3 : -d_{x_2} + d_y \leq 0, d_{x_2} + d_y = 0, -d_{x_1} + d_y \leq 0\} \\
T_3(x_1^*, x_2^*, y^*) &= \{(d_{x_1}, d_{x_2}, d_y) \in \Re^3 : -d_{x_2} + d_y = 0, d_{x_2} + d_y = 0, -d_{x_1} + d_y \leq 0\} \\
T_4(x_1^*, x_2^*, y^*) &= \{(d_{x_1}, d_{x_2}, d_y) \in \Re^3 : d_{x_2} + d_y \leq 0, -d_{x_2} + d_y \leq 0, -d_{x_1} + d_y = 0\}
\end{aligned}$$

3. Computing the Convex Cones

In this section we discuss how to compute the constraints

$$(U_l x + V_l y - w_l)_j \leq 0, \quad j \in \mathcal{J}_l \tag{4}$$

associated with a cone T_l when $n_l > 0$, $P(\Psi_l) \cap ri(F_l) \neq \emptyset$ and $(x^*, y^*) \in P(\Psi_l)$. For this purpose consider the dual constraints

$$B_{\mathcal{A}_l}^T \lambda = -Rx - Qy - r,\ \lambda \geq 0 \tag{5}$$

and suppose that $nl < ny$ or that $n_l \geq ny$ but a basis in (5) is not given explicitly. In these cases one can take a subset of the equalities in (5) and replace each equality of this subset by two inequalities of the opposite sense that can be rewritten as equalities by incorporating slack variables. Using this procedure we can rewrite the dual constraints (5) in the form

$$\tilde{B}_{\mathcal{A}_l}^T \tilde{\lambda} = -\tilde{R}x - \tilde{Q}y - \tilde{r},\ \tilde{\lambda} \geq 0 \tag{6}$$

where $\tilde{\lambda}$ includes λ and the additional slack variables, and $\tilde{B}$, $\tilde{R}$, $\tilde{Q}$ and $\tilde{r}$ include, respectively, B, R, Q and r as well as the coefficients of the added equalities. With this modification a basis can easily be extracted from $\tilde{B}_{\mathcal{A}_l}^T$.

The set of constraints (6) can be interpreted as the feasible set of a parametric linear program with $nx + ny$ parameters on the right and consequently we can compute the constraints (4) by using parametric linear programming techniques ([12]

and [15]). Although a parametric linear program might be a hard problem even with a single parameter [14], the computational effort required to compute the constraints (4) is significantly reduced when compared with the solution of multiparametric linear programs.

We introduce now a particular class of points in Υ, for which the computation of the convex cones of induced region directions is a much less involved task. For this purpose we recall the definition of extreme induced region points and extreme induced region directions [18]. $(\bar{x}, \bar{y}) \in \Upsilon$ is called an extreme induced region point if there exists $\bar{\lambda} \in \Re^{nc}$ such that $(\bar{x}, \bar{y}, \bar{\lambda})$ is an extreme point of the polyhedral set

$$\{(x, y, \lambda) : x \in \Re^{nx},\ y \in \Re^{ny},\ \lambda \in \Re^{nc},\ Qy + Rx + r + B^T\lambda = 0,\ Ax + By \leq c\}$$

and satisfies the complementarity condition $\bar{\lambda}^T(c - A\bar{x} - B\bar{y}) = 0$. As in linear programming, an extreme induced region point is called nondegenerate if the values of the basic variables are all positive. In addition, two extreme induced region points are said to be adjacent if their bases differ in exactly one column. Any direction that connects two adjacent extreme induced region points is called an extreme induced region direction.

Given a nondegenerate extreme induced region point $(\bar{x}, \bar{y})$, the set of convex cones of induced region directions at $(\bar{x}, \bar{y})$ is defined by extreme induced region directions. Hence its computation can be processed by performing all the possible pivot steps that keep the complementarity condition valid. These pivot steps produce the set of all adjacent extreme induced region points of $(\bar{x}, \bar{y})$ that in turn define the convex cones of induced region directions.

4. Number of Convex Cones

Consider $(x^*, y^*) \in \Upsilon$ and let $\mathcal{A}$ represent the set of active constraints at (x^*, y^*).

An upper bound for the number of convex cones of induced region directions at (x^*, y^*) is given by the number of faces that include (x^*, y^*), and is equal to $2^{|\mathcal{A}|}$,

Unfortunately this shows that the number of convex cones in the worst case might be very large. However this number can be significantly reduce in certain situations.

As in section 2, let $\{A_l\}_{l \in L}$ and $\{F_l\}_{l \in L}$ represent, respectively, all subsets of $\mathcal{A}$ and the corresponding faces of Ω. It is a simple matter to see that if $\mathcal{A}_{l_1} \subset \mathcal{A}_{l_2}$ for $l_1, l_2 \in L$, then $P(\Psi_{l_1}) \subset P(\Psi_{l_2})$. This follows since

$$P(\Psi_l) = \{(x, y) : x \in \Re^{nx},\ y \in \Re^{ny} \text{ and } Qy + Rx + r + B^T_{\mathcal{A}_l}\lambda = 0 \text{ for some } \lambda \geq 0\}.$$

Consequently, if a subset $\mathcal{A}_{\bar{l}}$ is found for which $P(\Psi_{\bar{l}}) = \emptyset$, then all faces F_l with $\mathcal{A}_l \subset \mathcal{A}_{\bar{l}}$ need not be considered since $P(\Psi_l) = \emptyset$. Hence the computation of the convex cones might begin with the faces F_l corresponding to sets $\mathcal{A}_l$ having large cardinalities.

Another special case is when the lower level objective function is linear (i.e., $Q = 0$ and $R = 0$). In this case the dual constraints are given by

$$r + B^T\lambda = 0,\ \lambda \geq 0$$

where neither upper nor lower level variables appear. Thus, given a face F_l, either $ri(F_l) \subset \Upsilon$ (and by continuity of the lower level solution, $F_l \subset \Upsilon$) or $ri(F_l) \cap \Upsilon = \emptyset$. Furthermore, given $(x^*, y^*) \in \Upsilon$ and the corresponding set of active constraints $\mathcal{A}$, if a face $F_{\bar{l}}$ (associated with a subset $\mathcal{A}_{\bar{l}}$ of $\mathcal{A}$) satisfies $ri(F_{\bar{l}}) \subset \Upsilon$, then all faces F_l with $\mathcal{A}_{\bar{l}} \subset \mathcal{A}_l$ satisfy $ri(F_l) \subset \Upsilon$. Therefore in this situation (and in contrast to the quadratic strictly convex case) the computation of the convex cones might begin with the faces F_l corresponding to sets $\mathcal{A}_l$ having small cardinalities.

5. Stationary Points and Local Minima

We have defined the set of induced region directions that emanate from a given induced region point (x^*, y^*). This set is composed of the union of a finite number of convex cones. As shown in theorem 1, each convex cone $T_l(x^*, y^*)$, $l \in \mathcal{L}$, is a tangent cone associated with a particular polyhedral set. We are now able to state necessary and sufficient optimality conditions for problem (1). We begin by defining stationarity.

Definition 1 *Let $f : \Omega \rightarrow \Re$ be continuously differentiable at $(x^*, y^*) \in \Upsilon$. The point $z^* = (x^*, y^*)$ is said to be a stationary point of problem (1) if*

$$\nabla f(z^*)^T d \geq 0 \ \text{ for all } \ d \in T_l(z^*),\ l \in \mathcal{L}.$$

The next theorem states that the first order necessary conditions for problem (1) are simple generalizations of the corresponding conditions for smooth linearly constrained programs.

Theorem 2 *Let $f : \Omega \longrightarrow \Re$ be continuously differentiable at (x^*, y^*). If (x^*, y^*) is a local minima of problem (1) then it is a stationary point.*

Stationary points can also be defined in terms of the polars of the tangent cones. The polar $T_l^0(x^*, y^*)$ of the tangent cone $T_l(x^*, y^*)$, $l \in \mathcal{L}$, is the set of all vectors $d^0 \in \Re^k$ satisfying $d^T d^0 \leq 0$ for all $d \in T_l(x^*, y^*)$. By Farkas lemma the polar of $T_l(x^*, y^*)$, $l \in \mathcal{L}$, is given by

$$T_l^0(x^*, y^*) = \{d^0 \in \Re^k : d^0 = C^T \alpha_l + W_l^T \beta_l,\ \alpha_l^c \geq 0,\ \beta_l \geq 0\}$$

when $n_l > 0$ and

$$T_l^0(x^*, y^*) = \{d^0 \in \Re^k : d^0 = C^T \alpha_l + S^T \gamma_l,\ \alpha_l \geq 0\}$$

when $\mathcal{A} = \emptyset$ or $n_l = 0$, where C is the row partition of $[\, A \ \ B \,]$ corresponding to the indices in $\mathcal{A}$, W_l is the row partition of $[\, U_l \ \ V_l \,]$ corresponding to the indices in $\mathcal{J}_l$, $S = [\, Q \ \ R \,]$, α_l^c is the partition of α_l corresponding to the indices in $\mathcal{A} \backslash \mathcal{A}_l$ and $\gamma_l \in \Re^{ny}$.

Using this result a stationary point (x^*, y^*) is an induced region point that satisfies

$$-\nabla f(x^*, y^*) \in T_l^0(x^*, y^*),\ l \in \mathcal{L}.$$

As in [10] we can also introduce the concept of nondegeneracy using the relative interior operator.

Definition 2 *A stationary point (x^*, y^*) of problem (1) is nondegenerate if*

$$-\nabla f(x^*, y^*) \in ri(T_l^0(x^*, y^*)),\ l \in \mathcal{L}.$$

If (x^*, y^*) is nondegenerate then (see [6]) the multiplier inequalities in the expression defining $T_l^0(x^*, y^*)$, $l \in \mathcal{L}$, can be made strict (i.e., strict complementarity holds).

The following theorem extends the second order sufficient conditions of smooth linearly constrained problems to problem (1).

Theorem 3 *Let $f : \Omega \to \Re$ be continuously differentiable over Ω and twice continuously differentiable at (x^*, y^*). If $z^* = (x^*, y^*)$ is a stationary point of problem (1) and $d^T \nabla^2 f(z^*) d > 0$ for all nonzero d satisfying $\nabla f(z^*)^T d = 0,\ d \in T_l(z^*),\ l \in \mathcal{L}$, then (x^*, y^*) is a strict local minimizer of problem (1).*

Note that the assumptions in theorems 2 and 3 can be relaxed by instead assuming, for each point $(x^*, y^*) \in \Upsilon$, that f is defined and continuously differentiable (theorem 2), or twice continuously differentiable (theorem 3), at (x^*, y^*) over each closed convex set $(x^*, y^*) \oplus T_l(x^*, y^*)$, $l \in \mathcal{L}$.

The following example shows a situation where the function f is not differentiable at $(x^*, y^*) \in \Upsilon$ when defined over Ω but differentiable over each close convex set $(x^*, y^*) \oplus T_l(x^*, y^*)$, $l \in \mathcal{L}$. Consider the following bilevel program defined in $\Re^2$

$$\min_{x,y} \quad f(x,y) = |x + y|$$
$$\text{subject to} \quad y \in \text{argmin} \{\frac{1}{2} y^2 : x - y \leq 0\}.$$

In this bilevel program Ω is the subset of $\Re^2$ defined by $\{(x,y) \in \Re^2 : x - y \leq 0\}$, and the induced region is $\Upsilon = \{(x,y) \in \Omega : y = x,\ x \geq 0\} \cup \{(x,y) \in \Omega : y = 0,\ x < 0\}$. At the point $(0,0) \in \Upsilon$ there are two convex cones of induced region directions $T_1(0,0) = \{(-\alpha, 0),\ \alpha \geq 0\}$ and $T_2(0,0) = \{(\alpha, \alpha),\ \alpha \geq 0\}$.

Whereas f is not differentiable at $(0,0)$ when defined over Ω it is infinitely differentiable at $(0,0)$ over the sets $\{(0,0)\} \oplus T_l(0,0) = T_l(0,0)$, $l = 1, 2$.

The definition of a projected gradient can also be generalized from mathematical programming to bilevel programs of the form (1). Given an induced region point $z^* = (x^*, y^*)$, the projected gradient $\nabla_\Upsilon f$ of f at z^* onto the induced region Υ is defined by

$$\nabla_\Upsilon f(z^*) = \text{argmin} \{||d + \nabla f(z^*)|| :\ d \in T_l(z^*),\ l \in \mathcal{L}\}.$$

If, for $l \in \mathcal{L}$, we let

$$d_l^0(\alpha_l, \beta_l, \gamma_l) = \begin{cases} C^T \alpha_l + W^T \beta_l & \text{when } n_l > 0 \\ C^T \alpha_l + S^T \gamma_l & \text{when } \mathcal{A} = \emptyset \text{ or } n_l = 0, \end{cases}$$

where $\mathcal{A}$, C, W_l, S, α_l, β_l, γ_l and n_l are defined as before, then (see [7]) this projected gradient can be computed using

$$\nabla_\Upsilon f(z^*) = -\nabla f(z^*) + d_\iota^0(\alpha_\iota^*, \beta_\iota^*, \gamma_\iota^*)$$

where, for $l \in \mathcal{L}$ and α_l^c defined as before, $(\alpha_l^*, \beta_l^*, \gamma_l^*)$ is a solution to the linear least-squares problem

$$\min\{||\nabla f(z^*)+d_l^0(\alpha_l, \beta_l, \gamma_l)|| : \alpha_l^c \geq 0 \text{ and } \beta_l \geq 0 \text{ when } n_l > 0,\ \alpha_l \geq 0 \text{ when } n_l = 0\}$$

and ι is the index of the minimum term in the set $\{||\nabla f(z^*)+d_l^0(\alpha_l^*, \beta_l^*, \gamma_l^*)||,\ l \in \mathcal{L}\}$.

6. Conclusions and Future Work

We have introduced necessary and sufficient optimality conditions for a particular class of bilevel programming problems. The main concepts involved were geometric and hinged on the fact that at each induced region point there exists a finite number of convex cones of induced region directions.

Our future work includes the extension of this work to bilevel programming problems with nonlinear convex lower level problems. These problems can be stated as

$$\begin{array}{rl} \min_{x,y} & f(x,y) \\ \text{subject to} & (x,y) \in \{(x,y) :\ y \in \ \text{argmin}\,\{q(x,y) :\ g(x,y) \leq 0\}\} \end{array}$$

where $q : \Re^{nx+ny} \to \Re$ is a convex function in y for all values of x defined on a closed convex set and $g : \Re^{nx+ny} \to \Re^{nc}$ is a vector function composed of convex functions in y for all values of x and defined on the same closed convex set.

We also consider the development of trust region algorithms and the identification of optimal active constraints based on the concept of a projected gradient.

References

1. F. Al-Khayyal, R. Horst and P. Pardalos, *Global optimization of concave functions subject to quadratic constraints: an application in nonlinear bilevel programming*, Annals of Operations Research 34 (1992) 125–147.
2. G. Anandalingam and T. Friesz, *Hierarchical optimization: an introduction*, Annals of Operations Research 34 (1992) 1–11.
3. J. Bard, *Optimality conditions for the bilevel programming problem*, Naval Research Logistics Quarterly 31 (1984) 13–26.
4. J. Bard, *Convex two-level optimization*, Mathematical Programming 40 (1988) 15–27.
5. Z. Bi, P. Calamai and A.R. Conn, *Optimality conditions for a class of bilevel programming problems*, Technical Report #191-O-191291, Department of Systems Design Engineering, University of Waterloo, 1991.
6. J. Burke and J. Moré, *On the identification of active constraints*, SIAM Journal on Numerical Analysis 25 (1988) 1197–1211.
7. P. Calamai and J. Moré, *Projected gradient methods for linearly constrained problems*, Mathematical Programming 39 (1987) 93–116.
8. P. Calamai and L. Vicente, *Generating quadratic bilevel programming problems*, ACM Transactions on Mathematical Software, 1994 (to appear).
9. P. Clarke and A. Westerberg, *A note on the optimality conditions for the bilevel programming problem*, Naval Research Logistics 35 (1988) 413–418.
10. J.C. Dunn, *On the convergence of projected gradient processes to singular critical points*, Journal of Optimization Theory and Applications 55 (1987) 203–216.
11. T. Edmunds and J. Bard, *Algorithms for nonlinear bilevel mathematical programming*, IEEE Transactions on Systems, Man and Cybernetics 21 (1991) 83–89.
12. T. Gal, *Postoptimal analysis, parametric programming and related topics*, McGraw-Hill, New York, 1979.

13. J. Gauvin and G. Savard, *The steepest descent direction method for the nonlinear bilevel programming problem*, École Polytechnique, Université de Montréal, 1991.
14. K. Murty, *Computational complexity of parametric linear programming*, Mathematical Programming 19 (1980) 213–219.
15. K. Murty, *Linear Programming*, John Wiley & Sons, New York, 1983.
16. K. Shimizu and E. Aiyoshi, *Optimality conditions and algorithms for parameter design problems with two-level structure*, IEEE Transactions on Automatic Control 30 (1985) 986–993.
17. L. Vicente and P. Calamai, *Bilevel and multilevel programming: a bibliography review*, Technical Report #180-O-666693, Department of Systems Design Engineering, University of Waterloo, 1993.
18. L. Vicente, G. Savard and J. Júdice, *Descent approaches for quadratic bilevel programming*, Journal of Optimization Theory and Applications, 1994 (to appear).
19. T. Rockafellar, *Convex Analysis*, Princeton University Press, Princeton, 1970

THE SPHERICAL ONE-CENTER PROBLEM*

GUOLIANG XUE
Department of Computer Science and Electrical Engineering, College of Engineering and Mathematics, University of Vermont, Burlington, VT 05405, USA. Email: xue@cs.uvm.edu

and

SHANGZHI SUN
Lattice Semiconductor Corporation, 1820 McCarthy Blvd, Milpitas, CA 95035, USA. Email: ssun@lattice.com

Abstract. In this paper we study the spherical one-center problem, i.e., finding a point on a sphere such that the maximum of the geodesic distances from this point to n given points on the sphere is at minimum. We show that this problem can be solved in $O(n)$ time using the multidimensional search technique developed by Megiddo[9] and Dyer[5] when the n given points all lie within a spherical circle of radius smaller than $\frac{\pi}{2}$ times the radius of the given sphere. We also show that the spherical one-center problem may have multiple solutions when the above condition is not satisfied.

1. Introduction

Let $a_1, a_2, \ldots, a_n$ be n points on the 3-dimensional sphere $S = \{x | x \in R^3, ||x|| = 1\}$, where $|| \bullet ||$ is the 2-norm. We want to find a point $x \in S$ that minimizes the maximum of the geodesic distances from x to all of the n given points, i.e.,

$$\min_{x \in S} \max_{1 \leq j \leq n} \cos^{-1}(a_j^T x), \tag{1}$$

where $\cos^{-1}(\bullet)$ is the inverse function of $\cos(\bullet)$. This is called the spherical one-center problem. Notice that $\cos^{-1}(a_j^T x)$ is the `geodesic distance` or `great circle distance` between x and a_j. Also notice that the spherical one-center problem on any given 3-dimensional sphere is equivalent to (1) (by applying a translation and a contraction).

Definition 1 ([2, 3, 7, 10]) A `spherical circle` with a given `center` and `radius` is defined on a sphere by the locus of all points whose geodesic distance from the center is equal to that radius. A spherical circle divides the sphere into two parts; a point is said to be `within` a spherical circle if and only if the point and the center of the spherical circle is included in the same part.

Therefore, in the spherical one-center problem, we are trying to find the smallest enclosing spherical circle of a given set of n points on a sphere.

* This research was supported in part by National Science Foundation grants No. NSF ASC-9409285 and NSF OSR-9350540.

D.-Z. Du and P. M. Pardalos (eds.), Minimax and Applications, 153–156.

The Euclidean one-center problem (weighted as well as unweighted) has a long history. See, for example, [1, 5, 6, 8, 9]. Although various algorithms have been designed for solving the problem in as early as in the 18th century (see [6]), optimal algorithms have been discovered only in the eighties. Megiddo [8] first presented a linear-time algorithm for the unweighted Euclidean one-center problem. Dyer [5] then presented a linear-time algorithm for the weighted Euclidean one-center problem.

A typical application of the one-center problem model is the location of an emergency service center for a given group of clients. This Euclidean model is very useful for small region location problems. However, when the region is very large, the spherical model is more appropriate. Min-sum spherical location problems have been considered by [3, 7, 10]. To the best of our knowledge, min-max spherical location problems have not been considered in the literature.

In the next section, we will show that finding the minimum spherical circle enclosing a given set of n points on the sphere S is equivalent to the minimization of the maximum of n linear functions in 3 variables subject to a single convex quadratic constraint involving the 3 variables, provided that the n given points all lie within a spherical circle of radius $\frac{\pi}{2}$. Therefore this problem can be solved in $O(n)$ time by the multidimensional search technique developed by Megiddo[9] and Dyer[5]. We will also show that the spherical one-center problem may have multiple solutions when the above condition is not satisfied.

2. Main Result

We are only interested in the case where the smallest enclosing spherical circle is unique. When $n = 3$ and the 3 given points lie on a great circle and at the same time sit at the vertices of an equilateral triangle, there will be two smallest enclosing spherical circles with radius equal to $\frac{\pi}{2}$. When $n = 4$ and the 4 given points sit at the vertices of a tetrahedron, there will be four smallest enclosing spherical circles.

In the rest of this paper, we will assume that the n given points all lie within a spherical circle of radius smaller than $\frac{\pi}{2}$. Under this assumption, we will prove that the smallest enclosing spherical circle is unique and can be computed in linear time.

Let us consider the following maximization problem:

$$\begin{aligned} \max\ & r \\ \text{s. t. }\ & a_j^T x \geq r,\ j = 1, 2, \cdots, n, \\ & \|x\|^2 \leq 1. \end{aligned} \tag{2}$$

We will prove that (1) and (2) are equivalent in the following sense.

Theorem 1 *Assume that the n given points $a_1, \cdots, a_n$ all lie within a spherical circle of radius smaller than $\frac{\pi}{2}$. Then (2) has an unique optimal solution, with a* *positive objective function value. In addition, the following assertions are true.*

1. *Let $(\overline{x}, \overline{r})$ be the unique optimal solution of (2). Then $\overline{x}$ is an optimal solution of (1), whose corresponding objective function value is $\cos^{-1}(\overline{r})$.*
2. *Let $\underline{x}$ be an optimal solution of (1). Then its corresponding objective function value is $\cos^{-1}(\underline{r})$, where $\underline{r} = \min_{1 \le j \le n} a_j^T \underline{x}$. In addition, $(\underline{x}, \underline{r})$ is the optimal solution of (2).*

Proof. In (2), the objective function is a linear function and the constraints are either linear or convex quadratic functions. Therefore the set of feasible solutions of (2) is a closed convex set. In addition, the linear objective function r to be maximized is bounded from above. Therefore there exists at least one optimal solution of (2). Since $a_1, \cdots, a_n$ all lie within a spherical circle of radius smaller than $\frac{\pi}{2}$, the optimal objective function value of (2) must be positive.

Now suppose that (x_1, r_1) and (x_2, r_2) are two different optimal solutions of (2). We will show that this is impossible. Since both (x_1, r_1) and (x_2, r_2) are optimal solutions of (2), we have $r_1 = r_2 > 0$. Therefore $x_1 + x_2 \neq 0$ (since $a_1^T(x_1 + x_2) \ge r_1 + r_2 > 0$). Let $x_3 = \frac{x_1 + x_2}{2}$ and $r_3 = \frac{r_1 + r_2}{2}$. We can verify that (x_3, r_3) is a feasible solution of (2). Now let $x_4 = \frac{x_3}{||x_3||}$ and $r_4 = \frac{r_3}{||x_3||}$. Then (x_4, r_4) is also a feasible solution (2). However, $r_4 > r_3 = r_1 = r_2$ because $||x_3|| < 1$. Therefore (x_4, r_4) is a `better` solution than (x_1, r_1) and (x_2, r_2). This contradiction proves the uniqueness of the optimal solution of (2).

Now we will prove the first assertion of the theorem. Let $(\overline{x}, \overline{r})$ be the optimal solution of (2). It is clear that $||\overline{x}|| \le 1$. We claim that $||\overline{x}|| = 1$, since $(\frac{\overline{x}}{||\overline{x}||}, \frac{\overline{r}}{||\overline{x}||})$ will be a better solution if $||\overline{x}|| < 1$. Therefore $\overline{x}$ is a point on the sphere S whose largest geodesic distance to the n given points is $\cos^{-1}(\overline{r})$. Therefore $\overline{x}$ is a feasible solution of (1) with an objective function value of $\cos^{-1}(\overline{r})$. For any point $x \in S$, let $r(x) = \min_{1 \le j \le n} a_j^T x$. Then $(x, r(x))$ is a feasible solution of (2). Therefore $r(x) \le \overline{r}$. Since $\cos(\bullet)$ is a monotonically decreasing function in the interval $[0, \pi]$, we must have $\cos^{-1}(r(x)) \ge \cos^{-1}(\overline{r})$ which means that $\cos^{-1}(\min_{1 \le j \le n} a_j^T x) \ge \cos^{-1}(\overline{r})$. Notice that $\cos^{-1}(\min_{1 \le j \le n} a_j^T x) = \max_{1 \le j \le n} \cos^{-1}(a_j^T x)$. Therefore $\overline{x}$ is an optimal solution of (1).

Last, we will prove the second assertion of the theorem. Let $\underline{x}$ be an optimal solution of (1). Clearly, its corresponding objective function value is $\max_{1 \le j \le n} \cos^{-1}(a_j^T \underline{x})$. Since $\cos(\bullet)$ is a decreasing function in $[0, \pi]$, we have

$$\max_{1 \le j \le n} \cos^{-1}(a_j^T \underline{x}) = \cos^{-1}(\min_{1 \le j \le n} a_j^T \underline{x}) = \cos^{-1}(\underline{r}), \tag{3}$$

where $\underline{r} = \min_{1 \le j \le n} a_j^T \underline{x}$.

Since $||\underline{x}|| = 1$ and $a_j^T \underline{x} \ge \underline{r}$, $(\underline{x}, \underline{r})$ is a feasible solution of (2). Since both $\cos^{-1}(\underline{r})$ and $\cos^{-1}(\overline{r})$ are the optimal objective function value of (1), we have

$$\cos^{-1}(\underline{r}) = \cos^{-1}(\overline{r}). \tag{4}$$

Therefore $\underline{r} = \overline{r}$ and hence $(\underline{x}, \underline{r})$ is an optimal solution of (2). This completes the proof. □

Notice that (2) is equivalent to *the maximization of the minimum of a set of n linear functions subject to a single convex quadratic inequality constraint*, where the total number of variables is 3 (independent of n). By introducing a negative sign into the linear functions, we find that (2) is also equivalent to *the minimization of the maximum of a set of n linear functions subject to a single convex quadratic inequality constraint*, where the total number of variables is 3 (independent of n). This kind of problems can be solved in $O(n)$ time using the multidimensional search technique developed by Megiddo[9] and Dyer[5]. Therefore we have

Corollary 1 *The spherical one-center problem* (1) *can be solved in* $O(n)$ *time.* □

3. Conclusions

In this paper, we have studied the unweighted spherical one-center problem and proved that it can be solved in linear time using the multidimensional search technique developed by Megiddo[9] and Dyer[5]. A more challenging problem is the weighted spherical one-center problem. However, we can not prove that the weighted spherical one-center problem is equivalent to a minimization problem like (2). Using the method of [4], one can prove that the weighted spherical one-center for n points on a given sphere is the same as one of the C_n^3 weighted spherical one-centers involving only 3 of the given points. This will lead to an algorithm with time complexity $O(n^3)$. We do not know whether there exists an algorithm for the weighted spherical one-center problem with a time complexity better than $O(n^3)$.

References

1. R. Chandrasekaran, The Weighted Euclidean 1-Center Problem, *Operations Research Letters*, Vol. 1(1982), pp. 111-112.
2. J.D.H. Donnay, *Spherical Trigonometry*, Interscience Publishers, Inc., 1945.
3. Z. Drezner and G.O. Wesolowsky, Facility Location on a Sphere, *Journal of the Operational Research Society*, Vol. 29(1978), pp. 997-1004.
4. Z. Drezner and G.O. Wesolowsky, Single Facility l_p-Distance Minmax Location, *SIAM Journal of Algebraic and Discrete Mathematics*, Vol. 1(1980), pp. 315-321.
5. M.E. Dyer, On a Multidimensional Search Technique and Its Applications to the Euclidean One-Centre Problem, *SIAM Journal on Computing*, Vol. 15(1986), pp. 725-738.
6. D. Hearn and J. Vijay, Efficient Algorithms for the (Weighted) Minimum Circle Problem, *Operations Research*, Vol. 30(1982), pp. 777-795.
7. I.N. Katz and L. Cooper, Optimal Location on a Sphere, *Computers and Mathematics with Applications*, Vol. 6(1980), pp. 175-196.
8. N. Megiddo, Linear-Time Algorithms for Linear Programming in R^3 and Related Problems, *SIAM Journal on Computing*, Vol. 12(1983), pp. 759-776.
9. N. Megiddo, Linear Programming in Linear Time When the Dimension Is Fixed, *Journal of the Association of Computing Machinery*, Vol. 31(1984), pp. 114-127.
10. G.L. Xue, A Globally Convergent Algorithm for Facility Location on a Sphere, *Computers and Mathematics with Applications*, Vol. 27-6(1994), pp. 37–50.

ON MIN-MAX OPTIMIZATION OF A COLLECTION OF CLASSICAL DISCRETE OPTIMIZATION PROBLEMS

GANG YU*
Department of Management Science and Information Systems
The University of Texas at Austin
Austin, TX 78712, USA.

and

PANAGIOTIS KOUVELIS
The Fuqua School of Business
Duke University
Durham, NC 27706, USA.

Abstract. In this paper, we study discrete optimization problems with min-max objective functions. This type of optimization has long been the attention of researchers, and it has direct applications in the recent development of robust optimization. The following well-known classes of problems are discussed: 1) the minimum spanning tree problem, 2) the resource allocation problem with separable cost functions, and 3) the production control problem. Computational complexities of the corresponding min-max version of the above-mentioned problems are analyzed. Pseudo-polynomial algorithms for these problems under certain conditions are provided.

1. Introduction

In this paper, we study discrete optimization problems of the following form:

$$(P) \qquad z_P = \min_{x \in X} \max_{s \in S} f^s(x) \tag{1}$$

where S is a discrete index set, and x is the decision variable restricted in the constraint set X. The integrality requirement on x is also included in X. This type of min-max discrete optimization has generated considerable research interest in the past few years. A major motivation comes from the recent development of robust optimization under uncertainties. In [30], Yu and Kouvelis defined S as the set of scenarios describing all possible outcomes, each of which occurs with positive but perhaps unknown probability, and the function $f^s(x)$ as the deviation (or percentage deviation) of the objective value with decision x from the optimal objective value under scenario $s \in S$. The scenario dependence of the objective function is due to the input data uncertainty of our decision model. Thus, with these definitions, the min-max optimization intends to minimize the maximum deviation from optimality for all decisions over all possible data scenarios. Throughout this paper, the set S will be referred to as scenario set and each index $s \in S$ will be said to correspond to a scenario.

* Gang Yu's research is supported in part by ONR grant N00014-91-J-1241, ONR grant N00014-92-J-1536, a URI research grant, and a CBA research award from The University of Texas at Austin.

D.-Z. Du and P. M. Pardalos (eds.), Minimax and Applications, 157–171.

The general min-max (or max-min) optimization problem in various forms has long had the attention of researchers. The continuous min-max resource allocation problem of many different variations has been studied by Kaplan [12]; Czuchra [4]; Luss and Smith [18]; Pang and Yu [20]; and Klein, Luss, and Rothblum [14]. The discrete min-max allocation problem has been studied by Jacobsen [11], Porteus and Yormark [22], Ichimori [10], and Tang [26]. The min-max location problem has been studied by Drezner and Wesolowsky [5], and Rangan and Govindan [24]. The min-max partition on trees problem has been studied by Agasi, Becker and Perl [1]. The continuous max-min knapsack problem with GLB constraints has been studied by Eislet [7]. The max-min 0-1 knapsack problem has been studied by Yu [29]. Recently, Yu and Kouvelis [31] conducted research on the complexity of several min-max discrete optimization problems.

Min-max continuous optimization algorithms also have been studied extensively. One motivation is that a convex nonlinear function can be approximated by the upper envelope of a set of approximating linear functions. Thus, the minimization of a nonlinear function can be formulated as min-max of the set of linear functions. See Lemaréchal [16] for a good introduction. Luss [17] has studied separable nonlinear min-max problems. Posner and Wu [21], Bazaraa and Goode [3], and Ahuja [2] have given efficient algorithms for linear min-max programming problems. Dutta and Vidyasagar [6]; Madsen and Jacobsen [19]; and Vincent, Goh, and Teo [28] have provided efficient algorithms for different classes of nonlinear min-max problems.

In this paper, we study min-max versions of some well-known classes of discrete optimization problems. The problems considered are: 1) the minimum spanning tree problem, 2) the resource allocation problem with separable cost functions, and 3) the production control problem. All the problems under consideration share a common feature in that their original optimization problems (i.e., equivalent to the single scenario case by setting $|S| = 1$ in program (P)) can be effectively solved by polynomial algorithms. We have found that the min-max versions of these problems become NP-hard even for very restricted cases. In the case when the set S is bounded (i.e., $|S|$ does not grow with problem size), pseudo-polynomial algorithms can be found for the listed problems under certain conditions. However, when S becomes unbounded, all the problems are strongly NP-hard.

In the complexity proofs of subsequent sections, we frequently refer to the well-known 2-partition and 3-partition problems. For clarity and completeness, we define these two problems here.

The 2-partition problem:

Instance: Finite set I and a size $a_i \in Z_+$ for $i \in I$.

Question: Is there a subset $I' \subseteq I$ such that $\sum_{i\in I'} a_i = \sum_{i \in I \setminus I'} a_i$?

It is well known that the 2-partition problem is NP-hard even when $|I'| = |I|/2$ (see Karp [13]).

The 3-partition problem:

Instance: A finite set I of $3l$ elements, a bound $B \in Z_+$, and a size $a_k \in Z_+$ for $k \in I$, such that each a_k satisfies $B/4 < a_k < B/2$ and such that $\sum_{k\in I} a_k = lB$.

Question: Can I be partitioned into l disjoint sets $I_1, I_2, ..., I_l$ such that, for $1 \leq i \leq l, \sum_{k \in I_i} a_k = B$?

The 3-partition problem is strongly NP-hard (see Garey and Johnson [8]).

Following this introduction, Section 2 studies the min-max spanning tree (MMST) problem. We show that the MMST problem is NP-hard even for grid graphs and with only two scenarios. A pseudo-polynomial algorithm is given for MMST in a special class of grid graphs with bounded number of scenarios, and the strong NP-hardness is proved for cases when S is unbounded. Section 3 discusses the min-max resource allocation (MMRA) problem with separable convex cost functions. We show that the MMRA problem is NP-hard even for linear decreasing cost functions with only two scenarios. A pseudo-polynomial algorithm is given for the MMRA problem with linear decreasing cost functions. When S is unbounded, the MMRA problem is shown strongly NP-hard. In Section 4, the min-max production control (MMPC) problem is investigated. We prove that the MMPC problem is NP-hard even in the case of two scenarios. A pseudo-polynomial algorithm is given for MMPC with bounded S, and strong NP-hardness of MMPC is shown for unbounded S. Finally, Section 5 summarizes and extends the results of this paper.

2. The Min-max Spanning Tree Problem

Given an undirected connected graph $G = (V, E)$, a spanning tree T is defined as a connected subgraph of G without cycles. In another words, a spanning tree is a minimally-connected subgraph of G. If every edge of the graph is associated with a cost, a minimum spanning tree refers to a spanning tree with minimum total edge cost. A minimum spanning tree can be easily found by the Prim's algorithm [23] in $O(\min\{|V|^2, |E| \log |V|\})$ time or by the Kruskal's algorithm [15] in $O(|E| \log |E|)$ time. To define the min-max version, we associate a nonnegative cost c_e^s to each edge $e \in E$ under each scenario $s \in S$. Thus, the min-max spanning tree (MMST) problem is defined as

$$(MMST) \qquad z_{MMST} = \min_T \max_{s \in S} \sum_{e \in T} c_e^s$$

subject to

$$T \text{ is a spanning tree.}$$

The MMST problem has many applications in designing telecommunication networks where edge costs are uncertain. The min-max formulation hedges against the worst possible contingency.

For further discussion, we now define a special class of graphs – the grid graph. A grid graph of order (m, n) is defined by the vertex set

$$V = \{v_{ij} : i = 1, ..., m; j = 1, ..., n\}$$

and edge set $E = E_r \cup E_c$ with

$$E_r = \{(v_{ij}, v_{i,j+1}) : i = 1, ..., m; j = 1, ..., n-1\}$$

and

$$E_c = \{(v_{ij}, v_{i+1,j}) : i = 1, ..., m-1; j = 1, ..., n\}.$$

Edges in E_r are called "row" edges, and edges in E_c are called "column" edges. A grid graph of order (m, n) has $m \cdot n$ nodes and $2mn - m - n$ edges.

The following theorem gives the complexity result for MMST.

Theorem 1 *The MMST problem is NP-hard even under the following restrictions:*
i) G is a grid graph with only two rows, i.e., $m = 2$;
ii) $c_e^s = 0, e \in E_c, s \in S$;
iii) $|S| = 2$.

Proof: We reduce the 2-partition problem described in Section 1 to the MMST problem with the specified restrictions.

Given a 2-partition problem with a set I and elements of I having sizes $a_j, j \in I$, construct a grid graph with $m = 2, n = |I| + 1$. Define a 2-scenario (i.e., $|S| = 2$) MMST problem with cost of the edges as:

$$\begin{aligned} c_e^s &= 0 & & e \in E_c; s \in S, \\ c^1_{v_{1j}, v_{1,j+1}} &= a_j & & j = 1, ..., n-1, \\ c^1_{v_{2j}, v_{2,j+1}} &= 0 & & j = 1, ..., n-1, \\ c^2_{v_{1j}, v_{1,j+1}} &= 0 & & j = 1, ..., n-1, \\ c^2_{v_{2j}, v_{2,j+1}} &= a_j & & j = 1, ..., n-1. \end{aligned}$$

We claim that there exists a 2-partition if and only if the MMST has an optimal objective value $z_{MMST} = \frac{1}{2}\sum_{j \in I} a_j$.

To prove the *only if* part, suppose that there exists a 2-partition, i.e., a subset $I' \subset I$ can be found with $\sum_{j \in I'} a_j = \sum_{j \in I \setminus I'} a_j$. We construct an MMST by selecting the following edges: all edges in E_c; edges $(v_{1j}, v_{1,j+1}), j \in I'$; and edges $(v_{2j}, v_{2,j+1}), j \in I \setminus I'$. The constructed spanning tree will give an objective value $z_{MMST} = \frac{1}{2}\sum_{j \in I} a_j$.

To prove the *if* part, let $z_{MMST} = \frac{1}{2}\sum_{j \in I} a_j$. Due to the nonnegativity of the row edge costs, there always exists an optimal min-max spanning tree with all edges in E_c selected. Here, we are only interested in optimal solutions of MMST containing all edges of E_c together with some edges in E_r. Assume in the first row of an MMST, only edges $(v_{1j}, v_{1,j+1}), j \in I'$ are included in the spanning tree, then edges $(v_{2j}, v_{2,j+1}), j \in I \setminus I'$ in row 2 must also be selected in order to form a tree. Thus, under scenario $s = 1$, we have total cost $z^1 = \sum_{j \in I'} a_j$, and under $s = 2$, we have total cost $z^2 = \sum_{j \in I \setminus I'} a_j$. By assumption of the proof, $z_{MMST} = \frac{1}{2}(z^1 + z^2)$. By definition, $z_{MMST} = \max\{z^1, z^2\}$. This implies $\frac{1}{2}(z^1 + z^2) = \max\{z^1, z^2\}$, which leads to the desired conclusion.

∎

Since the 2-partition problem is only weakly NP-hard, we may expect to solve the MMST problem in pseudo-polynomial time. In fact, this is true at least for the case with bounded scenario set S and with certain conditions.

Theorem 2 *A pseudo-polynomial time algorithm exists for the MMST problem satisfying the following conditions:*

i) G is a grid graph;

ii) $c_e^s = 0, e \in E_c, s \in S$;

iii) S is bounded, i.e., $|S|$ is bounded by a constant as the graph size (m, n) increases.

Proof: We prove the theorem by providing a pseudo-polynomial algorithm based on dynamic programming. Note that by Theorem 1, the MMST problem with the described restrictions is NP-hard for $|S| \geq 2$. An MMST can be constructed by selecting all edges in E_c and those edges in E_r determined by a recursive procedure described below.

Define:

$g_j(\alpha_1, ..., \alpha_{|S|})$ = min-max value of the partial spanning tree with all and only vertices in columns j through n are spanned, and when an additional cost of α_s is augmented for scenario $s \in S$.

With the above definition, the initial condition can be easily specified as:

$$g_n(\alpha_1, ..., \alpha_{|S|}) = \max_{s \in S} \alpha_s.$$

A recursive relation can be written as:

$$g_j(\alpha_1, ..., \alpha_{|S|}) = \min_{i=1,...,m} g_{j+1}(\alpha_1 + c^1_{v_{ij},v_{i,j+1}}, ..., \alpha_{|S|} + c^{|S|}_{v_{ij},v_{i,j+1}}).$$

The MMST objective value can be found by $z_{MMST} = g_1(0, ..., 0)$. The edges in the MMST include E_c and exactly one edge in each column satisfying the condition that $g_j(\alpha_1, ..., \alpha_{|S|}) = g_{j+1}(\alpha_1 + c^1_{v_{ij},v_{i,j+1}}, ..., \alpha_{|S|} + c^{|S|}_{v_{ij},v_{i,j+1}})$ for all α values. In the case where two or more edges in a column satisfy this condition, we can choose any one of them with arbitrary tie breaking. We now present the complete algorithm for finding an MMST.

```
procedure MMST(G = (V, E): grid graph; c: edge costs);
begin
  Initialization: for each scenario s ∈ S, compute a spanning tree value L_s by
      including all edges in E_c together with edges
      {(v_ij, v_i,j+1)|i = argmax_{i=1,...,m} c^s_{v_ij,v_i,j+1}; j = 1, ..., n − 1};
  for α_1 = 0 to L_1 do
  ...
      for α_|S| = 0 to L_|S| do
          g_n(α_1, ..., α_|S|) = max_{s∈S} α_s;

  for j = n − 1 downto 1 do
      for α_1 = 0 to L_1 do
      ...
          for α_|S| = 0 to L_|S| do
              g_j(α_1, ..., α_|S|) = min_{i=1,...,m} {g_{j+1}(α_1 + c^1_{v_ij,v_i,j+1}, ..., α_|S| + c^|S|_{v_ij,v_i,j+1})};
```

output $z_{MMST} = g_1(0, ..., 0)$ as the optimal objective value of MMST ;
end.

In the algorithm above, the parameters $L_s, s \in S$ is used to limit the range of α values. To analyze the complexity of the algorithm, notice that the initialization step takes $O(mn|S|)$ time; the main loop takes $O(mn \prod_{s \in S} L_s)$. Thus, the overall complexity of the algorithm is $O(mn \prod_{s \in S} L_s)$, which is bounded by $O(|E| L_{max}^{|S|})$, where $L_{max} = \max_{s \in S} L_s$. Thus, if $|S|$ is bounded by a constant, this algorithm runs in pseudo-polynomial time.

∎

Note that although the above algorithm runs in pseudo-polynomial time, the complexity increases as the number of scenarios grows. It is natural to conjecture that the MMST problem becomes strongly NP-hard when $|S|$ is an increasing function of the problem size (m, n). This fact is formally stated in the following theorem.

Theorem 3 *The MMST problem with unbounded number of scenarios is strongly NP-hard even for grid graphs.*

Proof: The above theorem is proven by reducing the strongly NP-hard 3-partition problem to the MMST problem defined on grid graphs.

Given a 3-partition problem described in Section 1 with a set I of $3l$ elements of sizes $a_i, i \in I$ and a constant B, construct a grid graph with $m = l, n = 3l + 1$. Define an l-scenario (i.e., $|S| = l$) MMST problem with cost of the edges as:

$$c_e^s = 0 \qquad e \in E_c; s \in S,$$

$$c_{v_{ij}, v_{i,j+1}}^s = \begin{cases} a_j & \text{if } s = i, \\ 0 & \text{otherwise,} \end{cases} \qquad i = 1, ..., m; j = 1, ..., n-1; s \in S.$$

We claim that there exists a 3-partition if and only if the MMST has an optimal objective value $z_{MMST} = B$.

To prove the above assertion, assume that an optimal solution of MMST includes edges $T = E_c \cup \left(\cup_{i=1}^l \{(v_{ij}, v_{i,j+1}) | j \in I_i\}\right)$. By the property of a spanning tree, $I_i, i = 1, ..., l$ defines a natural partition of set I, i.e., $I_i \cap I_{i'} = \emptyset, i \neq i'$ and $\cup_{i=1}^l I_i = I$. By the cost specification of the graph, the total spanning tree cost under scenario $s \in S$ is $z^s = \sum_{k \in I_s} a_k$. From the partition property, we have $\sum_{s \in S} \sum_{k \in I_s} a_k = \sum_{k \in I} a_k = lB$. By definition, $z_{MMST} = \max_{s \in S} z^s$. These, together with the fact that $|I| = l$, lead to the conclusion that there exists a 3-partition if and only if $z^s = B, s \in S$, i.e., $z_{MMST} = B$.

∎

3. The Min-max Resource Allocation Problem

The Resource Allocation (RA) problem with separable cost functions can be defined as follows. N units of a given resource are to be allocated to n activities. The operation of each activity incurs a cost. Let x_i be the amount of the resource allocated to activity i. Let $c_i(x_i)$ be the cost incurred from activity i by allocating x_i units of the resource to activity i. It is desirable to find an optimal allocation of the resource to minimize the total cost. The RA problem can then be defined by the following nonlinear integer program:

$$(RA) \qquad z_{RA} = \min \sum_{i=1}^{n} c_i(x_i)$$

subject to

$$\sum_{i=1}^{n} x_i \leq N$$

$$x_i \in Z_+ \qquad i = 1, ..., n.$$

In many applications, the functions $c_i(\cdot), i = 1, ..., n$, are decreasing and convex to reflect the fact that the more resources we allocate to an activity, the less cost will be incurred and the marginal decrease in cost diminishes. One such an application allocates workers to production lines to minimize total production time.

For decreasing convex cost functions $c_i(\cdot), i = 1, ..., n$, the RA problem can be solved in polynomial time by a simple greedy algorithm in $O(n^2)$ time (see Ibaraki and Katoh [9]).

The min-max resource allocation problem (MMRA) is defined as follows.

$$(MMRA) \qquad z_{MMRA} = \min_{x} \max_{s \in S} \sum_{i=1}^{n} c_i^s(x_i)$$

subject to

$$\sum_{i=1}^{n} x_i \leq N$$

$$x_i \in Z_+ \qquad i = 1, ..., n.$$

The complexity of the min-max extension of the classical resource allocation problem is significantly increased as indicated by the following theorem.

Theorem 4 *The MMRA problem is NP-hard even with the following restrictions:*
i) all the functions $c^s(\cdot)$ are linear decreasing;
ii) x is restricted to take only binary values; and
iii) $|S| = 2$.

Proof: We reduce the 2-partition problem to MMRA. It is known that the 2-partition problem is NP-hard even if the set I contains an even number of elements and the two partitioned subsets are restricted to have equal cardinality, i.e., $|I'| = |I|/2$.

Given a 2-partition problem as specified in Section 1, construct the MMRA problem with $|S| = 2$ as follows. Let the cost functions be:

$$c_i^1(x_i) = 2b - (2b - a_i)x_i,$$

$$c_i^2(x_i) = b + a_i - a_i x_i.$$

Clearly, all the cost functions are linear and decreasing if a large enough number b is selected. In fact, we only need to choose $b > \frac{1}{2}\max_{i=1,\dots,n} a_i$. Let $N = n/2$. Due to decreasing cost functions and binary restriction on the decision variables, exactly $n/2$ activities will be selected with one unit of resource allocated to each. If an optimal solution to MMRA has $x_i^* = 1, i \in I'; 0$ otherwise with $|I'| = n/2$. The total cost derived from scenario one is: $z^1 = nb + \sum_{i \in I'} a_i$; and the total cost obtained from scenario two is: $z^2 = nb + \sum_{i \in I \setminus I'} a_i$. By definition $z_{MMRA} = \max\{z^1, z^2\}$. We conclude that there exists a 2-partition with $|I'| = |I|/2$ if and only if the MMRA has an optimal objective value $z_{MMRA} = nb + \frac{1}{2}\sum_{i \in I} a_i$.

■

In the following, we show that in the case when the number of scenarios is bounded by a constant, a pseudo-polynomial algorithm based on dynamic programming can be devised to optimally solve the MMRA problem.

Theorem 5 *MMRA problem with linear decreasing cost functions can be solved by a pseudo-polynomial algorithm if the scenario set S is bounded.*

Proof: To prove the above theorem, we just need to provide an algorithm that runs in pseudo-polynomial time, and that solves the MMRA problem with linear decreasing cost functions to optimality.

First, consider the case with $x_i, i = 1, \dots, n$ restricted to be binary variables. Let the cost function be

$$c_i^s(x_i) = a_i^s - b_i^s x_i \qquad a_i^s \geq b_i^s \geq 0; i = 1, \dots, n; s \in S.$$

Define:

$$g_k(d; \alpha_1, \dots, \alpha_{|S|}) = \min_x \max_{s \in S} \left\{ \sum_{i=1}^{k} c_i^s(x_i) + \alpha_s \,\middle|\, \sum_{i=1}^{k} x_i = d, x_i \in \{0, 1\}, i = 1, \dots, n \right\};$$

i.e., $g_k(d; \alpha_1, \dots, \alpha_{|S|})$ is the min-max value for allocating d units of resource to the first k activities when each scenario $s \in S$ is augmented with a cost α_s.

Clearly,

$$g_1(d; \alpha_1, \dots, \alpha_{|S|}) = \begin{cases} \max_{s \in S}\{\sum_{i=1}^{n} a_i^s + \alpha_s\} & d = 0, \\ \max_{s \in S}\{\sum_{i=1}^{n} a_i^s - b_1^s + \alpha_s\} & d = 1, \\ \infty & d > 1. \end{cases}$$

We have the following general recursive formula:

$$g_{k+1}(d; \alpha_1, \dots, \alpha_{|S|}) = \min\left\{ g_k(d; \alpha_1, \dots, \alpha_{|S|}), g_k(d-1; \alpha_1 - b_{k+1}^1, \dots, \alpha_{|S|} - b_{k+1}^{|S|}) \right\}$$

The desired quantity is $z_{MMRA} = g_n(N; 0, \dots, 0)$. To construct an optimal solution for MMRA, if $g_{k+1}(d; \alpha_1, \dots, \alpha_{|S|}) = g_k(d-1; \alpha_1 - b_{k+1}^1, \dots, \alpha_{|S|} - b_{k+1}^{|S|})$ for all α values, then $x_k^* = 1$; else $x_k^* = 0$. The detailed procedure is listed below.

procedure MMRA(linear cost function coefficients a and b);
begin

Initialization: for each scenario $s \in S$, compute $L_s = \sum_{i=1}^{n} a_i^s$;
for $\alpha_1 = 0$ **to** L_1 **do**
...
  for $\alpha_{|S|} = 0$ **to** $L_{|S|}$ **do**
    for $d = 0$ **to** N **do**
      if $d \leq 1$ **then**
        $g_1(d; \alpha_1, ..., \alpha_{|S|}) = \max_{s \in S}\{\sum_{i=1}^{n} a_i^s - db_1^s + \alpha_s\}$;
      else
        $g_1(d; \alpha_1, ..., \alpha_{|S|}) = \infty$;

for $j = 1$ **to** $n-1$ **do**
  for $\alpha_1 = 0$ **to** L_1 **do**
  ...
    for $\alpha_{|S|} = 0$ **to** $L_{|S|}$ **do**
      for $d = 0$ **to** N **do**
        $g_{j+1}(d; \alpha_1, ..., \alpha_{|S|}) =$
        $\min\left\{g_j(d; \alpha_1, ..., \alpha_{|S|}), g_j(d-1; \alpha_1 - b_{j+1}^1, ..., \alpha_{|S|} - b_{j+1}^{|S|})\right\}$;
  output $z_{MMRA} = g_n(N; 0, ..., 0)$ as the optimal objective value of MMRA ;
end.

Both the initialization and the main loop take $O(nN\prod_{s \in S} L_s)$ time. Let $L_{max} = \max_{s \in S} L_s$. The overall complexity for the dynamic programming procedure is $O(nNL_{max}^{|S|})$. Since the complexity relates to the objective coefficients, when $|S|$ is bounded by a constant, the proposed dynamic programming algorithm runs in pseudo-polynomial time.

We now consider the general nonnegative integer case. To retain the linear decreasing property of the cost function, we assume $a_i^s \geq Nb_i^s, i = 1, ..., n; s \in S$. Again, define:

$$g_k(d; \alpha_1, ..., \alpha_{|S|}) = \min_x \max_{s \in S} \left\{ \sum_{i=1}^{k} f_i^s(x_i) + \alpha_s | \sum_{i=1}^{k} x_i = d, x_i \in Z_+, i = 1, ..., n \right\}.$$

Then,

$$g_1(d; \alpha_1, ..., \alpha_{|S|}) = \begin{cases} \max_{s \in S}\{\sum_{i=1}^{n} a_i^s - db_1^s + \alpha_s\} & d \leq N, \\ \infty & d > N. \end{cases}$$

The recursion becomes:

$$g_{k+1}(d; \alpha_1, ..., \alpha_{|S|}) = \min_{m=0,1,...,d} g_k(d-m; \alpha_1 - mb_{k+1}^1, ..., \alpha_{|S|} - mb_{k+1}^{|S|})$$

The optimal objective value of MMRA is given by $g_n(N; 0, ..., 0)$. To construct an optimal solution for the general integer case, if $g_{k+1}(d; \alpha_1, ..., \alpha_{|S|}) = g_k(d-m; \alpha_1 - mb_{k+1}^1, ..., \alpha_{|S|} - mb_{k+1}^{|S|})$ for all α values, then $x_k^* = m$.

A procedure similar to the one described in this section can be devised for the general integer case. For each k and a given d, $O(NL_{max}^{|S|})$ operations are needed to compute $g_{k+1}(d; \alpha_1, ..., \alpha_{|S|})$, where $L_{max} = N\max_{s \in S} a_s$. Thus, the overall complexity of the algorithm is $O(nN^2 L_{max}^{|S|})$, an $O(N)$ higher than the binary case. For scenario set S with cardinality bounded by a constant, the dynamic programming procedure remains pseudo-polynomial.

■

Again the complexity of the dynamic programming algorithm is an increasing function of the number of scenarios. The limiting situation is stated in the following theorem.

Theorem 6 *The MMRA problem is strongly NP-hard for unbounded scenario set.*

Proof: We reduce the strongly NP-hard Set Covering (SC) problem to the MMRA problem. Define the set-element incidence matrix for the SC problem as: $a_{is} = 1$ if element s is covered by (included in) set i; 0 otherwise. The SC problem tries to answer the question of whether there exists a solution x such that:

$$\begin{aligned} &\textstyle\sum_{i=1}^{n} x_i \leq N \\ &\textstyle\sum_{i=1}^{n} a_{is} x_i \geq 1 \quad s \in S \\ &x_i \in Z_+ \qquad i = 1, ..., n. \end{aligned}$$

This formulation can be rephrased as finding no more than N sets in a given collection of sets such that all elements in the space are covered. Note that the extension of the domain of the x variables from $\{0, 1\}$ to general nonnegative integers will not change the yes/no answer to the problem. This is due to the fact that elements of the set-element incidence matrix can only take values 0 or 1. For a given instance of SC problem, we define the following reduction:

$$f_i^s(x_i) = 1/n - a_{is} x_i$$

Thus $f(\cdot)$ is linear and decreasing. The corresponding MMRA problem is:

$$z_{MMRA} = \min y$$

subject to

$$1 - \sum_{i=1}^{n} a_{is} x_i \leq y \qquad s \in S$$

$$\sum_{i=1}^{n} x_i \leq N$$

$$x_i \in Z_+ \qquad i = 1, ..., n.$$

There exists a solution with $z_{MMRA} \leq 0$ for the MMRA problem if and only if there exists a feasible solution for SC, i.e., a set cover with no more than N covering sets.

■

4. The Min-max Production Control Problem

Given a finite time horizon, a deterministic demand on a single product in each time period, a production capacity at each time period, and production/inventory costs, the production control problem searches for an optimal production plan to satisfy all the demands with minimum total production/inventory cost.

To formulate the min-max production control (MMPC) problem, we first define the following parameters and decision variables:

- c_t^s = production unit cost in period t under scenario s;
- h_t^s = inventory holding cost in period t under scenario s;
- K_t = production capacity in period t;
- x_t = production quantity in period t; and
- y_t = inventory quantity in period t.

The Min-max Production Control (MMPC) problem is defined as:

$$(MMPC) \qquad \min_{x,y} \max_{s \in S} \sum_{t=1}^{T} c_t^s x_t + \sum_{t=1}^{T} h_t^s y_t$$

subject to

$$y_t = y_{t-1} + x_t - d_t \qquad t = 1, ..., T,$$

$$0 \le x_t \le K_t \text{ and integral} \qquad t = 1, ..., T,$$

$$y_t \ge 0 \text{ and integral} \qquad t = 1, ..., T.$$

The MMPC problem with a single scenario becomes the classical production control problem which can be solved in $O(T^2)$ time by using a dynamic programming procedure (see Wagner and Whitin [27]). However, the multi-scenario extension significantly increases the complexity.

Theorem 7 *The MMPC problem is NP-hard even in the case of two scenarios.*

Proof: We reduce the 2-partition problem to the MMPC problem. Given a 2-partition problem with each element $i \in I$ having size a_i, construct the following MMPC problem:

- $T = 2|I|; |S| = 2;$
- $d_t = 1$ if $t = 2i, i = 1, ..., |I|$; 0 otherwise;
- $K_t = 1, t = 1, ..., T;$
- $h_t^s = b, t = 2i, i = 1, ..., |I|; s \in S$; 0 otherwise, where b is a very large number (e.g., $b > \max_{s \in S} \sum_{i=1}^{n} a_i$);
- $c_t^1 = a_i$ if $t = 2i - 1, i = 1, ..., |I|$; 0 otherwise. $c_t^2 = a_i$ if $t = 2i, i = 1, ..., |I|$; 0 otherwise.

Due to large inventory carry-over cost from period $t = 2i$ to period $2i + 1, i = 1, ..., |I|$, the demand incurred at period $t = 2i$ must be satisfied by production either in period $t = 2i - 1$ or in period $t = 2i$. Assume an optimal production plan schedules a positive production in periods $t = 2i - 1, i \in I'$; it must also schedule a positive production in periods $t = 2i, i \in I \setminus I'$. The total cost for scenario one in this case is $z^1 = \sum_{i \in I'} a_i$; while for scenario two, we have total cost $z^2 = \sum_{i \in I \setminus I'} a_i$. By definition, $z_{MMPC} = \max\{z^1, z^2\}$. Thus, there exists a 2-partition if and only if the MMPC gives an optimal objective value $z_{MMPC} = \frac{1}{2}\sum_{i\in I} a_i$.

■

The following theorem states that for a bounded number of scenarios, the MMPC problem is only weakly NP-hard.

Theorem 8 *The MMPC problem can be solved in pseudo-polynomial time for a bounded number of scenarios.*

Proof: Define:

$g_t(y_{t-1}; \alpha_1, ..., \alpha_{|S|})$ = min-max total production/inventory cost incurred from period t through period T when an additional cost of α_s is augmented for scenario $s \in S$, and when the leftover inventory from period $t-1$ is y_{t-1}.

With the above definition, an initial condition can be easily specified:

$$g_T(y_{T-1}; \alpha_1, ..., \alpha_{|S|}) = \begin{cases} \max_{s\in S}\{c_T^s(d_T - y_{T-1}) + \alpha_s\} & y_{T-1} \le d_T \\ \infty & y_{T-1} > d_T \text{ or } d_T - y_{T-1} > K_T. \end{cases}$$

The recursive relation is:

$$g_t(y_{t-1}; \alpha_1, ..., \alpha_{|S|}) = \min_{d_t - y_{t-1} \le x_t \le K_t} g_{t+1}(y_{t-1} + x_t - d_t; \alpha_1 + c_t^1 x_t, ..., \alpha_{|S|} + c_t^{|S|} x_t).$$

The optimal objective value of MMPC is given by $z_{MMPC} = g_1(0; 0, ..., 0)$. To construct an optimal solution, if $g_t(p; \alpha_1, ..., \alpha_{|S|}) = g_{t+1}(p+q-d_t; \alpha_1 + c_t^1 q, ..., \alpha_{|S|} + c_t^{|S|} q)$ for all α values, we have $x_t^* = q, y_t^* = p + q - d_t$.

A detailed procedure for optimally solving MMPC is as follows.

procedure MMPC(c: production cost; h: holding cost; K: production capacity);
begin
  Initialization: for each scenario $s \in S$,
    compute $L_s = (\max_{t=1,...T} c_t^s + \max_{t=1,...T} h_t^s)\sum_{t=1}^T d_t$;
    also compute $d_{max} = \max_{t=1,...,T} d_t$;
  for $\alpha_1 = 0$ **to** L_1 **do**
  ...
    for $\alpha_{|S|} = 0$ **to** $L_{|S|}$ **do**
      for $p = 0$ **to** K_T **do**
        if $p > d_T$ **or** $p < d_T - K_T$ **then**
          $g_1(p; \alpha_1, ..., \alpha_{|S|}) = \infty$;
        else
          $g_1(p; \alpha_1, ..., \alpha_{|S|}) = \max_{s\in S}\{c_T^s(d_T - p) + \alpha_s\}$;

  for $t = T - 1$ **downto** 1 **do**
    for $\alpha_1 = 0$ **to** L_1 **do**
    ...
      for $\alpha_{|S|} = 0$ **to** $L_{|S|}$ **do**
        for $p = 0$ **to** d_{max} **do**

$$g_t(p;\alpha_1,...,\alpha_{|S|}) = \min_{d_t-p\le q\le K_t} g_j(p+q-d_t;\alpha_1+c_t^1 q,...,\alpha_{|S|}+c_t^{|S|}q);$$

output $z_{MMPC} = g_1(0;0,...,0)$ as the optimal objective value of MMPC ;

end.

The parameters $L_s, s \in S$ and d_{max} are used to limit the range of x and y values. The initialization takes $O(K_T \prod_{s\in S} L_s)$ time. The main loop takes $O(T \cdot d_{max} \max_{t=1,...,T} K_t \cdot \prod_{s\in S} L_s)$ time. Thus, the overall complexity is $O(Td_{max}K_{max}L_{max}^{|S|})$, where $K_{max} = \max_{t=1,...,T} K_t$ and $L_{max} = \max_{s\in S} L_s$. If the set S is bounded, the dynamic programming algorithm is pseudo-polynomial.

■

As the scenario set grows unbounded, the MMPC problem becomes strongly NP-hard as indicated by the following theorem.

Theorem 9 *The MMPC problem is strongly NP-hard if the scenario set S is unbounded.*

Proof: We reduce the 3-partition problem described in Section 1 to the MMPC problem.

Given a 3-partition problem with parameters defined as in Section 1, construct an MMPC problem as follows:

- $T = 3l^2; |S| = l;$
- $d_t = 1$ if $t = il, i = 1, ..., 3l$; 0 otherwise;
- $K_t = 1, t = 1, ..., T;$
- $h_t^s = b, t = l \cdot i, i = 1, ..., 3l; s \in S$; 0 otherwise, where b is a very large number (say, $b > \max_{s\in S} \sum_{t=1}^{T} a_i^s$);
- $c_t^s = a_i$ if $t = (i-1)l + s, i = 1, ..., l; s \in S$; 0 otherwise.

Similar to the proof of Theorem 9, due to large inventory carry-over cost from period $t = il$ to period $il + 1, i = 1, ..., 3l$, the demand incurred at period $t = il$ must be satisfied by production in period $t = (i-1)l + 1$ through period $t = il$. Let the optimal production plan be x^*. We can partition the indices into disjoint subsets by using x^* as follows:

$$I_s = \{i | x^*_{(i-1)l+s} = 1, i = 1, ..., 3l\} \qquad s = 1, ..., l.$$

By unit demand and unit production capacity restriction, we have $I_s \cap I_{s'} = \emptyset, s \neq s'$ and $\cup_{s\in S} I_s = I$. Thus, $I_s, s \in S$ is a partition of the set I. The total production/inventory cost under scenario $s \in S$ is $z^s = \sum_{i\in I_s} a_i$. By definition $z_{MMPC} = \max_{s\in S} z^s$. By the partition property, we have $\sum_{s\in S} z^s = \sum_{i\in I} a_i = lB$. Thus, there exists a 3-partition if and only if the optimal objective value for MMPC is $z_{MMPC} = B$.

■

5. Summary and Extensions

In this paper, we have investigated the min-max version of three well-known classes of discrete optimization problems. All these classical problems can be solved efficiently in their original form. However, the min-max extension remarkably increased their complexity. All the problems under consideration become NP-hard even in very restricted cases. We have shown that the complexity increases when the number of scenarios grows. However, as long as the number of scenarios is bounded by a constant (i.e., not an increasing function of the problem size), all the problems discussed under certain restrictions can be solved in pseudo-polynomial time. Pseudo-polynomial algorithms based on dynamic programming are provided. When the number of scenarios grows with the problem size, the min-max problems discussed in this paper become strongly NP-hard. The results obtained in this paper can be easily extended. Since the MMPC problem is a special case of the minimum cost network problem, the complexity result for MMPC implies that the min-max cost discrete network flow problem is NP-hard.

Although the min-max problems discussed in this paper can be solved in pseudo-polynomial time for some special cases with bounded scenario set, the high order of complexity for large $|S|$ prohibits realistic computation by applying the described algorithms. Finding optimal solutions of the min-max discrete problems becomes a challenging and important issue. A surrogate relaxation and decomposition approach might be appropriate. Further investigation in this direction is under way.

References

1. E. Agasi, R.I. Becker and Y. Perl, "A Shifting Algorithm for Constrained Min-max Partition on Trees," *Discrete Applied Mathematics* **45** (1993)1-28.
2. R.K. Ahuja, "Minimax Linear Programming Problem," *Operations Research Letters* **4** (1985) 131-134.
3. M.S. Bazaraa and J.J. Goode, "An Algorithm for Solving Linearly Constrained Minimax Problems," *European Journal of Operational Research* **11** (1982) 158-166.
4. W. Czuchra, "A Graphical Method to Solve a Maximin Allocation Problem," *European Journal of Operational Research* **26** (1986) 259-261.
5. Z. Drezner and G.O. Wesolowsky, "A Maximin Location Problem with Maximum Distance Constraints," *IIE Transactions* **12** (1980) 249-252.
6. R.S.K. Dutta and M. Vidyasagar, "New Algorithm for Constrained Minimax Optimization," *Mathematical Programming* **13** (1977) 140-155.
7. H.A. Eislet, "Continuous Maximin Knapsack Problems with GLB Constraints," *Mathematical Programming* **36** (1986) 114-121.
8. M.R. Garey and D.S. Johnson, *Computers and Intractability,* (W.H. Freeman, San Francisco, 1979).
9. T. Ibaraki and N. Katoh, *Resource Allocation Problems: Algorithmic Approaches,* (MIT Press, Cambridge, Massachusetts, 1988).
10. . T. Ichimori, "On Min-max Integer Allocation Problems," *Operations Research* **32** (1984) 449-450.
11. S. Jacobsen, "On Marginal Allocation in Single Constraint Min-max Problems," *Management Science* **17** (1971) 780-783.
12. S. Kaplan, "Application of Programs with Maximin Objective Functions to Problems of Optimal Resource Allocation," *Operations Research* **22** (1974) 802-807.
13. R.M. Karp, "Reducibility among Combinatorial Problems," in R.E. Miller and J.W. Thatcher, ed., *Complexity of Computer Communications,* (Plenum Press, NY, 1972) pp. 85-103.
14. R.S. Klein, H. Luss and U.G. Rothblum, "Minimax Resource Allocation Problems with Resource-Substitutions Represented by Graphs," *Operations Research* **41** 5 (1993) 959-971.
15. J.B. Kruskal, "On the Shortest Spanning Subtree of a Graph and the Traveling Salesman Problem," in *Proceedings of the American Mathematical Society* **7** (1956) pp. 48-50.
16. C. Lemaréchal, "Nondifferentiable Optimization," in: G.L. Nemhauser, A.H.G. Rinnooy Kan and M.J. Todd, eds., *Handbooks in Operations Research and Management Science*, Vol. 1 (Elsevier Science Publishers B.V., 1989) pp. 529-572.

17. H. Luss, "An Algorithm for Separable Nonlinear Minimax Problems," *Operations Research Letters* **6**, (1987) 159-162.
18. H. Luss and D.R. Smith, "Resource Allocation among Competing Activities: A Lexicographic Minimax Approach," *Operations Research Letters* **5** (1986) 227-231.
19. K. Madsen and H. Schjaer-Jacobsen, "Linearly Constrained Minimax Optimization," *Mathematical Programming* **14** (1978) 208-223.
20. J.S. Pang and C.S. Yu, "A Min-max Resource Allocation Problem with Substitutions," *European Journal of Operational Research* **41** (1989) 218-223.
21. M.E. Posner and C.T. Wu, "Linear Max-min Programming," *Mathematical Programming* **20** (1981) 166-172.
22. E.L. Porteus and J.S. Yormark, "More on the Min-max Allocation," *Management Science* **18** (1972) 520-527.
23. R.C. Prim, "Shortest Connection Networks and Some Generalizations," *Bell System Technical Journal* **36** (1957) 1389-1401.
24. C.P. Rangan and R. Govindan, "An $O(n \log n)$ Algorithm for a Maxmin Location Problem," *Discrete Applied Mathematics* **36** (1992) 203-205.
25. J. Rosenhead, M. Elton, and S.K. Gupta, "Robustness and Optimality as Criteria for Strategic Decisions," *Operational Research Quarterly* **23**, 4 (1972) 413-430.
26. C.S. Tang, "A Max-min Allocation Problem: Its Solutions and Applications," *Operations Research* **36** (1988) 359-367.
27. H.M. Wagner and T. Whitin, "Dynamic Problems in the Theory of the Firm," in T. Whitin, ed., *Theory of Inventory Management*, 2nd edition (Princeton University Press, Princeton, N.J., 1957).
28. T.L. Vincent, B.S. Goh and K.L. Teo, "Trajectory-Following Algorithms for Min-max Optimization Problems," *Journal of Optimization Theory and Applications* **75**, 3 (1992) 501-519.
29. G. Yu, "On the Max-min Knapsack Problem with Robust Optimization Applications," *Operations Research*, forthcoming.
30. G. Yu and P. Kouvelis, "Robust Optimization Problems are Hard Problems to Solve," Working paper, 92/93-3-6, Department of Management Science and Information Systems, Graduate School of Business, University of Texas at Austin (Austin, 1993).
31. G. Yu and P. Kouvelis, "Complexity Results for a Class of Min-Max Problems with Robust Optimization Applications," in: P.M. Pardalos, ed., *Complexity in Numerical Optimization* (World Scientific Publishing Co., 1993) pp. 503-511.

HEILBRONN PROBLEM FOR SIX POINTS IN A PLANAR CONVEX BODY

ANDREAS W.M. DRESS
Fakultät für Mathematik, Universität Bielefeld, D-33615 Bielefeld 1

and

LU YANG and ZHENBING ZENG
Chengdu Institute of Computer Applications, Academia Sinica, 610041 Chengdu, People's Republic of China

Abstract For any six points in a planar convex body K there must be at least one triangle, formed by three of these points, with area not greater than 1/6 of the area of K. This upper bound 1/6 is best possible.

1. Introduction

Let K be a planar convex body (that means a compact convex set with non-empty interior), $|K|$ the area of K; for any triangle $r_1r_2r_3$, by $(r_1r_2r_3)$ denote its area; and let

$$(r_1r_2\cdots r_n) := \min\{(r_ir_jr_k) \mid 1 \le i < j < k \le n\};$$

$$H_n(K) := \frac{1}{|K|}\sup\{(r_1r_2\cdots r_n) \mid r_i \in K, i = 1,\cdots,n\}.$$

The values $H_n(K), n = 3, 4, \cdots$ defined as above are called *Heilbronn numbers.* Obviously, Heilbronn numbers do not change under affine transformations. If $\{r_1, r_2, \cdots, r_n\}$ is a subset of K such that

$$\frac{1}{|K|}(r_1r_2\cdots r_n) = H_n(K),$$

we say that $\{r_1, r_2, \cdots, r_n\}$ or $r_1r_2\cdots r_n$ is a *Heilbronn arrangement* of n points in K, simply, an *H-arrangement* in K.

Usually, we drop the K in $H_n(K)$ and write H_n to denote the Heilbronn numbers for K, when K is a square or parallelogram. There has been a lot of work concerning these numbers, see [2-11]. Even if n is a small integer, it is not easy to compute the exact value of H_n. In his paper [1], M. Goldberg considered exact values of the first several Heilbronn numbers. Besides the trivial cases, $H_3 = H_4 = 1/2$, he asserted that, for $n < 8$, H_n can be reached by some affine regular n-gon contained in the square, i.e. there must be an affine regular n-gon $r_1r_2\cdots r_n$ in K such that $\frac{1}{|K|}(r_1r_2\cdots r_n) = H_n$. And he listed these values as follows:

$$H_5 = \frac{3-\sqrt{5}}{4} = 0.1909\cdots,$$

D.-Z. Du and P. M. Pardalos (eds.), Minimax and Applications, 173–190.

$$H_6 = \frac{1}{8} = 0.125,$$

$$H_7 = 0.0794\cdots.$$

But he didn't give any proof.

The above-mentioned assertions were examined in [12-16] where it was shown that only one of the three is true, that is $H_6 = 1/8$. By a careful and detailed analysis, Yang, Zhang and Zeng proved

$$H_5 = \frac{\sqrt{3}}{9} = 0.1924\cdots;$$

$$H_6 = \frac{1}{8} = 0.125.$$

The first disproves Goldberg's conjecture for $n = 5$, the latter confirms it for $n = 6$. In addition, they showed by a simple example that

$$H_7 \geq \frac{1}{12} = 0.08333\cdots > 0.0794\cdots,$$

disproving Goldberg's conjecture for $n = 7$. From these discussions we know that, in general, Heilbronn arrangements in a square are not necessarily affine regular n-gons even if n is small. As to the problem for a triangular region, please refer to [17] where it was proved that

$$H_5(\triangle) = 3 - 2\sqrt{2}, \quad H_6(\triangle) = \frac{1}{8}.$$

So far as we know the above results concerning H_5, $H_5(\triangle)$, H_6, and $H_6(\triangle)$ give the first exact values of Heilbronn numbers. No further results appear to be known, in particular none concerning general planar convex bodies.

In this paper, we prove the following

Theorem 1 *For any six points in a planar convex body K there must be at least one triangle, formed by three of these points, with area not greater than $1/6$ of $|K|$. This upper bound $1/6$ is best possible.*

To the corresponding problem for seven points, we conjecture that for any planar convex body K, it holds

$$H_7(K) \leq \frac{1}{9},$$

and the upper bound 1/9 is best possible.

One could easily find examples to check that, in general, $H_n(K)$ are not necessarily bounded by $H_n(K_n)$ for $n \geq 7$, where by K_n denote the regular n-gons.

2. Prerequisites

The lemmata stated in this section, except Lemma 3, all are known and can be found in [16]. We will give the arguments here, because [16] has not been published formally.

Given n points $r_1, \cdots, r_n$ in the plane, we define a triangle $r_i r_j r_k$, formed by three of these points, to be *tight* relative to $\{r_1, \cdots, r_n\}$ if $(r_i r_j r_k) = (r_1 \cdots r_n)$, and we define it to be *loose* if $(r_i r_j r_k) > (r_1 \cdots r_n)$. Given a convex polygon with vertices $r_1, r_2, \cdots, r_n$, a triangle $r_i r_j r_k$ is said to be *peripheral* if r_i, r_j, r_k are three consecutive adjacent vertices in the polygon.

Lemma 1 *Given a convex polygon $r_1 r_2 \cdots r_n$, the triangle $r_i r_j r_k$ must be peripheral if it is tight relative to $\{r_1, r_2, \cdots, r_n\}$.*

This is easily established. □

Lemma 2 *Given a convex n-gon $r_1 r_2 \cdots r_n$ with two loose peripheral triangles sharing an edge and a positive number δ, there is another convex n-gon $r'_1 \cdots r'_n$, contained in the former, such that*

$$(r'_1 \cdots r'_n) > (r_1 \cdots r_n)$$

and $|r'_i - r_i| < \delta$ for $i = 1, \cdots, n$.

Proof. Let $r_1 r_2 r_3$ and $r_2 r_3 r_4$ be two loose peripheral triangles sharing the edge $r_2 r_3$. Put $r'_3 := (1-\varepsilon) r_3 + \varepsilon r_2$, (as shown in Fig. 1), where $\varepsilon > 0$ small enough, which keeps

$$(r_1 r_2 r'_3) > (r_1 r_2 \cdots r_n), \quad (r_2 r'_3 r_4) > (r_1 r_2 \cdots r_n)$$

and makes $(r'_3 r_4 r_5) > (r_1 r_2 \cdots r_n)$ because

$$(r'_3 r_4 r_5) = (1-\varepsilon)(r_3 r_4 r_5) + \varepsilon (r_2 r_4 r_5)$$

and $r_2 r_4 r_5$ is not tight. Replacing r_3 by r'_3, we have a new n-gon with 3 peripheral triangles of area greater than $(r_1 r_2 \cdots r_n)$. Repeating this procedure, one finally gets an n-gon which is contained in the first one and every of its peripheral triangles has an area greater than $(r_1 r_2 \cdots r_n)$. □

Lemma 3 *Given points $B, C \in \mathbf{R}^2$ with coordinates $B = (1,0)$, $C = (0,1)$, let D, E be the intersections of the line connecting points $(\xi, 0)$ and $(0, \eta)$ with the lines $x = \rho$, $y = \tau$, respectively, where $0 < \rho < \xi < 1$, $0 < \tau < \eta < 1$. Then*

$$(BDE) = \frac{1}{2}\eta(1-\xi)\left(1 - \frac{\rho}{\xi} - \frac{\tau}{\eta}\right),$$

$$(CDE) = \frac{1}{2}\xi(1-\eta)\left(1 - \frac{\rho}{\xi} - \frac{\tau}{\eta}\right). \qquad (1)$$

Proof. A direct computation gives

$$D = (\rho, \quad \eta(1 - \frac{\rho}{\xi})), \quad E = (\xi(1 - \frac{\tau}{\eta}), \quad \tau)$$

as shown in Fig. 2, hence,

$$(BDE) = \frac{1}{2} \begin{vmatrix} 1 & 0 & 1 \\ \rho & \eta(1 - \frac{\rho}{\xi}) & 1 \\ \xi(1 - \frac{\tau}{\eta}) & \tau & 1 \end{vmatrix} = \frac{1}{2}\eta(1 - \xi)(1 - \frac{\rho}{\xi} - \frac{\tau}{\eta}),$$

$$(CDE) = \frac{1}{2} \begin{vmatrix} 0 & 1 & 1 \\ \rho & \eta(1 - \frac{\rho}{\xi}) & 1 \\ \xi(1 - \frac{\tau}{\eta}) & \tau & 1 \end{vmatrix} = \frac{1}{2}\xi(1 - \eta)(1 - \frac{\rho}{\xi} - \frac{\tau}{\eta}).$$

□

Remark 1 Obviously, (1) remains true at least up to sign, if the conditions $0 < \rho < \xi < 1$, $0 < \tau < \eta < 1$ are not fulfilled.

Corollary 1 *For $\rho \in (0, 1)$ let $X(\rho)$ denote the set of all pairs of distinct points $D, E \in \mathbf{R}^2$ whose coordinates are at least as large as ρ and for which the line connecting D and E intersects x-axis and y-axis at points $(\xi, 0)$ and $(0, \eta)$,respectively, with $\xi, \eta \in (0, 1)$. Then*

$$\sup_{(D,E)\in X(\rho)} \min\{(BDE), (CDE)\} = \frac{1}{2}(\frac{1}{2} - \rho)^2, \tag{2}$$

where $B = (1, 0)$, $C = (0, 1)$.

Proof. With ξ and η as above and

$$D_0 := (\rho, \quad \eta(1 - \frac{\rho}{\xi})), \quad E_0 := (\xi(1 - \frac{\eta}{\rho}), \quad \rho)$$

one has

$$(BDE) \leq (BD_0E_0) = \frac{1}{2}\eta(1 - \xi)(1 - \frac{\rho}{\xi} - \frac{\rho}{\eta}),$$

$$(CDE) \leq (CD_0E_0) = \frac{1}{2}\xi(1 - \eta)(1 - \frac{\rho}{\xi} - \frac{\rho}{\eta}),$$

for all $(D, E) \in X(\rho)$. Without loss of generality one can assume that $\eta \leq \xi$, hence

$$\begin{aligned} \sup_{(D,E)\in X(\rho)} \min\{(BDE), (CDE)\} &\leq (BD_0E_0) \\ &\leq \frac{1}{2}(\xi - 2\rho)(1 - \xi) \leq \frac{1}{2}(\frac{1}{2} - \rho)^2. \end{aligned}$$

□

Corollary 2 *Given points r_4, r_5 in a triangle $r_1r_2r_3$, one has*

$$(r_1r_2\cdots r_5) \le \frac{1}{4+2\sqrt{3}}(r_1r_2r_3), \tag{3}$$

where the coefficient $1/(4+2\sqrt{3})$ is best possible.

Proof. Assume that the line through r_4, r_5 intersects the edges r_1r_2 and r_1r_3 of the triangle $r_1r_2r_3$. Choose an affine coordinate system such that

$$r_1 = (0,0), \quad r_2 = (1,0), \quad r_3 = (0,1).$$

as shown in Fig. 3. Let $\rho_0 := \frac{1}{4+2\sqrt{3}}$. If the conclusion does not hold, that is, if

$$(r_ir_jr_k) > \frac{1}{4+2\sqrt{3}}(r_1r_2r_3), \text{ whenever } 1 \le i < j < k \le 5,$$

then $(r_4, r_5) \in X(\rho_0)$. In view of Corollary 1 we have

$$\min\{(r_2r_4r_5), (r_3r_4r_5)\} \le \frac{1}{2}(\frac{1}{2} - \rho_0)^2.$$

On the other hand,

$$(r_2r_4r_5) > \frac{1}{4+2\sqrt{3}}(r_1r_2r_3), \quad (r_3r_4r_5) > \frac{1}{4+2\sqrt{3}}(r_1r_2r_3),$$

hence

$$\frac{1}{4+2\sqrt{3}} < 2\min\{(r_2r_4r_5), (r_3r_4r_5)\} \le (\frac{1}{2} - \rho_0)^2 = \frac{1}{4+2\sqrt{3}},$$

a contradiction! □

Lemma 4 *Given a convex quadrilateral $r_1r_2r_3r_4$ with*

$$(r_2r_3r_4) = (r_1r_2r_3r_4)$$

and a point $r_5 \in \triangle r_2r_3r_4$, one has

$$(r_1r_2\cdots r_5) \le \frac{1}{8}|r_1r_2r_3r_4|,$$

where equality holds if and only if $r_1r_2r_3r_4$ is a parallelogram and either

$$(r_2r_3r_5) = (r_3r_1r_5) = (r_4r_2r_5) = \frac{1}{8}|r_1r_2r_3r_4|$$

or

$$(r_3r_4r_5) = (r_1r_3r_5) = (r_4r_2r_5) = \frac{1}{8}|r_1r_2r_3r_4|.$$

Proof. We may assume that

$$(r_1 r_2 \cdots r_5) \geq \frac{1}{8}|r_1 r_2 r_3 r_4|$$

and $r_5 \in \Delta r_1 r_2 r_3$. Choose an affine coordinate system such that

$$r_1 = (0,0), \quad r_2 = (1,0), \quad r_4 = (0,1)$$

and let

$$r_3 = (a,b), \quad r_5 = (x,y),$$

as shown in Fig. 4. $(r_2 r_3 r_4) = (r_1 r_2 r_3 r_4)$ obviously implies $a \leq 1$ and $b \leq 1$, while $r_3 \notin \Delta r_1 r_2 r_4$ implies $a + b \geq 1$.

Since $(r_3 r_1 r_5), (r_4 r_2 r_5)$ and $(r_2 r_3 r_5)$ are not less than $\frac{1}{8}|r_1 r_2 r_3 r_4| = \frac{1}{8}(a+b)$, we have

$$\begin{cases} bx - ay \geq \frac{1}{8}(a+b), \\ x + y - 1 \geq \frac{1}{8}(a+b), \\ -bx + (a-1)y + b \geq \frac{1}{8}(a+b). \end{cases} \tag{4}$$

If this system of linear inequalities for x and y has a solution at all, these must form a triangle, the intersection (x_0, y_0) of the lines

$$bx - ay = \frac{1}{8}(a+b), \quad x + y - 1 = \frac{1}{8}(a+b),$$

being one of its vertices, that is, we must have

$$-bx_0 + (a-1)y_0 + b \geq \frac{1}{8}(a+b). \tag{5}$$

Substituting the values of x_0 and y_0, we get

$$-2a^2 + 3ab + 5b^2 + a - 7b \geq 0. \tag{6}$$

Put $g(a,b) := -2a^2 + 3ab + 5b^2 + a - 7b$, which has a unique critical point $(\frac{31}{49}, \frac{25}{49})$ with the critical value

$$g(\frac{31}{49}, \frac{25}{49}) = -\frac{72}{49} < 0.$$

Thus, what we only need to compute the maximum on the boundary of the compact region:

$$0 \leq a \leq 1, \quad 0 \leq b \leq 1, \quad a + b \geq 1.$$

Now

$$\begin{aligned} &1) \quad g(1,b) = (5b+1)(b-1); \\ &2) \quad g(a,1) = -2(1-a)^2; \\ &3) \quad g(1-b,b) = -1-b. \end{aligned}$$

Therefore, $g(a,b) \leq 0$ for all admissable choices of a and b, and $g(a,b) = 0$ if and only if $a = b = 1$ and in which case $r_5 = (3/4, 1/2)$. That means, $r_1 r_2 r_3 r_4$ is a parallelogram and

$$(r_2 r_3 r_5) = (r_3 r_1 r_5) = (r_4 r_2 r_5) = \frac{1}{8}|r_1 r_2 r_3 r_4|.$$

Analogously, when $r_5 \in \Delta r_1 r_3 r_4$, that $(r_1 \cdots r_5) = \frac{1}{8}|r_1 r_2 r_3 r_4|$ if and only if $r_1 r_2 r_3 r_4$ is a parallelogram and

$$(r_3 r_4 r_5) = (r_1 r_3 r_5) = (r_4 r_2 r_5) = \frac{1}{8}|r_1 r_2 r_3 r_4|.$$

□

Lemma 5 *If the convex hull of six points $r_1 r_2 \cdots r_6$, which belong to a convex region K, is a quadrilateral, then*

$$(r_1 r_2 \cdots r_6) \leq \frac{1}{8}|K|.$$

Proof. Let $r_1 r_2 r_3 r_4$ be the convex hull of $r_1 r_2 \cdots r_6$ and and assume that $(r_2 r_3 r_4) = (r_1 r_2 r_3 r_4)$. If either r_5 or r_6 belongs to $\Delta r_2 r_3 r_4$, the conclusion holds in view of Lemma 4. If r_5 and r_6 both belong to $\Delta r_1 r_2 r_4$, by Corollary 2, we have

$$(r_1 r_2 r_4 r_5 r_6) \leq \frac{1}{4+2\sqrt{3}}(r_1 r_2 r_4) < \frac{1}{7}(r_1 r_2 r_4),$$

hence

$$|K| \geq |r_1 r_2 r_3 r_4| = (r_1 r_2 r_4) + (r_2 r_3 r_4) > 8(r_1 r_2 r_3 r_4 r_5 r_6).$$

□

3. Proof of the Main Theorem

In order to prove Theorem 1, we must consider the three distinct cases in which the convex hull of the six points is a 4-gon, a 5-gon, or a 6-gon, separately. The case of a 4-gon has been settled already in Lemma 5.

As we shall see, however, if the convex hull is a pentagon, things are much more complicated. Several further definitions, theorems and lemmata are needed.

Definition 1 *Given $S = \{r_1, \cdots, r_n\} \subset \mathbf{R}^2$, let*

$$h(S) := \frac{(r_1 \cdots r_n)}{|\text{conv}(S)|}$$

where $|\text{conv}(S)|$ is the area of the convex hull of S.

Definition 2 *Given $S = \{r_1, \cdots, r_6\} \subset \mathbf{R}^2$, we say S belongs to class Σ_k $(k = 0, 1, 2)$ if*

i. *the convex hull of S is a pentagon, say, $r_1 \cdots r_5$;*

ii. *r_6 is at neither the edges nor the diagonals of $r_1 \cdots r_5$;*

iii. *r_6 belongs to the intersection of k peripheral triangles of $r_1 \cdots r_5$, exactly.*

Lemma 6 *Given* $\{r_1, \cdots, r_6\} \in \Sigma_2$ *with* $r_6 \in \triangle r_3r_4r_5 \cap \triangle r_4r_5r_1$, *and*

$$h(\{r_1, \cdots, r_6\}) = \sup_{S \in \Sigma_2} h(S),$$

then $r_1r_2r_3$ *is tight relative to* $\{r_1, \cdots, r_6\}$.

Proof. Otherwise, let $r_2' := (1-\varepsilon)r_2 + \varepsilon r_6$ where $\varepsilon > 0$ small enough, that keeps the area of every triangle formed by three of $r_1, r_2', r_3, r_4, r_5, r_6$ not less than $(r_1 \cdots r_6)$ and makes

$$(r_1r_2'r_3) = (1-\varepsilon)(r_1r_2r_3) - \varepsilon(r_1r_3r_6) < (r_1r_2r_3),$$

hence

$$(r_1r_2'r_3r_4r_5r_6) \geq (r_1 \cdots r_6), \quad |r_1r_2'r_3r_4r_5| < |r_1 \cdots r_5|.$$

This leads to

$$h(\{r_1, r_2', r_3, r_4, r_5, r_6\}) > h(\{r_1, \cdots, r_6\}),$$

a contradiction! □

Lemma 7 *With the notations and assumptions as in the last lemma, there is a point* r_2' *such that when replacing* r_2 *by* r_2' *the following holds:*

i. $r_1r_2'r_3r_4r_5$ *is still a convex pentagon;*

ii. $h(\{r_1, r_2', r_3, r_4, r_5, r_6\}) = \sup_{S \in \Sigma_2} h(S)$;

iii. *besides* $r_1r_2'r_3$, *at least one of the two triangles* $r_5r_1r_2'$ *and* $r_2'r_3r_4$ *is tight relative to* $\{r_1, r_2', r_3, r_4, r_5, r_6\}$.

Proof. Let

$$r_2' := r_2 + \frac{(r_2r_3r_4) - (r_1r_2r_3)}{(r_1r_3r_4)}(r_3 - r_1). \tag{7}$$

Then, we have

$$(r_1r_2'r_3) = (r_1r_2r_3), \quad (r_2'r_3r_4) = (r_1r_2r_3), \quad (r_5r_1r_2') \geq (r_5r_1r_2)$$

and $(r_2'r_jr_k) \geq (r_2r_jr_k)$ for j, k else, hence

$$(r_1r_2'r_3r_4r_5r_6) = (r_1 \cdots r_6),$$

$$|r_1r_2'r_3r_4r_5| = |r_1 \cdots r_6|.$$

That means

$$h(\{r_1, r_2', r_3, r_4, r_5, r_6\}) = h(\{r_1, \cdots, r_6\}) = \sup_{S \in \Sigma_2} h(S).$$

□

Theorem 2 *Given a convex pentagon $r_1 \cdots r_5$, if a point r_6 belongs to the intersection of two peripheral triangles of the pentagon, then*

$$(r_1 \cdots r_6) \leq \mu_2 |r_1 \cdots r_5| \tag{8}$$

where $\mu_2 = 0.1397365 \cdots$ is the smallest root of the equation

$$25\mu^3 - 31\mu^2 + 11\mu - 1 = 0; \tag{9}$$

and μ_2 is best possible, that is, there exist $r_1, \cdots, r_5, r_6$ as above with $(r_1 \cdots r_6) = \mu_2 |r_1 \cdots r_5|$.

Proof. By Lemmata 6-7, we may assume that both $r_1r_2r_3$ and $r_2r_3r_4$ are tight relative to $\{r_1, \cdots, r_6\}$. Choose an affine coordinate system such that

$$r_1 = (0,1), \quad r_2 = (0,0), \quad r_3 = (1,0),$$

and let

$$r_4 = (a,1), \quad r_5 = (u,b), \quad r_6 = (x,y),$$

as shown in Fig. 5. In this case,

$$|r_1 \cdots r_5| = \frac{1}{2}(ab+1),$$

hence

$$h(\{r_1, \cdots, r_6\}) = \frac{1}{ab+1}.$$

And $(r_1r_2r_4) \geq (r_1 \cdots r_6) = (r_1r_2r_3) = 1/2, (r_2r_3r_5) \geq (r_1 \cdots r_6) = (r_1r_2r_3) = 1/2$ obviously implies $a \geq 1, b \geq 1$.

Since $(r_6r_1r_4), (r_6r_5r_3)$ and $(r_6r_4r_5)$ are not less than $(r_1r_2r_3)$, we have

$$\begin{cases} a(y-1) \geq 1, \\ bx - (u-1)y - b \geq 1, \\ (1-b)x + (u-a)y + ab - u \geq 1. \end{cases} \tag{10}$$

Similarly as above, the consistency of the system of inequalities leads to

$$a^2b^2 - ab(a+b) - ab + b - 1 \geq (ab - a - 1)u. \tag{11}$$

And $(r_5r_1r_2) \geq (r_1r_2r_3)$ implies $u \geq 1$. Noting $b \geq y \geq (a+1)/a$, hence $ab - a - 1 \geq 0$, by (11) we have

$$a^2b^2 - ab(a+b+2) + (a+b) \geq 0. \tag{12}$$

Observing the symmetry in a and b, put $p := ab$ and $s := a+b$. It follows that

$$p^2 - s(p-1) - 2p \geq 0. \tag{13}$$

Set $f(p,s) := p^2 - s(p-1) - 2p$. Since $2\sqrt{p} \leq s$ and $f(p,s)$ is monotone increasing with s while $ab \geq 1$, we have

$$f(p, 2\sqrt{p}) \geq f(p,s) \geq 0,$$

that is

$$p^2 - 2\sqrt{p}(p-1) - 2p \geq 0. \tag{14}$$

Let $\mu := h(\{r_1 \cdots, r_6\}) = \dfrac{1}{p+1}$, then (14) leads to

$$25\mu^3 - 31\mu^2 + 11\mu - 1 \leq 0 \tag{15}$$

under which and a trivial bound for μ,

$$\mu = h(\{r_1 \cdots, r_6\}) \leq \frac{1}{5},$$

the maximum of μ is

$$\mu_2 = 0.1397365 \cdots,$$

the smallest positive root of the equation

$$25\mu^3 - 31\mu^2 + 11\mu - 1 = 0.$$

Furthermore, μ_2 is best possible because the equality holds when

$$x = y = 1.403031 \cdots, \quad u = 1, \quad a = b = 2.481194 \cdots.$$

□

Lemma 8 *Given* $\{r_1, \cdots, r_6\} \in \Sigma_1$ *with* $r_6 \in \triangle r_4 r_5 r_1$*, if*

$$h(\{r_1, \cdots, r_6\}) = \sup_{S \in \Sigma_1} h(S),$$

then $r_6 r_1 r_4$ *is tight relative to* $\{r_1, \cdots, r_6\}$*.*

Proof In the case $\angle r_1 r_2 r_6 + \angle r_2 r_6 r_5 > \pi$ or $\angle r_5 r_6 r_3 + \angle r_6 r_3 r_4 > \pi$, say, the former holds, the triangles $r_1 r_2 r_5$ and $r_1 r_6 r_5$ are loose. If $r_1 r_4 r_6$ is also loose, let $r_1' := (1-\varepsilon) r_1 + \varepsilon r_5$ where $\varepsilon > 0$ small enough, that keeps $(r_1' r_2 r_5), (r_1' r_6 r_5), (r_1' r_4 r_6)$ still greater than $(r_1 \cdots r_6)$ and

$$(r_1' r_2 r_3 r_4 r_5 r_6) \geq (r_1 \cdots r_6),$$

while

$$|r_1' r_2 r_3 r_4 r_5| = |r_1 \cdots r_5| - \varepsilon (r_1 r_2 r_5) < |r_1 \cdots r_5|.$$

Thus,

$$h(\{r_1, \cdots, r_6\}) < h(\{r_1', r_2, r_3, r_4, r_5, r_6\}) \leq \sup_{S \in \Sigma_1} h(S)$$

which contradicts the hypothesis.

In the case $\angle r_1 r_2 r_6 + \angle r_2 r_6 r_5 \leq \pi$ and $\angle r_5 r_6 r_3 + \angle r_6 r_3 r_4 \leq \pi$, we have

$$(r_6 r_1 r_2) \geq (r_5 r_1 r_2), \quad (r_6 r_3 r_4) \geq (r_5 r_3 r_4).$$

If $(r_6r_1r_4)$ is loose, let

$$r_5' := r_5 + \varepsilon(r_6 - r_5), \quad r_6' := r_6 + \varepsilon(r_6 - r_5)$$

with $\varepsilon > 0$ small enough, that keeps $(r_6'r_1r_4) > (r_1 \cdots r_6)$ and the area of every triangle formed by three of $r_1, r_2, r_3, r_4, r_5', r_6'$ not less than $(r_1 \cdots r_6)$ and makes $(r_5'r_1r_4) < (r_5r_1r_4)$. Thus,

$$(r_1r_2r_3r_4r_5'r_6') \geq (r_1 \cdots r_6), \quad |r_1r_2r_3r_4r_5'| < |r_1 \cdots r_5|,$$

and therefore,

$$h(\{r_1, \cdots, r_6\}) < h(\{r_1, r_2, r_3, r_4, r_5', r_6'\}) \leq \sup_{S \in \Sigma_1} h(S)$$

which contradicts the hypothesis. □

Theorem 3 *Given a convex pentagon $r_1 \cdots r_5$, if a point r_6 belongs to one and only one peripheral triangle of the pentagon, then*

$$(r_1 \cdots r_6) \leq \mu_1 |r_1 \cdots r_5| \tag{16}$$

where $\mu_1 = 0.14860979 \cdots$ is the unique real root of the equation

$$11\mu^3 + 10\mu^2 + 5\mu - 1 = 0; \tag{17}$$

and μ_1 is best possible.

Proof. By Lemma 8, we may assume $r_6r_1r_4$ is tight relative to $\{r_1, \cdots, r_6\}$. Choose an affine coordinate system such that

$$r_6 = (0,0), \quad r_1 = (1,0), \quad r_4 = (0,1),$$

and let

$$r_2 = (a, v), \quad r_3 = (u, b), \quad r_5 = (-x, -y),$$

as shown in Fig. 6. It is easy to find the area of the pentagon:

$$|r_1 \cdots r_5| = \frac{1}{2}(ab + 1 + x + y - (u-1)(v-1)). \tag{18}$$

Since $(r_6r_3r_4), (r_6r_1r_2), (r_6r_3r_5)$ and $(r_6r_5r_2)$ are not less than $(r_6r_1r_4)$, we have

$$\begin{cases} u \geq 1, \\ v \geq 1, \\ bx - uy \geq 1, \\ ay - vx \geq 1. \end{cases} \tag{19}$$

The system leads to

$$\begin{cases} u - 1 \leq \dfrac{1}{y}(bx - y - 1), \\ v - 1 \leq \dfrac{1}{x}(ay - x - 1), \\ bx \geq y + 1, \\ ay \geq x + 1. \end{cases} \tag{20}$$

Substituting the former two of (20) into (18), we have

$$|r_1 \cdots r_5| \geq \frac{1}{2xy}(x^2y + xy^2 + bx(1+x) + ay(1+y) - x - y - 1); \tag{21}$$

and then, substituting the latter two of (20) into (21), we obtain

$$|r_1 \cdots r_5| \geq \frac{1}{2xy}(x^2y + xy^2 + 2xy + x + y + 1). \tag{22}$$

Set

$$g(x,y) := \frac{xy}{x^2y + xy^2 + 2xy + x + y + 1}, \tag{23}$$

then

$$\frac{(r_1 \cdots r_6)}{|r_1 \cdots r_5|} \leq g(x,y). \tag{24}$$

To find the critical values of $g(x,y)$, we need solve the following system:

$$\begin{cases} \mu(x^2y + xy^2 + 2xy + x + y + 1) - xy = 0, \\ x^2y - y - 1 = 0, \\ xy^2 - x - 1 = 0. \end{cases} \tag{25}$$

We employ an efficient algorithm, "successive resultant computation"[18], which is supported by current softwares for computer algebra such as MAPLE, MACSYMA, REDUCE or MATHEMATICA. The following program was written in REDUCE:

$g_0 := \mu(x^2y + xy^2 + 2xy + x + y + 1) - xy\$$
$g_1 := x^2y - y - 1\$$
$g_2 := xy^2 - x - 1\$$
$R_2 :=$ resultant$(g_1, g_2, y)\$$
$R_1 :=$ resultant$(g_0, g_1, y)\$$
$R_0 :=$ resultant$(R_1, R_2, x)\$$
$R^* :=$ factorize(R_0);
;end;

The result for R^* is the polynomial

$$(\mu - 1)^2(11\mu^3 + 10\mu^2 + 5\mu - 1) \tag{26}$$

whose roots consist of all the real and complex critical values of $g(x,y)$.

Now let us go to find the global maximum of $g(x,y)$ in region $\{x \geq 0, y \geq 0\}$. When $x \geq 6$ or $y \geq 6$, clearly $g(x,y) < 1/8 = 0.125$. Then we consider the compact region $D : \{0 \leq x \leq 6, 0 \leq y \leq 6\}$. Since $g(x,y) < 1/8$ when $x = 6$ or $y = 6$ and $g(x,y) = 0$ when $x = 0$ or $y = 0$, we have $g(x,y) < 1/8$ over the boundary of D. On the other hand, if $g(x,y)$ takes its maximum at some point belonging to the interior of D, that maximum has to be a real root of (26). Noting $g(x,y) < 1$ whenever $x > 0$ and $y > 0$, we can assert that the unique critical value of $g(x,y)$ over region D is the unique real root of the equation

$$11\mu^3 + 10\mu^2 + 5\mu - 1 = 0,$$

$$\mu_1 = 0.14860949\cdots,$$

so is the global maximum of $g(x, y)$ over region $\{x \geq 0, y \geq 0\}$. Then it follows (24) that

$$(r_1 \cdots r_6) \leq \mu_1 |r_1 \cdots r_5|.$$

And μ_1 is best possible because the equality holds when

$$x = y = 1.3247\cdots, \quad u = v = 1, \quad a = b = 1.7548\cdots.$$

□

Lemma 9 *Given* $\{r_1, \cdots, r_6\} \in \Sigma_0$ *with* r_6 *in the convex pentagon* $r_1 \cdots r_5$, *if*

$$h(\{r_1, \cdots, r_6\}) = \sup_{S \in \Sigma_0} h(S),$$

then there is at least one diagonal of the pentagon, its end points with r_6 *form a triangle tight relative to* $\{r_1, \cdots, r_6\}$.

Proof. Otherwise, we can prove that at most one of triangles $r_6 r_{i-1} r_i$ $(i = 1, \cdots, 5; r_0 := r_5)$ is tight relative to $\{r_1, \cdots, r_6\}$. If there are two, they have either one or no common edge. In the former case, say,

$$(r_6 r_1 r_2) = (r_6 r_2 r_3) = (r_1 \cdots r_6),$$

the triangle $r_6 r_1 r_3$ has to be tight because

$$(r_6 r_1 r_3) + (r_1 r_2 r_3) = 2(r_1 \cdots r_6).$$

In the latter case, say,

$$(r_6 r_1 r_2) = (r_6 r_3 r_4) = (r_1 \cdots r_6),$$

then

$$\angle r_6 r_1 r_2 + \angle r_1 r_2 r_3 \geq \pi, \quad \angle r_2 r_3 r_4 + \angle r_3 r_4 r_6 \geq \pi,$$

which makes the sum of the interior angles of quadrilateral $r_1 r_2 r_3 r_4$ greater than 2π, impossible!

Thus, we can assume that triangles $r_6 r_{i-1} r_i$ $(i = 2, 3, 4, 5)$ are loose relative to $\{r_1, \cdots, r_6\}$. Let $r_6' := r_6 + \varepsilon(r_3 - r_6)$ where $\varepsilon > 0$ small enough such that

$$(r_6' r_i r_j) \geq (r_1 \cdots r_6), \qquad \text{for} \ \ 1 \leq i < j \leq 5.$$

On the other hand, at most two peripheral triangles are tight relative to $\{r_1, \cdots, r_5, r_6'\}$. Otherwise, there must be two of them with a common edge, say, $r_1 r_2 r_3$ and $r_2 r_3 r_4$ are tight. In this case,

$$(r_6' r_2 r_3) < \max\{(r_1 r_2 r_3), (r_2 r_3 r_4)\} = (r_1 \cdots r_5 r_6'),$$

that is impossible. And therefore, there are two loose peripheral triangles with a common edge, that leads to a contradiction in view of Lemma 2. □

Theorem 4 *Given a convex pentagon $r_1 \cdots r_5$, if a point r_6 in the pentagon belongs to none of the peripheral triangles, then*

$$(r_1 \cdots r_6) \le \mu_0 |r_1 \cdots r_5| \tag{27}$$

where $\mu_0 = (\sqrt{5}-1)/10 = 0.1236\cdots$ is best possible.

Proof. By Lemma 9, we may assume $r_6 r_4 r_1$ is tight relative to $\{r_1, \cdots, r_6\}$. Choose an affine coordinate system such that

$$r_6 = (0,0), \quad r_4 = (1,0), \quad r_1 = (0,1),$$

and let

$$r_2 = (-a,-v), \quad r_3 = (-u,-b), \quad r_5 = (x,y),$$

as shown in Fig. 7. It is easy to find the area of the pentagon:

$$|r_1 \cdots r_5| = \frac{1}{2}(ab + a + b + x + y - uv). \tag{28}$$

Since $(r_6 r_1 r_3), (r_6 r_2 r_4), (r_6 r_3 r_5)$ and $(r_6 r_5 r_2)$ are not less than $(r_6 r_4 r_1)$, we have

$$\begin{cases} u \ge 1, \\ v \ge 1, \\ bx - uy \ge 1, \\ ay - vx \ge 1. \end{cases} \tag{29}$$

The latter two of the system leads to

$$\begin{cases} a \ge \dfrac{1}{y}(vx+1), \\ b \ge \dfrac{1}{x}(uy+1). \end{cases} \tag{30}$$

Substituting the former two of (30) into (28), we have

$$|r_1 \cdots r_5| \ge \frac{1}{2xy}(x^2y + xy^2 + vx^2 + uy^2 + (v+1)x + (u+1)y + 1); \tag{31}$$

and then, substituting the former two of (29) into (31), we obtain

$$|r_1 \cdots r_5| \ge \frac{1}{2xy}(x^2y + xy^2 + x^2 + y^2 + 2x + 2y + 1). \tag{32}$$

Set

$$g(x,y) := \frac{xy}{x^2y + xy^2 + +x^2 + y^2 + 2x + 2y + 1}, \tag{33}$$

then

$$\frac{(r_1 \cdots r_6)}{|r_1 \cdots r_5|} \le g(x,y). \tag{34}$$

To find the critical values of $g(x, y)$, we need solve the following system:

$$\begin{cases} \mu(x^2y + xy^2 + x^2 + y^2 + 2x + 2y + 1) - xy = 0, \\ y(y+1)(1+y-x^2) = 0, \\ x(x+1)(1+x-y^2) = 0. \end{cases} \tag{35}$$

Solving it we obtain 8 critical points of $g(x, y)$ but only one of them is in the first quadrant, that is, $(\frac{1}{2}(\sqrt{5}+1), \frac{1}{2}(\sqrt{5}+1))$.

Now let us go to find the global maximum of $g(x, y)$ in region $\{x \geq 0, y \geq 0\}$. When $x \geq 7$ or $y \geq 7$, clearly $g(x, y) < 1/9 = 0.1111\cdots$. Then we consider the compact region $D : \{0 \leq x \leq 7, 0 \leq y \leq 7\}$. Since $g(x, y) < 1/9$ when $x = 7$ or $y = 7$ and $g(x, y) = 0$ when $x = 0$ or $y = 0$, we have $g(x, y) < 1/9$ over the boundary of D. On the other hand, if $g(x, y)$ takes its maximum at some point belonging to the interior of D, that maximum has to be a critical value, i.e.

$$\mu_0 := g(\frac{1}{2}(\sqrt{5}+1), \frac{1}{2}(\sqrt{5}+1)) = \frac{\sqrt{5}-1}{10} = 0.1236\cdots,$$

so is the global maximum of $g(x, y)$ over region $\{x \geq 0, y \geq 0\}$. Then it follows (34) that

$$(r_1 \cdots r_6) \leq \mu_0 |r_1 \cdots r_5|.$$

And μ_0 is best possible because the equality holds when

$$a = b = x = y = \frac{1}{2}(\sqrt{5}+1), \quad u = v = 1.$$

□

Now we are ready to complete the proof of Theorem 1.

Proof of Theorem 1. We have pointed out that the conclusion holds when the convex hull of the six points is a quadrilateral. Also, by Theorems 2-4, the conclusion holds when the convex hull is a pentagon because of

$$(r_1 \cdots r_6) \leq \mu_1 |r_1 \cdots r_5| < \frac{1}{6}|K|.$$

It remains that points $r_1, \cdots, r_6$ form a convex hexagon. Let

$$p_1 := r_1r_4 \cap r_3r_6, \quad p_2 := r_2r_5 \cap r_3r_6, \quad p_3 := r_1r_4 \cap r_2r_5,$$

as shown in Fig. 8. Then it is easy to see that each of the six triangles $r_1r_2p_1$, $r_2r_3p_1, r_3r_4p_2, r_4r_5p_2, r_5r_6p_3, r_6r_1p_3$ is of area not less than $(r_1 \cdots r_6)$, e.g.

$$(r_1r_2p_1) \geq \min\{(r_1r_2r_6), (r_1r_2r_3)\} \geq (r_1 \cdots r_6).$$

And therefore,

$$\begin{aligned} |r_1 \cdots r_6| &= |r_1r_2p_1| + |r_2r_3p_1| + |r_3r_4p_2| \\ &\quad + |r_4r_5p_2| + |r_5r_6p_3| + |r_6r_1p_3| + |p_1p_2p_3| \\ &\geq 6(r_1 \cdots r_6), \end{aligned}$$

hence,

$$(r_1 \cdots r_6) \le \frac{1}{6}|r_1 \cdots r_6| \le \frac{1}{6}|K|.$$

And the constant $\frac{1}{6}$ is best possible because the equality holds when K is an affine regular hexagon and $r_1, \cdots, r_6$ are those vertices. □

References

1. Goldberg, M., Maximizing the smallest triangle made by points in a square, *Math. Mag.*, **45**(1972), 135-144.
2. Komlos, J. et al., On Heilbronn's problem, *J. London Math. Soc.* (2), **24**(1981), 385-396.
3. Komlos, J. et al., A lower bound for Heilbronn's problem, *J. London Math. Soc.* (2), **25**(1982), 13-14.
4. Moser, W., Problems on extremal properties of a finite set of points, in "Discrete Geometry and Convexity", *Ann. New York Acad. Sci.* , **440**(1985), 52-64.
5. Moser, W. & Pach,J., "100 Research Problem in Discrete Geometry", MacGill University, Montreal, Que. 1986.
6. Roth, K.F., On a problem of Heilbronn, *J. London Math. Soc.*, **26**(1951), 198-204.
7. Roth, K.F., On a problem of Heilbronn II, *Proc. London Math. Soc.*, **25**(1972), 193-212.
8. Roth, K.F., On a problem of Heilbronn III, *Proc. London Math. Soc.*, **25**(1974), 543-549.
9. Roth, K.F., Estimation of the area of the smallest triangle obtained by selecting three out of n points in a disc of unit area, *Amer. Math. Soc. Proc. Symp. Pure Math.*, **24**: 251-262, 1973.
10. Roth, K.F., Developments in Heilbronn's triangle problem, *Advances in Math.*, **22**(1976), 364-385.
11. Schmidt, W.M., On a problem of Heilbronn, *J. London Math. Soc.* (2), **4**(1971), 545-550.
12. Yang Lu & Zhang Jingzhong, The problem of 6 points in a square, in "Lectures in Math. (2)", 151-175, Sichuan People's Publishing House 1980, pp. 151-175. (in Chinese).
13. Yang Lu & Zhang Jingzhong, A conjecture concerning six points in a square, in "Mathematical Olympiad in China", Hunan Education Publishing House 1990.
14. Yang Lu, Zhang Jingzhong & Zeng Zhenbing, On the conjecture and computing for exact values of the first several Heilbronn numbers, *Chin. Ann. Math.*(A), **13**:4(1992),503-515. (in Chinese).
15. Yang Lu, Zhang Jingzhong & Zeng Zhenbing, On the first several Heilbronn numbers of a triangle, *Acta Math. Sinica*, **37**:5(1994), 678-689. (in Chinese).
16. Yang Lu, Zhang Jingzhong & Zeng Zhenbing, Heilbronn problem for five points, Preprint, International Centre for Theoretical Physics, 1991, IC/91/252.
17. Yang Lu, Zhang Jingzhong & Zeng Zhenbing, On Goldberg's conjecture: computing the first several Heilbronn numbers, Preprint, Universität Bielefeld, 1991, ZiF-Nr.91/29, SFB-Nr.91/074.
18. Yang Lu, Zhang Jingzhong & Zeng Zhenbing, Searching dependency between algebraic equations: an algorithm applied to automated reasoning, Preprint, International Centre for Theoretical Physics, 1991, IC/91/6.

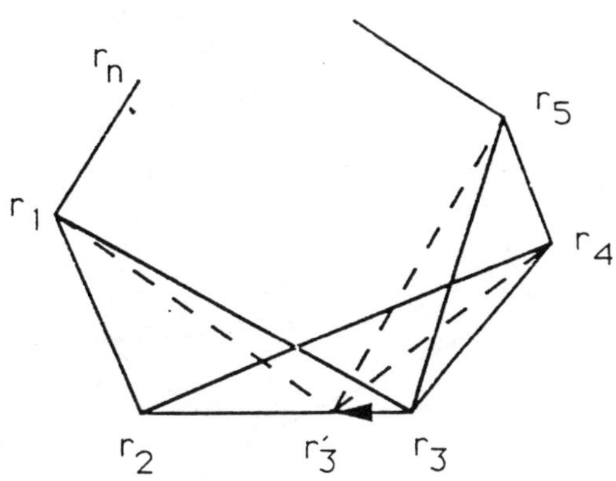

Figure 1

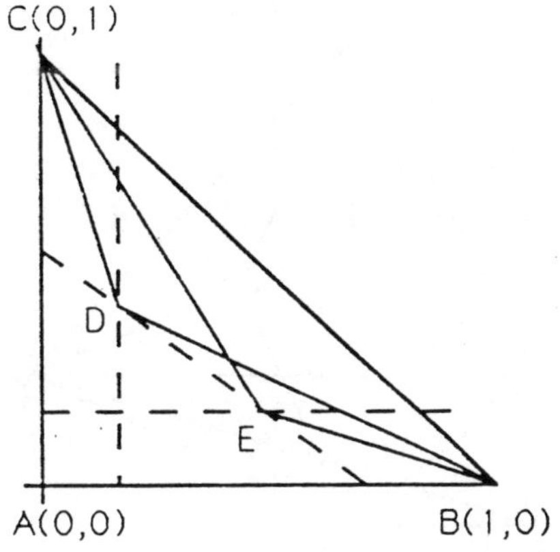

Figure 2

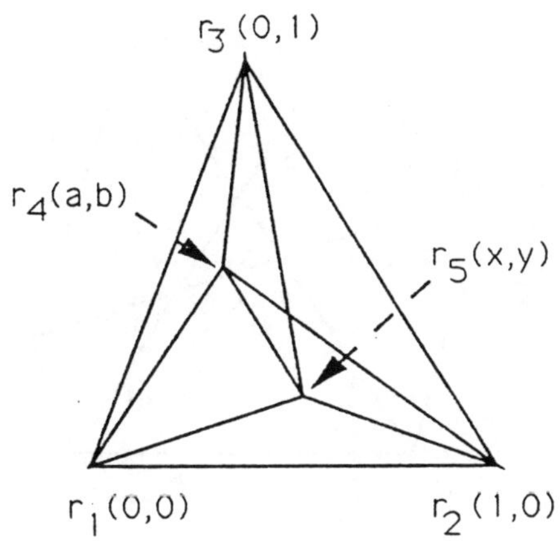

Figure 3

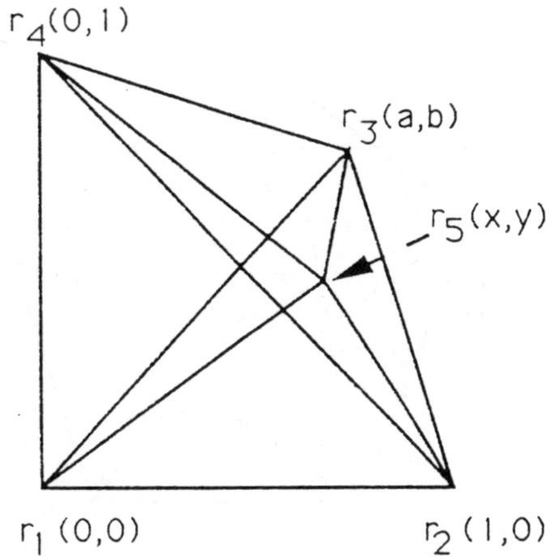

Figure 4

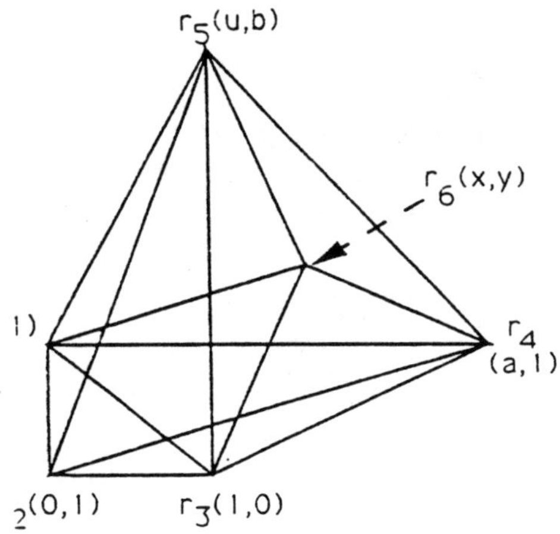

Figure 5

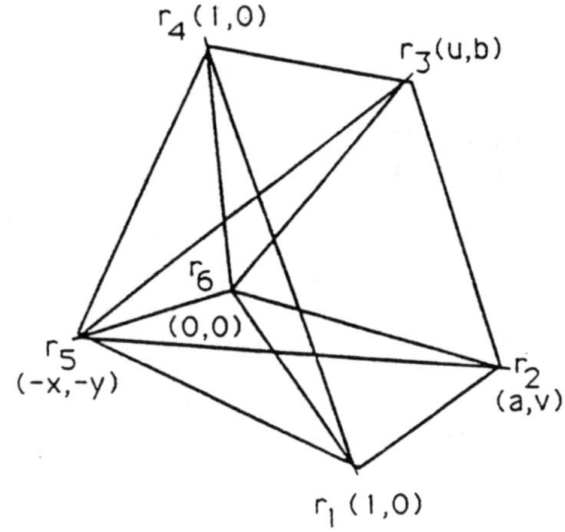

Figure 6

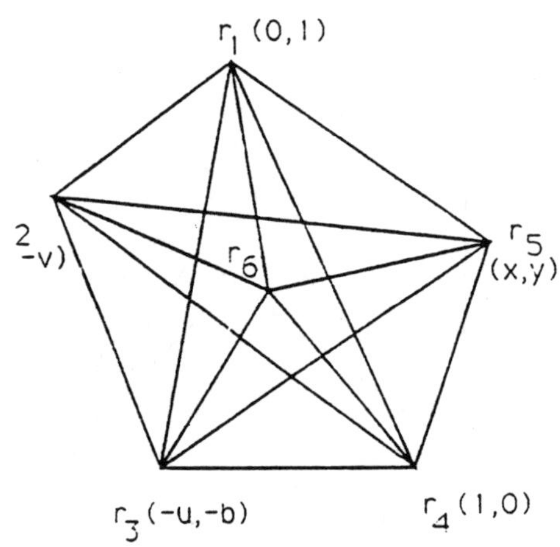

Figure 7

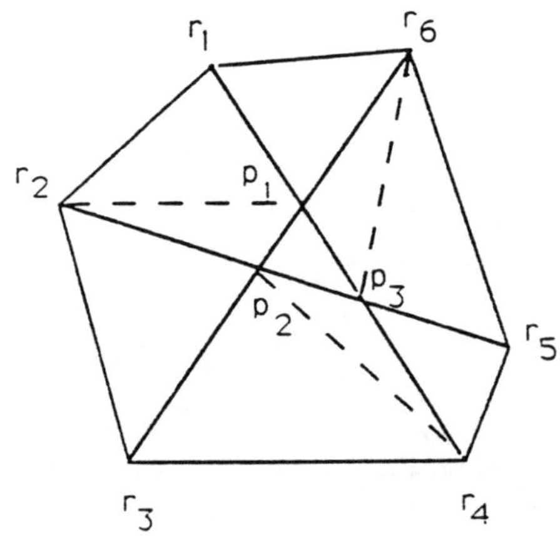

Figure 8

HEILBRONN PROBLEM FOR SEVEN POINTS IN A PLANAR CONVEX BODY

LU YANG and ZHENBING ZENG*
Chengdu Institute of Computer Applications, Academia Sinica, 610041 Chengdu, People's Republic of China

1. Introduction

Let K be a planar convex body (that means a compact convex set with non-empty interior), $|K|$ the area of K; for any triangle $r_1r_2r_3$, by $(r_1r_2r_3)$ denote its area; and let

$$(r_1r_2\cdots r_n) := \min\{(r_ir_jr_k)\ |1 \le i < j < k \le n\};$$
$$H_n(K) := \frac{1}{|K|}\sup\{(r_1r_2\cdots r_n)\ |r_i \in K, i = 1, \cdots, n\}.$$

The values $H_n(K), n = 3, 4, \cdots$ defined as above are called *Heilbronn numbers.* Obviously, Heilbronn numbers do not change under affine transformations. If $\{r_1, r_2, \cdots, r_n\}$ is a subset of K such that

$$\frac{1}{|K|}(r_1r_2\cdots r_n) = H_n(K),$$

we say that $\{r_1, r_2, \cdots, r_n\}$ or $r_1r_2\cdots r_n$ is a *Heilbronn arrangement* of n points in K, simply, an *H-arrangement* in K.

Usually, we drop the K in $H_n(K)$ and write H_n to denote the Heilbronn numbers for K, when K is a square or parallelogram. There has been a lot of work concerning these numbers, see [2-11]. It is not easy to compute the exact value of $H_n(K)$ even if n is a small integer.

In his paper [1], M. Goldberg considered exact values of the first several Heilbronn numbers. Besides the trivial cases, $H_3 = H_4 = 1/2$, he asserted that, for $n < 8$, H_n can be reached by some affine regular n-gon contained in the square, i.e. there must be an affine regular n-gon $r_1r_2\cdots r_n$ in K such that $\frac{1}{|K|}(r_1r_2\cdots r_n) = H_n$. And he listed these values as follows:

$$H_5 = \frac{3-\sqrt{5}}{4} = 0.1909\cdots, \quad H_6 = \frac{1}{8} = 0.125, \quad H_7 = 0.0794\cdots.$$

But he didn't give any proof.

* Supported in part by Sonderforschungsbereich 343 "Diskrete Strukturen in der Mathematik", Universität Bielefeld, Fakultät für Mathematik, 33615 Bielefeld 1, Germany

D.-Z. Du and P. M. Pardalos (eds.), Minimax and Applications, 191–218.

The above-mentioned assertions were examined in [13-18] where it was shown that only one of the three is true, that is $H_6 = 1/8$. By a careful and detailed analysis, Yang, Zhang and Zeng proved

$$H_5 = \frac{\sqrt{3}}{9} = 0.1924\cdots; \quad H_6 = \frac{1}{8} = 0.125.$$

The first disproves Goldberg's conjecture for $n = 5$, the latter confirms it for $n = 6$. In addition, they showed by an example that

$$H_7 \geq \frac{1}{12} = 0.08333\cdots > 0.0794\cdots,$$

disproving Goldberg's conjecture for $n = 7$. From these discussions we know that, in general, Heilbronn arrangements in a square are not necessarily affine regular n-gons even if n is small. As to the problem for a triangular region, please refer to [16, 19] where it was proved that

$$H_5(\triangle) = 3 - 2\sqrt{2}, \quad H_6(\triangle) = \frac{1}{8}.$$

About the bounds of $H_n(K)$ for general convex bodies, Dress, Yang and Zeng [20] have proved in recent that

$$H_6(K) \leq \frac{1}{6}.$$

Yet there are many left unsolved. Some open problems of P. Erdös (see [12]) are closely related to the lower and upper bounds of $H_n(K)$ for $n = 5, 6$. In this paper, we prove the following

Theorem. For any seven points in a planar convex body K there must be at least one triangle, formed by three of these points, with area not greater than $1/9$ of $|K|$. This upper bound $1/9$ is best possible.

2. Propositions and Proofs for Easier Cases

The goal of this section is prove the main theorem partly using some known propositions.

Given a configuration $r_1 \cdots r_7$ of seven points, we shall consider distinct cases according to its convex hull. If it is a triangle, or a quadrilateral, or a pentagon, then the main theorem is easier to prove. For the proofs of following propositions, see [?].

Proposition 2.1 Given a triangle $r_1r_2r_3$ and two points $r_4, r_5 \in r_1r_2r_3$, then

$$(r_1 \cdots r_5) \leq \frac{1}{4 + 2\sqrt{3}}(r_1r_2r_3).$$

□

The coefficient $1/(4 + 2\sqrt{3})$ in Proposition 2.1 is the best. The next three are about convex pentagon and a point contained in it.

For a convex polygon with vertices $r_1, r_2, \cdots, r_n$, a triangle is said to be peripheral if r_i, r_j, r_k are three consecutive adjacent vertices in the polygon.

Proposition 2.2 Given a convex pentagon $r_1 \cdots r_5$ and a point $r_6 \in r_1 \cdots r_5$, if r_6 is contained in none of peripheral triangles of the pentagon, then

$$(r_1 \cdots r_6) \leq \frac{\sqrt{5}-1}{10} |r_1 \cdots r_5|.$$

□

Proposition 2.3 Given a convex pentagon $r_1 \cdots r_5$ and a point $r_6 \in r_1 \cdots r_5$, if r_6 is contained in one and only one of peripheral triangles of the pentagon, then

$$(r_1 \cdots r_6) \leq \lambda_1 |r_1 \cdots r_5|,$$

where $\lambda_1 = 0.148609 \cdots$ is the unique real root of the equation

$$11\lambda^3 + 10\lambda^2 + 5\lambda - 1 = 0.$$

□

Proposition 2.4 Given a convex pentagon $r_1 \cdots r_5$ and a point $r_6 \in r_1 \cdots r_5$, if r_6 is contained in the intersection of two peripheral triangle of the pentagon, then

$$(r_1 \cdots r_6) \leq \lambda_2 |r_1 \cdots r_5|,$$

where $\lambda_2 = 0.139736 \cdots$ is the smallest root of the equation

$$25\lambda^3 - 31\lambda^2 + 11\lambda - 1 = 0.$$

□

The coefficients $(\sqrt{5}-1)/10, \lambda_1, \lambda_2$ in Propositions 2.2-4 are the best.

Now we prove the main theorem for the cases that the convex hull of given configuration are triangle, quadrilateral and pentagon, though it is trivial when the convex hull is a triangle since any 4 points in a triangle decompose the triangle into 9 smaller ones with disjoint interiors.

Theorem 2.1 If the convex hull of $r_1 \cdots r_7$ is a triangle, then

$$(r_1 \cdots r_7) \leq \frac{1}{8+2\sqrt{3}} |r_1 \cdots r_7|.$$

Proof. Assume $r_1 r_2 r_3$ is the convex hull of $r_1 \cdots r_7$. It is easy to prove that there must be $r_i \in \{r_4, \cdots, r_7\}$, say r_4, such that $r_4 r_1 r_2$, $r_4 r_2 r_3$, $r_4 r_3 r_1$ contains 0,1,2 (up to order) of r_5, r_6, r_7, respectively. Hence, by Proposition 2.1,

$$\begin{aligned} (r_1 r_2 r_3) &= (r_4 r_1 r_2) + (r_4 r_2 r_3) + (r_4 r_3 r_1) \\ &\geq (r_1 \cdots r_7) + 3(r_1 \cdots r_7) + (4+2\sqrt{3})(r_1 \cdots r_7) \\ &= (8+2\sqrt{3})(r_1 \cdots r_7). \end{aligned}$$

□

Theorem 2.2 If the convex hull of $r_1 \cdots r_7$ is a quadrilateral, then

$$(r_1 \cdots r_7) < 0.102786 \cdot |r_1 \cdots r_7|.$$

Proof. Assume $r_1r_2r_3r_4$ is the convex hull of $r_1 \cdots r_7$. We consider two cases as below:

CASE A: If $r_1r_2r_3$, $r_4r_1r_3$ (or $r_2r_3r_4$, $r_4r_1r_2$) contains 1,2 (up to order) of r_5, r_6, r_7, respectively, then by Proposition 2.1,

$$\begin{aligned}|r_1r_2r_3r_4| &\geq 3(r_1 \cdots r_7) + (4+2\sqrt{3})(r_1 \cdots r_7)\\ &> 10.464101 \cdot (r_1 \cdots r_7).\end{aligned}$$

CASE B: If not CASE A, then r_5, r_6, r_7 must be located in the same side of r_1r_3, as well as in the same side of r_2r_4. Without loss of generality, we might assume that $r_5, r_6, r_7 \in r_1r_2r_3 \cap r_4r_1r_2$. Then, if the convex hull of r_1, r_3, r_5, r_6, r_7 or of r_2, r_4, r_5, r_6, r_7 is a triangle, say, the convex hull of r_1, r_3, r_5, r_6, r_7 is $r_1r_6r_3$, then, we have

$$\begin{aligned}|r_1r_2r_3r_4| &= (r_4r_1r_3) + (r_1r_2r_6) + (r_2r_3r_6) + (r_1r_6r_3)\\ &\geq 3(r_1 \cdots r_7) + (4+2\sqrt{3})(r_1 \cdots r_7)\\ &> 10.464101 \cdot (r_1 \cdots r_7);\end{aligned}$$

if the convex hull of r_1, r_3, r_5, r_6, r_7 or of r_2, r_4, r_5, r_6, r_7 is a pentagon, say, the convex hull of r_1, r_3, r_5, r_6, r_7 is $r_1r_5r_6r_7r_3$, then, we have

$$\begin{aligned}|r_1r_2r_3r_4| &= (r_4r_1r_3) + (r_1r_2r_5) + (r_3r_1r_5) + (r_2r_3r_5)\\ &\geq 3(r_1 \cdots r_7) + (4+2\sqrt{3})(r_1 \cdots r_7)\\ &> 10.464101 \cdot (r_1 \cdots r_7);\end{aligned}$$

if the convex hulls of r_1, r_3, r_5, r_6, r_7 and of r_2, r_4, r_5, r_6, r_7 are both quadrilateral, say, the convex hull of r_1, r_3, r_5, r_6, r_7 is $r_1r_6r_7r_3$, and the convex hull of r_2, r_4, r_5, r_6, r_7 is $r_2r_4r_5r_6$, then r_5 is contained in exact one peripheral triangle $r_1r_6r_4$ of the pentagon $r_1r_6r_7r_3r_4$, according to Proposition 2.3, we have

$$\begin{aligned}|r_1r_2r_3r_4| &= (r_1r_2r_6) + (r_2r_7r_6) + (r_2r_3r_7) + |r_1r_6r_7r_3r_4|\\ &\geq 3(r_1 \cdots r_7) + \frac{1}{\lambda_1}(r_1 \cdots r_7)\\ &> 9.729031 \cdot (r_1 \cdots r_7).\end{aligned}$$

This completes the proof. □

Theorem 2.3 If the convex hull of $r_1 \cdots r_7$ is a pentagon, then

$$(r_1 \cdots r_7) < 0.109215 \cdot |r_1 \cdots r_7|.$$

Proof. Assume $r_1r_2r_3r_4r_5$ is the convex of $r_1 \cdots r_7$. If both of r_6, r_7 are contained in the same triangle $r_ir_jr_k, 1 \leq i < j < k \leq 5$, then, by Proposition 2.1,

$$\begin{aligned}|r_1 \cdots r_5| &\geq (r_ir_jr_k) + 2(r_1 \cdots r_7)\\ &\geq (4+2\sqrt{3})(r_1 \cdots r_7) + 2(r_1 \cdots r_7)\\ &> 9.464101 \cdot (r_1 \cdots r_7).\end{aligned}$$

If each triangle $r_ir_jr_k, 1 \leq i < j < k \leq 5$ contains at most one point of r_6, r_7, without loss of generality we might assume that either $r_6 \in r_4r_5r_1 \cap r_5r_1r_2, r_7 \in r_1r_2r_3 \cap r_2r_3r_4$, or $r_6 \in r_3r_4r_1 \cap r_5r_1r_2, r_7 \in r_1r_2r_3 \cap r_2r_3r_4$. In both cases, $r_2r_3r_4r_5r_6$ is a pentagon, $r_7 \in r_2r_3r_4 \cap r_6r_2r_3$, the intersection of two peripheral triangles of the $r_2r_3r_4r_5r_6$. According to Proposition 2.4, we have

$$\begin{aligned} |r_1 \cdots r_5| &\geq (r_1r_2r_6) + (r_6r_5r_1) + |r_2r_3r_4r_5r_6| \\ &\geq 2(r_1 \cdots r_7) + \frac{1}{\lambda_2}(r_1 \cdots r_7) \\ &> 9.156326 \cdot (r_1 \cdots r_7). \end{aligned}$$

This completes the proof. □

The proof for the case that the convex hull of $r_1 \cdots r_7$ is a convex hexagon is much more complicated. To make the expression clear, we divide the configuration of seven points which convex hull is a hexagon into four combinatorial classes $\Sigma_1, \Sigma_2, \Sigma_3, \Sigma_4$ as below.

A configuration $r_1 \cdots r_7 \in \Sigma_k$ $(k = 1, 2)$ if and only if it satisfies: (1) the convex hull of $r_1 \cdots r_7$ is a hexagon, say, $r_1 \cdots r_6$; (2) r_7 is contained in exactly k peripheral triangles of $r_1 \cdots r_6$.

A configuration $r_1 \cdots r_7 \in \Sigma_3$ if and only if it satisfies: (1) the convex hull of $r_1 \cdots r_7$ is a hexagon, say, $r_1 \cdots r_6$; (2) r_7 is contained in the triangle intersected by the three diagonal r_1r_4, r_2r_5, r_3r_6 of $r_1 \cdots r_6$.

A configuration $r_1 \cdots r_7 \in \Sigma_4$ if and only if its convex hull is a hexagon and $r_1 \cdots r_7 \notin \Sigma_1 \cup \Sigma_2 \cup \Sigma_3$.

At the end of this section, we prove the main theorem for $r_1 \cdots r_7 \in \Sigma_4$ by using Proposition 2.1.

Theorem 2.4 If a configuration of seven points $r_1 \cdots r_7 \in \Sigma_4$, then

$$(r_1 \cdots r_7) \leq 0.110009 \cdot |r_1 \cdots r_7|.$$

Proof. Without loss of generality, we might assume that either $r_7 \in r_1r_4r_5 \cap r_2r_3r_6$ or $r_7 \in r_1r_3r_4 \cap r_2r_3r_6$. In both cases, r_7 is contained in the convex pentagon $r_1r_2r_3r_5r_6$, but not in any peripheral of that. By Proposition 2.2, we have

$$\begin{aligned} |r_1 \cdots r_6| &= (r_3r_4r_5) + |r_1r_2r_3r_5r_6| \\ &\geq (r_1 \cdots r_7) + \frac{10}{\sqrt{5} - 1}(r_1 \cdots r_7) \\ &> 9.0901699 \cdot (r_1 \cdots r_7). \end{aligned}$$

□

3. Configurations with Stability

Call a configuration $r_1 \cdots r_n \subset K$ is stable (relative to K) if there is $\delta > 0$ such that

$$\frac{(r_1' \cdots r_n')}{|r_1' \cdots r_n'|} \leq \frac{(r_1 \cdots r_n)}{|r_1 \cdots r_n|}$$

for any configuration $r'_1 \cdots r'_n \subset K$ with $|r'_i - r_i| < \delta, i = 1, \cdots, n$. Particularly, call it is stable at $r_i (1 \le i \le n)$ (relative vto K) if there is $\delta > 0$ such that

$$\frac{(r_1 \cdots r_{i-1} r'_i r_{i+1} \cdots r_n)}{|r_1 \cdots r_{i-1} r'_i r_{i+1} \cdots r_n|} \le \frac{(r_1 \cdots r_n)}{|r_1 \cdots r_n|}$$

for all $r'_i \in K$ with $|r'_i - r_i| > \delta$. It is clear that if $r_1 \cdots r_n$ is stable, then it is stable at each point $r_i, i = 1, \cdots, n$.

It is obvious that any Heilbronn arrangement in the given convex body K must be stable relative to K. For simplify we don't specify the convex region when it means the whole plane.

Given a configuration $r_1 \cdots r_n$, we define a triangle $r_i r_j r_k$, formed by three of these points, to be tight (relative to $r_1 \cdots r_n$) if $(r_i r_j r_k) = (r_1 \cdots r_n)$, and define it to be loose if $(r_i r_j r_k) > (r_1 \cdots r_n)$.

Lemma 3.1 If $n > 5$ and convex n-gon $r_1 \cdots r_n$ is stable, then each peripheral triangle of it must be tight, that is,

$$(r_{i-1} r_i r_{i+1}) = (r_1 \cdots r_n)$$

for all $i, 1 \le i \le n, r_0 := r_n, r_{n+1} := r_1$.

Proof. It is easy to see that any tight triangle in a convex polygon must be peripheral. Given a convex n-gon $r_1 \cdots r_n$ $(n > 5)$, if not all its peripheral triangles are tight, say,

$$(r_1 r_2 r_3) > (r_1 \cdots r_n),$$

then there are only two triangles $r_2 r_3 r_4$, $r_n r_1 r_2$ which are incident with r_2 and possibly tight. Let $r'_2 = r_2 + \varepsilon(r_5 - r_2)$ with $0 < \varepsilon < \delta$, where

$$\delta := \frac{(r_1 r_2 r_3) - (r_1 \cdots r_n)}{(r_1 r_2 r_5) + (r_2 r_3 r_5)} > 0,$$

then

$$\begin{aligned} (r_1 r'_2 r_3) &= (r_1 r_2 r_3) - \varepsilon((r_1 r_2 r_5) + (r_2 r_3 r_5)) > (r_1 \cdots r_n), \\ (r'_2 r_3 r_4) &= (1 - \varepsilon)(r_2 r_3 r_4) + \varepsilon(r_3 r_4 r_5) \ge (r_1 \cdots r_n), \\ (r_n r_1 r'_2) &= (1 - \varepsilon)(r_n r_1 r_2) + \varepsilon(r_n r_1 r_5) \ge (r_1 \cdots r_n), \end{aligned}$$

hence,

$$(r_1 r'_2 r_3 \cdots r_n) \ge (r_1 r_2 r_2 \cdots r_n),$$

meanwhile

$$|r_1 r'_2 r_3 \cdots r_n| = |r_1 r_2 r_2 \cdots r_n| - \varepsilon((r_1 r_2 r_5) + (r_2 r_3 r_5)) < |r_1 r_2 r_3 \cdots r_n|,$$

this means that $r_1 r_2 r_3 \cdots r_n$ is not stable at vertex r_2. □

Lemma 3.2 If a configuration of seven points $r_1 \cdots r_7 \in \Sigma_1$ with $r_7 \in r_1 r_2 r_3 \cap r_2 r_4 r_5$ is stable, then

$$\begin{aligned} & (r_3 r_4 r_5) = (r_7 r_1 r_6) = (r_7 r_1 r_3) \\ = & (r_2 r_3 r_4) = (r_4 r_5 r_6) \\ = & (r_1 \cdots r_7). \end{aligned}$$

Proof. It is easy to observe that among all triangles in the given configuration, only the following ones can be tight:

$$r_2r_3r_4,\ r_3r_4r_5,\ r_4r_5r_6,\ r_5r_6r_1,\ r_6r_1r_2,\ \text{(not incident with } r_7);$$
$$r_7r_1r_2,\ r_7r_1r_3,\ r_7r_1r_6,\ r_7r_2r_3,\ r_7r_2r_5,\ r_7r_2r_6,\ \text{(incident with } r_7).$$

By making a small pertubation of r_4 along r_4r_7, it is easy to know that $r_1\cdots r_7$ is not stable (at r_4) if $r_3r_4r_5$ is loose. So we are going to prove that $r_7r_1r_6$ and $r_7r_1r_3$ are also tight under the assumption that $r_1\cdots r_7$ is stable.

It is clear that not all three triangles $r_7r_1r_6, r_5r_1r_6, r_2r_1r_6$ can be loose, and not both $r_5r_1r_6, r_2r_1r_2$ can be tight. Thus if $r_7r_1r_6$ is loose, then one has either CASE A: $r_2r_1r_6$ is loose, $r_5r_1r_6$ is tight or CASE B: $r_2r_1r_6$ is tight, $r_5r_1r_6$ is loose. Both cases lead to contradictions as in following way:

CASE A: $r_2r_1r_6$ is loose, $r_5r_1r_6$ is tight. In this case, it is easy to see that $r_7r_1r_2$ must be loose, too. Let $r_1' = r_1 + \varepsilon(r_4 - r_1)$ with $\varepsilon > 0$ small enough such that

$$(r_6r_1'r_2) > (r_1\cdots r_7),\quad (r_7r_1'r_2) > (r_1\cdots r_7),\quad (r_7r_1'r_6) > (r_1\cdots r_7);$$

and

$$(r_5r_6r_1') \geq (r_1\cdots r_7),\quad (r_7r_1'r_3) \geq (r_1\cdots r_7),$$

hence,

$$(r_1'r_2\cdots r_7) \geq (r_1\cdots r_7),\quad |r_1'r_2\cdots r_7| < |r_1\cdots r_7|,$$

contradicts to the stability of $r_1\cdots r_7$ at r_1.

CASE B: $r_2r_1r_6$ is tight, $r_5r_1r_6$ is loose. In this case, $r_7r_2r_6$ must be loose, too. Let $r_6' = r_6 + \varepsilon(r_7 - r_6)$ with $\varepsilon > 0$ small enough such that

$$(r_5r_6'r_1) > (r_1\cdots r_7),\quad (r_7r_1r_6') > (r_1\cdots r_7),\quad (r_7r_2r_6') > (r_1\cdots r_7);$$

and

$$(r_4r_5r_6') \geq (r_1\cdots r_7),\quad (r_6'r_1r_2) \geq (r_1\cdots r_7),$$

hence,

$$(r_1\cdots r_5r_6'r_7) \geq (r_1\cdots r_7),\quad |r_1\cdots r_5r_6'r_7| < |r_1\cdots r_7|,$$

contradicts to the stability of $r_1\cdots r_7$ at r_6.

Now we are going to prove that $r_7r_1r_3$ must be tight. If $r_7r_1r_3$ is loose, we consider two cases as below:

CASE A: $r_6r_1r_2$ is loose. In this case, it ie easy to know that $r_7r_1r_2$ must be loose, too. Let $r_1' = r_1 + \varepsilon(r_2 - r_1)$ with $\varepsilon > 0$ small enough such that

$$(r_6r_1'r_2) > (r_1\cdots r_7),\quad (r_7r_1'r_2) > (r_1\cdots r_7),$$
$$(r_7r_1'r_3) > (r_1\cdots r_7),\quad (r_7r_1'r_6) > (r_1\cdots r_7);$$

and

$$(r_5r_6r_1') > (r_1\cdots r_7),$$

hence,

$$(r_1'r_2\cdots r_7) \geq (r_1\cdots r_7),\quad |r_1'r_2\cdots r_7| < |r_1\cdots r_7|,$$

contradicts to the stability of $r_1 \cdots r_7$ at r_1.

CASE B: $r_6 r_1 r_2$ is tight. Let $r_2' = r_2 + \varepsilon(r_7 - r_2)$ and $r_7' = r_7 + \varepsilon(r_7 - r_2)$ with $\varepsilon > 0$ small enough such that

$$(r_7' r_1 r_3) > (r_1 \cdots r_7),$$

then

$$(r_2' r_3 r_4) > (r_1 \cdots r_7), \quad (r_6 r_1 r_2') > (r_1 \cdots r_7),$$
$$(r_7' r_1 r_2') = (r_7 r_1 r_2) = (r_1 \cdots r_7), \quad (r_7' r_1 r_3) = (r_7 r_1 r_3) = (r_1 \cdots r_7),$$
$$(r_7' r_2' r_6) = (r_7 r_2 r_6) = (r_1 \cdots r_7), \quad (r_7' r_2' r_6) = (r_7 r_2 r_6) = (r_1 \cdots r_7),$$

hence

$$(r_1 r_2' r_3 \cdots r_6 r_7') \geq (r_1 \cdots r_7), \quad |r_1 r_2' r_3 \cdots r_6 r_7'| < |r_1 \cdots r_7|,$$

contradicts to the stability of $r_1 \cdots r_7$.

Now we assume that $r_4 r_5 r_6$ is loose. Let $r_4' = r_4 + \varepsilon_1(r_4 - r_3)$ and $r_5' = r_5 + \varepsilon_2(r_5 - r_3)$ with $\varepsilon_1, \varepsilon_2 > 0$ small enough such that

$$(r_3 r_4' r_5') = (r_3 r_4 r_5) = (r_1 \cdots r_7)$$

and

$$(r_4' r_5' r_6) > (r_1 \cdots r_7),$$

then

$$(r_2 r_3 r_4') > (r_1 \cdots r_7), \quad (r_5' r_6 r_1) > (r_1 \cdots r_7), \quad (r_7 r_2 r_5') \geq (r_1 \cdots r_7),$$

hence

$$(r_1 r_2 r_3 r_4' r_5' r_6 r_7) \geq (r_1 \cdots r_7), \quad |r_1 r_2 r_3 r_4' r_5' r_6 r_7| < |r_1 \cdots r_7|,$$

contradicts to the stability of $r_1 \cdots r_7$.

To complete the proof, we prove that $r_2 r_3 r_4$ is also tight. If it is loose, we can choose $r_4^* = r_4 + \varepsilon(r_3 - r_5)$ with $\varepsilon > 0$ small enough such that $(r_2 r_3 r_4^*) > (r_1 \cdots r_7)$, then

$$(r_3 r_4^* r_5) = (r_3 r_4 r_5) = (r_1 \cdots r_7), \quad (r_4^* r_5 r_6) > (r_1 \cdots r_7).$$

Let $r_4' = r_4^* + \varepsilon_1(r_4^* - r_3)$ and $r_5' = r_5 + \varepsilon_2(r_3 - r_5)$ with $\varepsilon_1, \varepsilon_2 > 0$ small enough such that

$$(r_3 r_4' r_5') = (r_3 r_4^* r_5) = (r_3 r_4 r_5) = (r_1 \cdots r_7),$$

then

$$(r_4' r_5' r_6) > (r_1 \cdots r_7),$$

and

$$(r_2 r_3 r_4') > (r_1 \cdots r_7), \quad (r_5' r_6 r_1) > (r_1 \cdots r_7), \quad (r_7 r_2 r_5') \geq (r_1 \cdots r_7),$$

hence,

$$(r_1 r_2 r_3 r_4' r_5' r_6 r_7) \geq (r_1 \cdots r_7), \quad |r_1 r_2 r_3 r_4' r_5' r_6 r_7| < |r_1 \cdots r_7|,$$

contradicts to the stability of $r_1 \cdots r_7$, too. □

Lemma 3.3 If a configuration of seven points $r_1 \cdots r_7 \in \Sigma_1$ with $r_7 \in r_1r_2r_3 \cap r_2r_4r_5$ is stable, then it holds

$$\begin{aligned} & (r_5r_6r_1) = (r_7r_2r_3) = (r_1 \cdots r_7), \\ \text{or} \quad & (r_6r_1r_2) = (r_7r_1r_2) = (r_7r_2r_6) = (r_1 \cdots r_7), \\ \text{or} \quad & (r_7r_2r_3) = (r_7r_2r_6) = (r_1 \cdots r_7), \\ \text{or} \quad & (r_7r_2r_5) = (r_7r_2r_6) = (r_1 \cdots r_7). \end{aligned}$$

Proof. If both $r_5r_6r_1$ and $r_7r_2r_6$ are loose, then let $r_6' = r_6 + \varepsilon(r_3 - r_6)$ with $\varepsilon > 0$ small enough such that

$$(r_5r_6'r_1) > (r_1 \cdots r_7), \quad (r_7r_2r_6') > (r_1 \cdots r_7),$$

hence,

$$(r_4r_5r_6') \geq (r_1 \cdots r_7), (r_6'r_1r_2) > (r_1 \cdots r_7), \quad (r_7r_1r_6') \geq (r_1 \cdots r_7),$$

and therefore,

$$(r_1 \cdots r_5r_6'r_7) \geq (r_1 \cdots r_7), \quad |r_1 \cdots r_5r_6'r_7| < |r_1 \cdots r_7|,$$

contradicts to the stablity of $r_1 \cdots r_7$ at r_6.

Thus if $r_7r_2r_6$ is loose, then $r_5r_6r_1$ must be tight, and hence, $r_6r_1r_2, r_7r_1r_2$ must be loose. We prove that $r_7r_2r_3$ must be tight in this case. Otherwise, $(r_7r_2r_3) > (r_1 \cdots r_7)$, let $r_7' = r_7 + \varepsilon(r_7 - r_5)$ with $\varepsilon > 0$ small enough such that

$$(r_7'r_1r_2) > (r_1 \cdots r_7), \quad (r_7'r_2r_3) > (r_1 \cdots r_7), \quad (r_7'r_2r_6) > (r_1 \cdots r_7),$$

then it is easy to check that $(r_1 \cdots r_7') = (r_1 \cdots r_7)$. Consider the figuration $r_1 \cdots r_6r_7'$, which $r_6r_1r_2$,$r_7'r_1r_2$, $r_7'r_1r_3$, are loose (relative to $r_1 \cdots r_6r_7'$). Let $r_1' = r_1 + \varepsilon(r_2 - r_1)$ with $\varepsilon > 0$ small enough such that

$$(r_6r_1'r_2) > (r_1 \cdots r_7') = (r_1 \cdots r_7), \quad (r_7'r_1'r_2) > (r_1 \cdots r_7') = (r_1 \cdots r_7),$$

$$(r_7'r_1'r_3) > (r_1 \cdots r_7') = (r_1 \cdots r_7), \quad (r_7'r_1'r_6) > (r_1 \cdots r_7') = (r_1 \cdots r_7);$$

and

$$(r_5r_6r_1') > (r_1 \cdots r_7') = (r_1 \cdots r_7),$$

hence,

$$(r_1'r_2 \cdots r_6r_7') \geq (r_1 \cdots r_7') = (r_1 \cdots r_7), \quad |r_1'r_2 \cdots r_6r_7'| < |r_1 \cdots r_6r_7'| = |r_1 \cdots r_7|,$$

contradicts to the stability of $r_1 \cdots r_7$.

We are going to prove that if $r_7r_2r_6$ is tight, then at least one of $r_6r_1r_2, r_7r_2r_3$, $r_7r_2r_5$ must be tight. If all these three are loose, then let $r_2' = r_2 + \varepsilon(r_1 - r_2)$ with $\varepsilon > 0$ such that

$$(r_6r_1r_2') > (r_1 \cdots r_7), \quad (r_7r_1r_2') > (r_1 \cdots r_7),$$

$$(r_7 r_2' r_3) > (r_1 \cdots r_7), \quad (r_7 r_2' r_5) > (r_1 \cdots r_7),$$

hence

$$(r_2' r_3 r_4) \geq (r_1 \cdots r_7), \quad (r_7 r_2' r_6) \geq (r_1 \cdots r_7),$$

and

$$(r_1' r_2 \cdots r_7) \geq (r_1 \cdots r_7), \quad |r_1' r_2 \cdots r_7| < |r_1 \cdots r_7|,$$

contradicts to the stability of $r_1 \cdots r_7$ at r_1. □

Lemma 3.4 If a configuration of seven points $r_1 \cdots r_7 \in \Sigma_2$ with $r_7 \in r_1 r_2 r_3 \cap r_6 r_1 r_2$ is stable, then

$$\begin{aligned} & (r_2 r_3 r_4) = (r_3 r_4 r_5) = (r_4 r_5 r_6) = (r_5 r_6 r_1) \\ = & (r_7 r_1 r_2) = (r_7 r_1 r_3) = (r_7 r_2 r_6) \\ = & (r_1 \cdots r_7). \end{aligned}$$

Proof. It is easy to observe that among all triangles in the given configuration, only the following ones can be tight:

$$\begin{aligned} & r_2 r_3 r_4,\ r_3 r_4 r_5,\ r_4 r_5 r_6,\ r_5 r_6 r_1,\ \text{(not incident with } r_7); \\ & r_7 r_1 r_2,\ r_7 r_1 r_3,\ r_7 r_1 r_6,\ r_7 r_2 r_3,\ r_7 r_2 r_6,\ \text{(incident with } r_7). \end{aligned}$$

It is easy to prove that $r_2 r_3 r_4, r_3 r_4 r_5, r_4 r_5 r_6, r_5 r_6 r_1$ are tight, by making small perbation of r_3 along $r_3 r_6$, r_4 along $r_4 r_1$, r_5 along $r_5 r_2$, r_6 along $r_6 r_3$, respectively. For example, if $r_2 r_3 r_4$ is loose, then let $r_3' = r_3 + \varepsilon(r_6 - r_3)$ with $\varepsilon > 0$ small enough such that $(r_2 r_3' r_4) > (r_1 \cdots r_7)$, it holds

$$(r_1 r_2 r_3' r_4 r_5 r_6 r_7) \geq (r_1 \cdots r_7), \quad |r_1 r_2 r_3' r_4 r_5 r_6 r_7| < |r_1 \cdots r_7|,$$

which contradicts to the stability of $r_1 \cdots r_7$ at r_3.

To prove that $(r_7 r_1 r_2) = (r_7 r_1 r_6) = (r_7 r_2 r_6) = (r_1 \cdots r_7)$, we indicate the fact that not both $r_7 r_2 r_3$ and $r_7 r_1 r_6$ can be tight. Otherwise,

$$\angle r_7 r_2 r_3 + \angle r_2 r_3 r_6 \geq \pi, \quad \angle r_7 r_1 r_6 + \angle r_1 r_6 r_3 \geq \pi,$$

it leads to

$$\angle r_1 r_2 r_3 + \angle r_2 r_3 r_6 + \angle r_2 r_1 r_6 + \angle r_1 r_6 r_3 > \pi,$$

which is impossible. Thus, without loss of generality, assume that $r_7 r_2 r_3$ is loose. Then if $r_7 r_1 r_2$ is also loose, let $r'2 = r_2 + \varepsilon(r_5 - r_2)$ with $\varepsilon > 0$ small enough such that

$$(r_7 r_1 r_2') > (r_1 \cdots r_7), \quad (r_7 r_2' r_3) > (r_1 \cdots r_7),$$

it is to verify that

$$(r_1 r_2' r_3 \cdots r_7) \geq (r_1 \cdots r_7), \quad |r_1 r_2' r_3 \cdots r_7| < |r_1 \cdots r_7|$$

which contradicts to the stability of $r_1 \cdots r_7$ at r_2; if $r_7 r_2 r_6$ is loose under the assumption that $r_7 r_2 r_3$ is loose, then let $r_2' = r_2 + \varepsilon(r_1 - r_7)$ with $\varepsilon > 0$ small enough such that

$$(r_7 r_2' r_3) > (r_1 \cdots r_7), \quad (r_7 r_2' r_6) > (r_1 \cdots r_7),$$

it is easy to verify that

$$(r_1r_2'r_3\cdots r_7)\geq(r_1\cdots r_7),\quad |r_1r_2'r_3\cdots r_7|<|r_1\cdots r_7|$$

which contradicts to the stability of $r_1\cdots r_7$ at r_2, too; if $r_7r_1r_3$ is loose under the assumption that $r_7r_2r_3$ is loose, then let $r_3'=r_3+\varepsilon(r_2-r_4)$ with $\varepsilon>0$ small enough such that

$$(r_7r_1r_3')>(r_1\cdots r_7),\quad (r_7r_2r_3')>(r_1\cdots r_7),$$

it is to verify that $r_3'r_4r_5$ is loose relative to the configuration $r_1r_2r_3'r_4r_5r_6r_7$, and

$$(r_1r_2r_3'r_4\cdots r_7)\geq(r_1\cdots r_7),\quad |r_1r_2r_3'r_4\cdots r_7|=|r_1\cdots r_7|,$$

let $r_4'=r_4+\varepsilon(r_1-r_4)$ with $\varepsilon>0$ small enough such that $(r_3'r_4'r_5)>(r_1r_2r_3'r_4\cdots r_7)\geq(r_1\cdots r_7)$, it is to verify that

$$(r_2r_3'r_4')\geq(r_1r_2r_3'r_4\cdots r_7)\geq(r_1\cdots r_7),$$

$$(r_4'r_5r_6)\geq(r_1r_2r_3'r_4\cdots r_7)\geq(r_1\cdots r_7)$$

and hence

$$(r_1r_2r_3'r_4'r_5r_6r_7)\geq(r_1r_2r_3'r_4\cdots r_7)\geq(r_1\cdots r_7),$$

$$|r_1r_2r_3'r_4'r_5r_6r_7|<|r_1r_2r_3'r_4\cdots r_7|=|r_1\cdots r_7|,$$

which contradicts to the stability of $r_1\cdots r_7$, too. □

Lemma 3.5 If a configuration of seven points $r_1\cdots r_7\in\Sigma_3$ with r_7 contained in the triangle determined by three diagonal r_1r_4, r_2r_5, r_3r_6 of $r_1\cdots r_6$ is stable, then

$$\begin{aligned}&(r_7r_1r_2)=(r_7r_3r_4)=(r_7r_5r_6)\\ =\ &(r_7r_1r_4)=(r_7r_2r_5)=(r_7r_3r_6)\\ =\ &(r_1\cdots r_7).\end{aligned}$$

Proof. It easy to observe that the following triangles

$$r_7r_1r_3,\ r_7r_1r_5,\ r_7r_1r_6,\ r_7r_2r_3,\ r_7r_2r_4,$$
$$r_7r_2r_6,\ r_7r_3r_5,\ r_7r_4r_5,\ r_7r_4r_6$$

must be loose if configuration $r_1\cdots r_7\in\Sigma_3$.

Assume that $r_7r_3r_4$ is loose. Since not both $r_2r_3r_4, r_3r_4r_5$ can be tight, we consider two cases as below:

Case A: both $r_2r_3r_4$ and $r_3r_4r_5$ are loose, let $r_3'=r_3+\varepsilon(r_4-r_3)$ with $\varepsilon>0$ small enough such that

$$(r_2r_3'r_4)>(r_1\cdots r_7),\quad (r_3'r_4r_5)>(r_1\cdots r_7),\quad (r_7r_3'r_4)>(r_1\cdots r_7),$$

then it is easy to verify that

$$(r_1r_2r_3')>(r_1\cdots r_7),\quad (r_7r_3'r_6)>(r_1\cdots r_7),$$

and hence

$$(r_1r_2r'_3r_4\cdots r_7) \geq (r_1\cdots r_7), \quad |r_1r_2r'_3r_4\cdots r_7| < |r_1\cdots r_7|,$$

contradicts to the stability of $r_1\cdots r_7$ at r_3.

CASE B: One of $r_2r_3r_4, r_3r_4r_5$ is loose and the other one is tight. Without loss of generality, we can assume that $r_2r_3r_4$ is loose and $r_3r_4r_5$ is tight. In this case, let $r'_3 = r_3 + \varepsilon_1(r_5 - r_3), r'_4 = r_4 + \varepsilon_2(r_4 - r_5)$ with $\varepsilon_1, \varepsilon_2 > 0$ small enough such that

$$(r_2r'_3r'_4) > (r_1\cdots r_7), \quad (r'_3r'_4r_5) = (r_3r_4r_5) = (r_1\cdots r_7), \quad (r_7r'_3r'_4) > (r_1\cdots r_7),$$

then it is to verify that

$$(r_1r_2r'_3) > (r_1\cdots r_7), \quad (r'_4r_5r_6) > (r_1\cdots r_7),$$

$$(r_7r_1r'_4) > (r_1\cdots r_7), \quad (r_7r'_3r_6) \geq (r_1\cdots r_7),$$

and hence,

$$(r_1r_2r'_3r'_4r_5r_6r_7) \geq (r_1\cdots r_7), \quad |r_1r_2r'_3r'_4r_5r_6r_7| > |r_1\cdots r_7|,$$

which contradicts to the stability of $r_1\cdots r_7$.

Analogously, $r_7r_1r_2$, $r_7r_5r_6$ must be tight, too.

Now we are going to prove that $r_7r_1r_4, r_7r_2r_5, r_7r_3r_6$ are tight. Assume that $r_7r_2r_5$ is loose. Since not both $r_2r_3r_4$ and $r_3r_4r_5$ can be tight, without loss of generality we can assume that $r_2r_3r_4$ is loose. We consider two cases as below

CASE A: $r_1r_2r_3$ is loose, too. In this case, let $r'_2 = r_2 + \varepsilon(r_3 - r_2)$ with $\varepsilon > 0$ small enough such that

$$(r_1r'_2r_3) > (r_1\cdots r_7), \quad (r'_2r_3r_4) > (r_1\cdots r_7), \quad (r_7r'_2r_5) > (r_1\cdots r_7),$$

then, it ie easy to check that

$$(r_6r_1r'_2) > (r_1\cdots r_7), \quad (r_7r_1r'_2) > (r_1\cdots r_7),$$

therefore,

$$(r_1r'_2r_3\cdots r_7) \geq (r_1\cdots r_7), \quad |r_1r'_2r_3\cdots r_7| \geq |r_1\cdots r_7|,$$

which contradicts to the stability of $r_1\cdots r_7$ at r_2.

CASE B: $r_1r_2r_3$ is tight. In this case, let $r^*_2 = r_2 + \varepsilon(r_3 - r_1)$ with $\varepsilon > 0$ small enough such that

$$(r^*_2r_3r_4) > (r_1\cdots r_7), \quad (r_7r^*_2r_5) > (r_1\cdots r_7),$$

then, it ie easy to check that

$$(r_1r^*_2r_3) = (r_1r_2r_3) = (r_1\cdots r_7),$$

and

$$(r_6r_1r^*_2) > (r_1\cdots r_7), \quad (r_7r_1r^*_2) > (r_1\cdots r_7),$$

therefore,

$$(r_1 r_2^* r_3 \cdots r_7) = (r_1 \cdots r_7), \quad |r_1 r_2^* r_3 \cdots r_7| = |r_1 \cdots r_7|,$$

and $r_7 r_1 r_2^*$ is loose relative to $r_1 r_2^* r_3 \cdots r_7$.

On the other hand, we prove that $\varepsilon > 0$ can be chosen such that $r_1 r_2^* r_3 \cdots r_7$ is also stable at point r_2^* and at line $r_1 r_2^*$. Since $r_1 \cdots r_7$ is stable, there is $\delta > 0$ such that

$$\frac{(r_1' \cdots r_7')}{|r_1' \cdots r_7')|} \leq \frac{(r_1 \cdots r_7)}{|r_1 \cdots r_7|}$$

for any configuration $r_1' \cdots r_7'$ with $|r_i' - r_i| < \delta, i = 1, \cdots, 7$. Hence, if choose $r_2^* = r_2 + \varepsilon(r_3 - r_1)$ with $0 < \varepsilon < \delta/2$ such that

$$(r_2^* r_3 r_4) > (r_1 \cdots r_7), \quad (r_7 r_2^* r_5) > (r_1 \cdots r_7),$$

then for all $r_1' \cdots r_7'$ with $|r_i' - r_i| < \varepsilon, i = 1, \cdots, 7$ and $|r_2' - r_2^*| < \varepsilon$, one has

$$\frac{(r_1' \cdots r_7')}{|r_1' \cdots r_7')|} \leq \frac{(r_1 \cdots r_7)}{|r_1 \cdots r_7|} = \frac{(r_1 r_2^* r_3 \cdots r_7)}{|r_1 r_2^* r_3 \cdots r_7|},$$

hence $r_1 r_2^* r_3 \cdots r_7$ is stable, meanwhile $r_7 r_1 r_2^*$ is loose relative to $r_1 r_2^* r_3 \cdots r_7$. This is a contradiction in view of CASE A. We have proved that $r_7 r_2 r_5$ is tight whenever $r_1 r_2 r_3$ is tight or loose. Analogously, $r_7 r_1 r_4$, $r_7 r_3 r_6$ must be tight, too. □

4. Computing the Smallest Triangle

In this section, we shall compute the maximums of the smallest triangles in certain configurations of seven points. Equivalently, we fix the area of the smallest triangle, and compute the minimal area of the configuration. Without loss of generality, we assume the considered configurations are stable. The general routine is

(1): choose an affine coordinate system with coordinate triangle to be one of the tight triangles in the configuration,

(2): solve the constraint equations determined by tight triangles, and reduce some variables,

(3): solve the obtained programming problem. Usually, the constraint set is given by some polynomial inequalities. To find the minimum of the objective function, we calculate its critical values in the interior of the constraint set and interiors of the constraint polynomial curves, and its values at the vertices of the constraint set.

There are some technique in (1) and (2), but the most work is in (3) because of the massive computation comes from nonlinearity and differentiation. All these calculations are done by a computer algebra system MATHEMATICA in Unix and verified by another one REDUCE in VM/XA SP system.

Throughout this section, by $h(\cdot)$ and $k(\cdot)$ we denote

$$h(r_1 \cdots r_n) := \frac{(r_1 \cdots r_n)}{|r_1 \cdots r_n|}, \quad h(r_1 \cdots r_n) := \frac{|r_1 \cdots r_n|}{(r_1 \cdots r_n)},$$

for a given finite points configuration $r_1 \cdots r_n$.

Theorem 4.1 If $r_1 \cdots r_7$ is a convex heptagon, then

$$(r_1 \cdots r_7) \leq \mu_0 |r_1 \cdots r_7|,$$

where $\mu_0 = 4\sin^2(\pi/7)/7 = 0.107574\cdots$ is the best possible.

Proof. It is clear that if $r_1 \cdots r_7$ is convex and

$$h(r_1 \cdots r_7) = \sup\{h(S) \mid S \text{ is a convex heptagon}\},$$

then $r_1 \cdots r_7$ is stable. In view of Lemma 3.1 we might assume that all peripheral triangles of $r_1 \cdots r_7$ are of same area:

$$(t_{i-1} r_i r_{i+1}) = (r_1 \cdots r_7), \quad i = 1, \cdots, 7, r_0 := r_7, r_8 := r_1.$$

Choose an affine coordinate system such that

$$r_1 = (0,0), \quad r_2 = (1,0), \quad r_7 = (1,0)$$

and put

$$r_3 = (a,1), \quad r_4 = (x,v), \quad r_5 = (u,y), \quad r_6 = (1,b)$$

with $a, b, u, v, x, y > 1$. Then

$$(r_1 \cdots r_7) = \frac{1}{2}, \quad |r_1 \cdots r_7| = \frac{1}{2}(3 + ay).$$

Since $(r_2 r_3 r_4) = (r_3 r_4 r_5) = (r_4 r_5 r_6) = (r_5 r_6 r_7) = (r_1 \cdots r_7)$,

$$\begin{aligned} f_1 &= x + v - av = 0, \\ f_2 &= a(v - y) - (x - u) + xy - uv = 1, \\ f_3 &= (v - y) - b(x - u) + xy - uv = 1, \\ f_4 &= y + u - b = 0. \end{aligned}$$

Solving $f_1(x) = 0, f_4(y) = 0$ and substitute x, y into f_2, f_3 and $k = k(r_1 \cdots r_7)$ we get

$$x = (a-1)v, \quad y = (b-1)u,$$

and

$$\begin{aligned} k &= 3 + a(b-1)u, \\ f_2 &= u + (1 + b - ab)v + (ab - a - b)uv = 1, \\ f_3 &= (1 + a - ab)u + v + (ab - a - b)uv = 1. \end{aligned}$$

Solving $f_2(u,v) = f_3(u,v) = 0$ and $k = k(u)$ and substitute u, v into f_2 or f_3, we get

$$v = \frac{a(b-1)}{b(a-1)}u, \quad u = \frac{k-3}{a(b-1)},$$

and

$$f = (k-3)^2 - (ab-1)(k-3) - \frac{ab(a-1)(b-1)}{ab - a - b} = 0.$$

Thus we have transformed the problem into the following programming:

$$\begin{aligned} &\min\ k(a,b),\\ &\text{s.t.}\ \ f(k(a,b),a,b)=0. \end{aligned}$$

The natural boundary is

$$a \geq 1, \quad b \geq 1.$$

Noticed that if $a \longrightarrow 1^+, b > 1$, then

$$v = \frac{a(b-1)}{b(a-1)} \longrightarrow +\infty,$$

hence $k > v \longrightarrow +\infty$. On the other hand, if $a \geq 4, b \geq 4$, then

$$k > 3 + 2(r_7r_1r_3) + 2(r_6r_7r_3) > 3 + a + b \geq 11.$$

Therefore, the minimum point of $k(a,b)$ must be in the interior of

$$D := \{(a,b) \mid 1 < a, b < 4\},$$

hence, it must satisfy

$$\frac{\partial k(a,b)}{\partial a} = \frac{\partial k(a,b)}{\partial b} = 0,$$

that is equivalent to,

$$\frac{\partial f(k,a,b)}{\partial a} = 0, \quad \frac{\partial f(k,a,b)}{\partial b} = 0,$$

or

$$a^2 - b + 2ab - 2a^2b + b^2 - 2ab^2 + a^2b^2 + k(a^2 + 2ab - 2a^2b + b^2 - 2ab^2 + a^2b^2),$$
$$a^2 - a + 2ab - 2a^2b + b^2 - 2ab^2 + a^2b^2 + k(a^2 + 2ab - 2a^2b + b^2 - 2ab^2 + a^2b^2),$$

which leads to $a = b$, thus $u = v$, $x = y$, that is say, $r_1 \cdots r_7$ is an affine regular heptagon. It is easy to calculate that

$$(r_1 \cdots r_7) = \frac{4}{7}\sin^2(\frac{\pi}{7})|r_1 \cdots r_7|.$$

□

Theorem 4.2 If a configuration of seven points $r_1 \cdots r_7 \in \Sigma_1$, then

$$(r_1 \cdots r_7) \leq \frac{1}{6+2\sqrt{3}}|r_1 \cdots r_7|,$$

and the coefficient $1/(6+2\sqrt{3})$ is best possible.

Proof. It is clear that if $r_1 \cdots r_7 \in \Sigma_1$and $h(r_1 \cdots r_7) = \sup_{S \in \Sigma_1} h(S)$, then $r_1 \cdots r_7$ is stable. Without loss of generality, we might assume that $r_7 \in r_1r_2r_3 \cap r_2r_4r_5$, and

$$\begin{aligned} &(r_2r_3r_4) = (r_3r_4r_5) = (r_4r_5r_6)\\ =\ &(r_7r_1r_3) = (r_7r_1r_6)\\ =\ &(r_1 \cdots r_7), \end{aligned}$$

in view of Lemma 3.2.

Choose an affine coordinate system such that

$$r_4 = (0,0), \quad r_5 = (1,0), \quad r_3 = (0,1),$$

and put

$$r_6 = (a,1), \quad r_2 = (1,b), \quad r_1 = (u,v), \quad r_7 = (x,y),$$

with $a, b, u, v, x, y > 1$. Then

$$(r_1 \cdots r_7) = \frac{1}{2}, \quad |r_1 \cdots r_7| = \frac{1}{2}(av + bu - u - v + 2).$$

Since $(r_7 r_1 r_3) = (r_7 r_1 r_6) = (r_1 \cdots r_7)$, we have

$$x = \frac{uv - u - 1}{v - 1}, \quad y = v.$$

Let

$$\begin{aligned}
k &= av + bu - u - v + 2, \\
f_1 &:= 2(r_5 r_6 r_1) - 1 = -u - v + av, \\
f_2 &:= 2(r_6 r_1 r_2) - 1 = -u - v - ab + av + bu, \\
f_3 &:= 2(r_7 r_1 r_2) - 1 = 1 + b - 2v, \\
f_4 &:= 2(r_7 r_2 r_3) - 1 = 1 - b + u + v - bu - uv - v^2 + buv, \\
f_5 &:= 2(r_7 r_2 r_5) - 1 = 1 - v - bu - bv + buv, \\
f_6 &:= 2(r_7 r_2 r_6) - 1 = 1 + b - u - 3v - ab + bu + av + uv + v^2 \\
&\qquad + abv - buv - av^2,
\end{aligned}$$

then the programming problem is

$$\begin{aligned}
&\min\ k = av + bu - u - v + 2, \\
&\text{s.t.}\ \ f_1 \geq 0, f_2 \geq 0, \cdots f_6 \geq 0.
\end{aligned}$$

According to Lemma 3.3, when k obtains its minimum at (a, b, u, v), at least one of the following four equation systems must hold:

$$\begin{aligned}
&\text{(i)} \quad f_1(a,b,u,v) = f_4(a,b,u,v) = 0, \\
&\text{(ii)} \quad f_2(a,b,u,v) = f_3(a,b,u,v) = f_6(a,b,u,v) = 0, \\
&\text{(iii)} \quad f_5(a,b,u,v) = f_6(a,b,u,v) = 0, \\
&\text{(iv)} \quad f_4(a,b,u,v) = f_6(a,b,u,v) = 0.
\end{aligned}$$

Their solutions are respectively

$$\text{(i)} \quad a = \frac{u+v}{v}, \quad b = \frac{-1 - u - v + uv + v^2}{-1 - u + uv};$$

(ii) $u = \dfrac{1-a+b+ab}{-2+2b}, \quad v = \dfrac{1+b}{2};$

(iii) $a = \dfrac{-1-2u+u^2+7uv-2u^2v+3v^2-6uv^2+u^2v^2-v^3+uv^3}{(-1+v)(1-v-uv-v^2+uv^2)},$

$b = \dfrac{-1+v}{-u-v+uv};$

(iv) $u = \dfrac{4+2a-a^2-4ab+a^2b^2}{a(-1+b)(-2-a+ab)}, \quad v = \dfrac{-2+ab}{a}.$

Notice that

$$-u-v+uv>0, \quad -1-u+uv>0$$

in each case since $x=(uv-u-1)/(v-1)>1$,

$$1-v-uv-v^2+uv^2>0$$

in case (iii) since that then $b=(-1+v)/(-u-v+uv)<v$, and

$$-2-a+ab>0$$

in case (iv) since that then $v=(-2+ab)/a>1$. Hence the programming can be decomposed to the following four problems with less free variables:

I. min $k = k_1(u,v) = \dfrac{-2-3u-u^2+uv+u^2v+uv^2}{-1-u+uv}$

s.t. $f_2, f_3, f_5, f_6 \geq 0$:
$-1-u-v+uv+v^2 \geq 0,$
$2+2u+v-2uv \geq 0,$
$-1+u^2+2v+3uv-2u^2v+v^2-4uv^2+u^2v^2-v^3+uv^3 \geq 0,$
$-u-u^2+v+uv+2u^2v+uv^2-u^2v^2+v^3-uv^3 \geq 0.$

II. min $k = k_2(a,b) = 2+ab$

s.t. $f_1, f_4, f_5 \geq 0$:
$-1-a-b+ab \geq 0,$
$-4-a+ab \geq 0,$
$2-3b-ab-b^2+ab^2 \geq 0.$

III. min $k = k_3(u,v)$

$$= \frac{3u-u^2+2v-10uv+2u^2v-4v^2+10uv^2-u^2v^2+2v^3-4uv^3-v^4+uv^4}{(-1+v)(-u-v+uv)(1-v-uv-v^2+uv^2)},$$

s.t. $f_1, f_2, f_3, f_4 \geq 0$:
$u-4uv-2v^2+6uv^2+3v^3-3uv^3 \geq 0,$

$$-1-u+3uv+2v^2-2uv^2 \geq 0,$$
$$-1-u+3uv+2v^2-2uv^2 \geq 0,$$
$$1-u^2-2v-3uv+2u^2v-v^2+4uv^2-u^2v^2+v^3-uv^3 \geq 0.$$

$$\text{IV. } \min\ k = k_4(a,b) = -\frac{a\left(1+b+ab-ab^2\right)}{-2-a+ab}$$

$$\begin{aligned}\text{s.t. } & f_1, f_2, f_3, f_5 \geq 0: \\ & -4a-a^2-4b+2ab+5a^2b+a^3b+4ab^2 \\ & \quad -3a^2b^2-2a^3b^2-a^2b^3+a^3b^3 \geq 0, \\ & 4+a-ab \geq 0, \\ & 4+a-ab \geq 0, \\ & a-2b \geq 0.\end{aligned}$$

It is easy to calculate the critical points for each $k_i (1 \leq i \leq 4)$ and verify that each has no critical point in its feasible set, which is obviously contained in $D_0 := \{(a,b)\ |a,b>1\}$.

I. Deduce from $\frac{\partial k_1}{\partial u}=0, \frac{\partial k_1}{\partial v}=0$ and $u,v>1$,

$$1+2u+u^2+v-2uv-2u^2v-v^2+u^2v^2=0,$$
$$u+u^2-2uv-2u^2v+u^2v^2=0.$$

There is only one real critical points of $k_1(u,v)$ contained in D_0:

$$(u_0, v_0) = (2.07959\cdots, 2.32471\cdots).$$

It is easy to check that

$$f_3(u_0, v_0) = -1.1850\cdots < 0.$$

Hence there is no critical value of $k_1(u,v)$ in its feasible set.

II. The critical point of $f_2(a,b) = 2+ab$ is $(0,0)$, which is obviously not feasible.

III. Deduce from $\frac{\partial k_3}{\partial u}=0, \frac{\partial k_3}{\partial v}=0$ and $u,v>1$,

$$\begin{aligned}&-u^2+v-2uv+6u^2v-4v^2+8uv^2-14u^2v^2+5v^3-12uv^3+17u^2v^3 \\ &-2v^4+10uv^4-12u^2v^4+v^5-6uv^5+5u^2v^5-v^6+2uv^6-u^2v^6=0,\end{aligned}$$

$$\begin{aligned}&-u+2u^3-u^4+4uv+10u^2v-18u^3v+6u^4v+6uv^2-58u^2v^2+57u^3v^2 \\ &-14u^4v^2+4v^3-48uv^3+124u^2v^3-84u^3v^3+16u^4v^3-13v^4+76uv^4 \\ &-118u^2v^4+60u^3v^4-9u^4v^4+12v^5-42uv^5+46u^2v^5-18u^3v^5+2u^4v^5 \\ &-2v^6+5uv^6-4u^2v^6+u^3v^6=0,\end{aligned}$$

it can be reduced to

$$\begin{cases} u = 0, \\ v = 0, \end{cases} \quad \text{or} \quad \begin{cases} 1-9v+25v^2-25v^3+11v^4-2v^5=0, \\ 37-132u+58u^2+2u^3-6u^4+u^5=0. \end{cases}$$

Thus there is only one real critical points of $k_3(u,v)$ contained in D_0:

$$(u_0, v_0) = (3.55386\cdots, 1.47343\cdots),$$

which is, however, also not feasible, since

$$f_1(u_0, v_0) = -0.6507\cdots < 0.$$

Hence there is no critical value of $k_3(u,v)$ in its feasible set, too.

IV. Deduce from $\frac{\partial k_4}{\partial a} = 0, \frac{\partial k_4}{\partial b} = 0,$

$$2 + 2b + 4ab + a^2b - 4ab^2 - 2a^2b^2 + a^2b^3 = 0,$$

$$2a + 4a^2 + a^3 - 4a^2b - 2a^3b + a^3b^2 = 0,$$

it has unique root $a = 0, b = -1$. It is obviously not feasible.

Thus we know that for every k_i, the minimum is obtained on the boundary of the related feasible set. Since in each problem, the feasible set is a polygon with three or four curved edges, we could decompose I-IV into 13 univariate programmings. Let I_2 represent the problem I with one more tight constrain $f_2 = 0$, and I_3 represent I with $f_3 = 0$, so on. Because $r_5r_6r_1, r_6r_1r_2$ cannot be tight simultaneously, the feasible set of I_2 is empty. And so is that of II_1. The feasible set of I_3 is also empty also since $f_3 = 0 \Longrightarrow f_2 = 0$. Furthermore,

$$\mathrm{III}_2 = \mathrm{III}_3 = \mathrm{II}_5, \quad \mathrm{IV}_1 = \mathrm{I}_6, \quad \mathrm{IV}_2 = \mathrm{IV}_3 = \mathrm{II}_4, \quad \mathrm{IV}_5 = \mathrm{III}_4.$$

Hence we can reduce I-IV into the following six programmings with one free variate:

$$\begin{aligned}
\mathrm{I}_5 \ \min\ k &= \frac{-2 - 3u - u^2 + uv + u^2v + uv^2}{-1 - u + uv}, \\
\text{s.t.}\ & -1 + u^2 + 2v + 3uv - 2u^2v + v^2 - 4uv^2 + u^2v^2 - v^3 + uv^3 = 0; \\
& -1 - u - v + uv + v^2 > 0, \\
& 2 + 2u + v - 2uv > 0, \\
& -u - u^2 + v + uv + 2u^2v + uv^2 - u^2v^2 + v^3 - uv^3 \geq 0.
\end{aligned}$$

$$\begin{aligned}
\mathrm{I}_6 \ \min\ k &= \frac{-2 - 3u - u^2 + uv + u^2v + uv^2}{-1 - u + uv}, \\
\text{s.t.}\ & -u - u^2 + v + uv + 2u^2v + uv^2 - u^2v^2 + v^3 - uv^3 = 0; \\
& -1 - u - v + uv + v^2 > 0, \\
& 2 + 2u + v - 2uv > 0, \\
& -1 + u^2 + 2v + 3uv - 2u^2v + v^2 - 4uv^2 + u^2v^2 - v^3 + uv^3 \geq 0.
\end{aligned}$$

$$\begin{aligned}
\mathrm{II}_4 \ \min\ k &= 6 + a, \\
\text{s.t.}\ & -8 + 2a + a^2 \geq 0,
\end{aligned}$$

$$-8-2a+a^2\geq 0.$$

$$\text{II}_5\ \min\ k=\frac{-4+5b+b^2}{-1+b},$$
$$\text{s.t.}\ \ -2+b\geq 0.$$

$$\text{III}_1\ \min\ k=\frac{-2+3v^2}{-1+v},$$
$$\text{s.t.}\ \ 1-6v+13v^2-15v^3+6v^4\geq 0.$$

III_4 min k

$$=\frac{3u-u^2+2v-10uv+2u^2v-4v^2+10uv^2-u^2v^2+2v^3-4uv^3-v^4+uv^4}{(-1+v)(-u-v+uv)(1-v-uv-v^2+uv^2)}$$

$$\text{s.t.}\ \ 1-u^2-2v-3uv+2u^2v-v^2+4uv^2-u^2v^2+v^3-uv^3=0;$$
$$u-4uv-2v^2+6uv^2+3v^3-3uv^3\geq 0,$$
$$-1-u+3uv+2v^2-2uv^2\geq 0.$$

Now we are going to calculate the minimum of k for each problem. II_4, II_5 and III_1 are quite easy. Their solutions are given below.

II_4 : $\min k(a)=10$,
the minimum point is $a=4, b=2$.

II_5 : $\min k(b)=10$,
the minimum point is also $a=4, b=2$.

III_1 : $\min k(v)=6+2\sqrt{3}=9.46410\cdots$,
the minimum point is $u=2+\frac{2}{3}\sqrt{3}, v=1+\frac{\sqrt{3}}{3}$.

The other three need more computation. To each of them, we calculate its critical points in $D_0=\{(u,v)\mid u,v>1\}$ and decide if they are in the feasible set related to this problem, and calculate the values of k in the vertices of the feasible set.

I_5: Let $f=-1+u^2+2*v+3*u*v-2*u^2*v+v^2-4*u*v^2+u^2*v^2-v^3+u*v^3$. Then the critical points of

$$k=\frac{-2-3u-u^2+uv+u^2v+uv^2}{-1-u+uv}$$

under implicit function $f=0$ are given by

$$f=0,\quad \frac{\partial k}{\partial u}/\frac{\partial k}{\partial v}=\frac{\partial p}{\partial u}/\frac{\partial p}{\partial v},$$

which can reduced to

$$49 - 28u + 18u^2 - 18u^3 + 5u^4 = 0,$$
$$-4 - 2v^3 + v^4 = 0,$$

under condition $u, v > 1$. Solving this equation system, we know that $k|_{f=0}$ has unique critical points $(u_0, v_0) = (2.12895\cdots, 2.32023\cdots) \in D_0$. It is easy to check that this point is not in the feasible set.

The feasible set in this problem has a unique vertex (u_1, v_1) given by

$$f = -1 + u^2 + 2v + 3uv - 2u^2v + v^2 - 4uv^2 + u^2v^2 - v^3 + uv^3 = 0,$$
$$-u - u^2 + v + uv + 2u^2v + uv^2 - u^2v^2 + v^3 - uv^3 = 0.$$

and $u, v > 1$, hence

$$u_1 = 3.99671\cdots, \quad v_1 = 1.40373\cdots.$$

Thus the minimum of k in this problem is

$$\min k|_{f=0} = k(u_1, v_1) = 9.68806\cdots.$$

I_6: Let $f = -u - u^2 + v + uv + 2u^2v + uv^2 - u^2v^2 + v^3 - uv^3$. Then deduced from

$$f = 0, \quad \frac{\partial k}{\partial u} / \frac{\partial k}{\partial v} = \frac{\partial p}{\partial u} / \frac{\partial p}{\partial v},$$

and $u, v > 1$,

$$5 + 24u + 131u^2 + 336u^3 + 221u^4 + 16u^5 - 7u^6 - 28u^7 - 38u^8 - 8u^9 + 2u^{10},$$

$$1 - 13v^2 + 20v^3 - 58v^4 + 100v^5 - 120v^6 + 84v^7 - 35v^8 + 4v^9 + v^10.$$

Solving this equation system we know that $(u_0, v_0) = (1.66664\cdots, 2.36111\cdots)$ is the unique critical point of $k|_{f=0}$ in D_0. It is easy to check that (u_0, v_0) is not contained in the feasible set.

The feasible set in this problem has also a unique vertex

$$(u_1, v_1) = (3.99671\cdots, 1.40373\cdots),$$

given by the same equation system as that in I_5, and $u, v > 1$. Thus the minimum of $k|_{f=0}$ in this problem is also

$$\min k|_{f=0} = k(u_1, v_1) = 9.68806\cdots.$$

III_4: Let $f = 1 - u^2 - 2v - 3uv + 2u^2v - v^2 + 4uv^2 - u^2v^2 + v^3 - uv^3$. The critical points of k in this problem under the implicit function $f = 0$ are determined by

$$f = 0, \quad \frac{\partial k}{\partial u} / \frac{\partial k}{\partial v} = \frac{\partial p}{\partial u} / \frac{\partial p}{\partial v},$$

which can be reduced to

$$-169 - 383u - 170u^2 + 10u^3 - 13u^4 + 5u^5 = 0,$$
$$-3 + 42v - 96v^2 + 64v^3 - 16v^4 + 4v^5 = 0.$$

Hence we get the unique critical point $(u_0, v_0) = (4.61585\cdots, 1.32481\cdots)$ of k in D_0. By substitute u_0, v_0 into the constraints in this problem, we know that this critical point is in the feasible set, and

$$k_(u_0, v_0) = 9.58686\cdots.$$

It is easy to know that the feasible set in this problem has two vertices,

$$(u_1, v_1) = (3.99671\cdots, 1.40373\cdots), \quad (u_2, v_2) = (\frac{7}{2}, \frac{3}{2}),$$

and the values of k at vertices are

$$k(u_1, v_1) = 9.68806\cdots, \quad k(u_2, v_2) = 10.$$

Thus the minimum of $k|_{f=0}$ in this problem is

$$\min k|_{f=0} = f(u_0, v_0) = 9.58688\cdots.$$

So finally, we have proved that the minimum of $f = av + bu - u - v + 2$ under the constraint conditions $f_1, \cdots, f_6 \geq 0$ is

$$\min f = av + bu - u - v + 2 = 6 + 2\sqrt{3}.$$

□

Theorem 4.3 If a configuration of seven points $r_1 \cdots r_7 \in \Sigma_2$, then

$$(r_1 \cdots r_7) \leq \mu_2 |r_1 \cdots r_7|,$$

where $\mu_2 = 0.091084\cdots$ is the largest root of the equation

$$1 - 15\mu^1 + 49\mu^2 - 55\mu^3 + 19\mu^4 = 0,$$

and it is best possible.

Proof. It is clear that if $r_1 \cdots r_7 \in \Sigma_2$ and $h(r_1 \cdots r_7) = \sup_{S \in \Sigma_2} h(S)$, then $r_1 \cdots r_7$ must be stable. Without loss of generality, we might assume that $r_7 \in r_1r_2r_3 \cap r_6r_1r_2$, and

$$\begin{aligned} &(r_2r_3r_4) = (r_3r_4r_5) = (r_4r_5r_6) = (r_5r_6r_1) \\ = \; &(r_7r_1r_2) = (r_7r_1r_3) = (r_7r_2r_6) \\ = \; &(r_1 \cdots r_7), \end{aligned}$$

in view of Lemma 3.4.

Choose an affine coordinate system such that

$$r_7 = (0,0), \quad r_1 = (1,0), \quad r_2 = (0,1)$$

and put

$$r_3 = (-a,-1), \quad r_4 = (-u,-y), \quad r_5 = (-x,-v), \quad r_6 = (-1,-b).$$

with $a, b \geq 1$, $u, v, x, y > 0$. Then,

$$(r_1 \cdots r_7) = \frac{1}{2}, \quad |r_1 \cdots r_7| = \frac{1}{2}(3 + a + v + av - x).$$

Since $(r_2r_3r_4) = (r_3r_4r_5) = (r_4r_5r_6) = (r_5r_6r_1) = (r_1 \cdots r_7)$,

$$\begin{aligned} f_1 &= -1 + a - 2u + ay = 0, \\ f_2 &= -1 + b - 2v + bx = 0, \\ f_3 &= -1 - bu - v + uv + bx + y - xy = 0, \\ f_4 &= -1 - u - av + uv + x + ay - xy = 0. \end{aligned}$$

Solving $f_1(u,v) = 0, f_2(u,v) = 0$ and substituting u, v into f_3, f_4, h, we get

$$u = \frac{1}{2}(-1 + a + ay), \quad v = \frac{1}{2}(-1 + b + bx),$$

and

$$\begin{aligned} k &= \frac{1}{2}(5 + a + b + ab - 2x + bx + abx), \\ f_3 &= -1 - a - b - ab + bx + abx + 4y - ay - aby - 4xy + abxy, \\ f_4 &= -1 - a - b - ab + 4x - bx - abx + ay + aby - 4xy + abxy. \end{aligned}$$

Solving $f_3(x,y) = f_4(x,y) = 0$ and $k = k(x)$ and substituting x, y into f_3 or f_4, we get

$$y = \frac{2k - ab - a - b - 5}{ab + a - 2}, \quad x = \frac{2k - ab - a - b - 5}{ab + b - 2},$$

and

$$\begin{aligned} f &= 16 + 11a + a^2 + 11b + 9ab + b^2 - a^2b^2 \\ &\quad + k\left(-16 - 5a - 5b - ab + a^2b + ab^2 + a^2b^2\right) \\ &\quad + k^2\left(4 - ab\right) = 0. \end{aligned}$$

Thus we have transformed the problem into the following programming:

$$\begin{aligned} &\min\ k(a,b), \\ &\text{s.t.}\ \ f(k(a,b),a,b) = 0. \end{aligned}$$

The natural boundary is given by

$$a \geq 1, \quad b \geq 1.$$

To make the feasible set compact, we notice that if $a \geq 4$ or $b \geq 4$, $k = ab+a+b+2 \geq 11$, hence put the following upper bounds:

$$a \leq 4, \quad b \leq 4.$$

Another condition for a, b comes from the fact that the discriminant of the quadratic equation $f(k, a, b) = 0$ is non-negative whenever k is real:

$$\begin{aligned}\Delta \ &:= -16a + 9a^2 - 16b + 2ab + 22a^2b - 6a^3b + 9b^2 + 22ab^2 + a^2b^2 \\ &-12a^3b^2 + a^4b^2 - 6ab^3 - 12a^2b^3 - 4a^3b^3 + 2a^4b^3 + a^2b^4 + 2a^3b^4 + a^4b^4 \\ &\geq 0.\end{aligned}$$

It can be check that $\Delta(1,1) = 0$, and it is a local maximum, thus there exists $\delta > 0$ such that $\Delta(a,b) < 0$ whenever $0 < (a-1)^2 + (b-1)^2 \leq 1 + \delta$. On the other hand, it is easy to know that $a = 1, b = 1$ is not feasible, that means it is impossible that both $r_7r_2r_3$ and $r_7r_1r_6$ are tight. Hence we put an additional lower bound:

$$(a-1)^2 + (b-1)^2 \geq 1 + \delta.$$

By D we denote the region

$$D := \{(a,b) \mid 1 \leq a, b \leq 4, (a-1)^2 + (b-1)^2 \geq 1 + \delta\},$$

which is clearly containing the feasible set of (a, b).

Now we are going to compute the minimums of $k(a, b)$, limited in the interior and each edge of D.

First we consider the interior of D. If the minimum of $f(a, b)$ is got in the interior of D, it must be a critical value. Since $f(k(a,b), a, b) = 0$,

$$\frac{\partial k(a,b)}{\partial a} = \frac{\partial f}{\partial a} = 11 + 2a + 9b - 2ab^2 - 5k - bk + 2abk + b^2k + 2ab^2k - bk^2 = 0,$$

$$\frac{\partial k(a,b)}{\partial b} = \frac{\partial f}{\partial b} = 11 + 9a + 2b - 2a^2b - 5k - ak + a^2k + 2abk + 2a^2bk - ak^2 = 0.$$

To solve the equation system

$$f = f(a,b,k) = 0, \quad f_a := \frac{\partial f}{\partial a} = 0, \quad f_b := \frac{\partial f}{\partial a} = 0,$$

let

$$\begin{aligned}R_2 &:= Resultant[f, f_a, b], \\ R_1 &:= Resultant[f, f_b, b], \\ R_0 &:= Resultant[R_2, R_1, a],\end{aligned}$$

then the critical values of $k(a, b)$ must be the roots of

$$\begin{aligned}R_0 \ = \ & k(-5+2k)^4(-4+k)^4(-4+3k)(-2+k)^21(-1+k)^5 \\ & (-44 + 109k - 84k^2 + 20k^3)(-8 + 21k - 16k^2 + 4k^3) \\ & (19 - 55k + 49k^2 - 15k^3 + k^4)^2.\end{aligned}$$

This equation has 14 distinct real roots, among which only one is large than 4, that is $k_0 = 10.978796\cdots$, the largest root of the equation

$$19 - 55k + 49k^2 - 15k^3 + k^4 = 0.$$

It is easy to verify that k_0 is the unique feasible critical value in the interior of D. It is the largest root of

Second, we consider the interior of each edge of D. It is known if the minimum of k is obtained in the interior of some edge, it must be a critical value of k limited on this edge. By the choice of δ, it is obvious that k is not real in the curve

$$(a-1)^2 + (b-1)^2 = 1 + \delta, \quad (a, b > 1),$$

and hence there in no real critical value of k in this edge.

To compute the critical value in the boundary of $a = 1, 1+\sqrt{\delta} \le b \le 4$, let $a = 1$, then

$$f(k,a,b)|_{a=1} = 28 + 20b - 21k - 5bk + 2b^2k + 4k^2 - bk^2 = 0.$$

Hence

$$\frac{k(1,b)}{\partial b} = 20 - 5k + 4bk - k^2.$$

Solving the equation system

$$28 + 20b - 21k - 5bk + 2b^2k + 4k^2 - bk^2 = 0. \quad 20 - 5k + 4bk - k^2 = 0,$$

we get

$$(-4+k)^2(25 - 14k + k^2) = 0,$$

the unique root of this equation which is larger than 4 is $k_1 = 7+2\sqrt{6} = 11.898979\cdots$, it is the largest root of equation $k^2 - 14k + 25 = 0$. It is easy to check that k_1 is the unique critical value of $k(1,b)$ in the interior of the edge $a = 1, 1+\sqrt{\delta} \le b \le 4$.

In the same way, $k_1 = 7+2\sqrt{6}$ is also the unique critical value of $k(a,1)$ in the interior of the edge $b = 1, 1+\sqrt{\delta} \le a \le 4$.

As we have pointed out, the minimum of $k(a,b)$ on the edge $a = 4, 1 \le b \le 4$ or $b = 4, 1 \le a \le 4$ is not less than 11.

At last, we indicate that at all vertices of D, k is either not real or not less than 11.

Therefore, we have proved the global minimum of $k(a,b)$ in the feasible set is $k_0 = 10.978796\cdots$, the largest root of the equation $19 - 55k + 49k^2 - 15k^3 + k^4 = 0$. It is best possible because it is feasible. This completes the proof of theorem. □

Theorem 4.4 If a configuration of seven points $r_1, \cdots r_7 \in \Sigma_3$, then

$$(r_1 \cdots r_7) \le \frac{1}{9}|r_1 \cdots r_7|,$$

where the coefficient 1/9 is best possible.

Proof. It is clear that if $r_1 \cdots r_7 \in \Sigma_7$ and $h(r_1 \cdots r_7) = \sup_{S \in \Sigma_3} h(S)$, then $r_1 \cdots r_7$ must be stable. In view of Lemma 3.5, we might assume that the convex hull of $r_1 \cdots r_7$ is $r_1 \cdots r_6$, r_7 is contained in the triangle determined by three diagonal of r_1r_4, r_2r_5, r_3r_6 and

$$\begin{aligned} & (r_7r_1r_2) = (r_7r_3r_4) = (r_7r_5r_6) \\ & (r_7r_1r_4) = (r_7r_2r_5) = (r_7r_3r_6) \\ = \; & (r_1 \cdots r_7). \end{aligned}$$

Choose an affine coordinate system such that

$$r_7 = (0,0), \quad r_1 = (1,0), \quad r_4 = (0,1)$$

and put

$$r_2 = (a,1), \quad r_3 = (1,b), \quad r_5 = (-u,-y), \quad r_6 = (-x,-v),$$

with $a, b, x, y \geq 1$ and $u = 2(r_7r_6r_1) > 1, v = 2(r_7r_4r_5) > 1$. Then

$$(r_1 \cdots r_7) = \frac{1}{2}, \quad |r_1 \cdots r_7| = \frac{1}{2}(2 + ab + u + v).$$

Since $(r_7r_2r_5) = (r_7r_3r_6) = (r_7r_5r_6) = (r_1 \cdots r_7)$,

$$\begin{aligned} f_1 &= -1 - u + ay = 0, \\ f_2 &= -1 - v + bx = 0, \\ f_3 &= -1 + uv - xy = 0. \end{aligned}$$

Solving $f_1(y) = 0, f_2(x) = 0$ and substitute x, y into f_3 we get

$$x = \frac{1+v}{b}, \quad y = \frac{1+u}{a},$$

and

$$f_3 = -1 + uv + \frac{(1+u)(1+v)}{ab} = 0,$$

hence

$$ab = \frac{(1+u)(1+v)}{uv-1}.$$

Substitute it into h we get

$$k := \frac{1}{h(r_1 \cdots r_7)} = \frac{uv(u+v) + 3uv - 1}{uv - 1}.$$

Thus we have transformed the problem into the programming:

$$\begin{aligned} & \min_{.} \, k(u,v) = \frac{uv(u+v) + 3uv - 1}{uv - 1}, \\ & \text{s.t.} \quad u, v > 1. \end{aligned}$$

Noticed that $u+v \geq 2\sqrt{uv}$, let $s=\sqrt{uv}$, then

$$k(u,v) \geq k_0(s) := \frac{2s^3+3s^2-1}{s^2-1}.$$

It is easy to calculate that

$$\min_{s>1} k_0(s) = 9,$$

and the minimum point is $s=2$, and hence

$$\min_{u,v>1} k(u,v) = 9$$

with minimum point $u=2, v=2$. The related values of a,b,x,y are

$$a=b=x=y=\sqrt{3}.$$

This completes the proof. □

Finally, the main theorem is proved immediately.

5. Open Problems

As mentioned in §1, we knew that for a square or parallelogram, $H_7 \geq 1/12$ by a special configuration (see [15, 18]). We didn't find any configuration with better result yet. There are strong evidence to expect equality there.

Problem 1 Prove $H_7 = \dfrac{1}{12}$.

Observed from Theorems proved in §§2,4, we have seen that if $r_1 \cdots r_7 \notin \Sigma_3$, then

$$(r_1 \cdots r_7) < \frac{1}{9.09}|r_1 \cdots r_7|,$$

so if K is sufficiently close to the "optimal" convex body K_0 which Heilbronn number is $1/9$, then the Heilbronn arrangement of 7 points in K must be in Σ_3. This suggests particularly that $H_7(\triangle)$ can be determined by the largest K_0 inscribed in a triangle.

Problem 2 Prove $H_7(\triangle) = \dfrac{\sqrt{3}}{18}$.

For disks, by Theorems we have obtained, one can prove that

$$H_7(D) = (\sin\frac{2\pi}{7} - \sin\frac{2\pi}{7}\cos\frac{2\pi}{7})/\pi.$$

To the problem about the upper bound of $H_8(K)$, the optimal configuration is likely an affine regular heptagon with its center. We conjecture

Problem 3 Prove or disprove $H_8(K) \leq \dfrac{1}{14\cos(\pi/7)}$.

So far we know no result about the exact lower bound of $H_n(K)$, even for $n = 4, 5$. We conjecture that for $n = 5$, the smallest Heilbronn number is given by $H_5(\triangle) = 3 - 2\sqrt{2}$.

Problem 4 Prove or disprove that $H_5(K) \geq 3 - 2\sqrt{2}$.

One could expect to find the lower bound for $H_4(K)$ easier. If these can be solved, one could get the answer to a ten dollar open problem (in [12]) of P. Erdös'.

More unsolved problems about Heilbronn numbers for square and disk can be found in [1].

References

1. Goldberg, M., Maximizing the smallest triangle made by points in a square, *Math. Mag.*, **45**(1972), 135-144.
2. Komlos, J. et al., On Heilbronn's problem, *J. London Math. Soc.* (2), **24**(1981), 385-396.
3. Komlos, J. et al., A lower bound for Heilbronn's problem, *J. London Math. Soc.* (2), **25**(1982), 13-14.
4. Moser, W., Problems on extremal properties of a finite set of points, in "Discrete Geometry and Convexity", *Ann. New York Acad. Sci.*, **440**(1985), 52-64.
5. Moser, W. & Pach,J., "100 Research Problem in Discrete Geometry", MacGill University, Montreal, Que. 1986.
6. Roth, K.F., On a problem of Heilbronn, *J. London Math. Soc.*, **26**(1951), 198-204.
7. Roth, K.F., On a problem of Heilbronn II, *Proc. London Math. Soc.*, **25**(1972), 193-212.
8. Roth, K.F., On a problem of Heilbronn III, *Proc. London Math. Soc.*, **25**(1974), 543-549.
9. Roth, K.F., Estimation of the area of the smallest triangle obtained by selecting three out of n points in a disc of unit area, *Amer. Math. Soc. Proc. Symp. Pure Math.*, **24**: 251-262, 1973.
10. Roth, K.F., Developments in Heilbronn's triangle problem, *Advances in Math.*, **22**(1976), 364-385.
11. Schmidt, W.M., On a problem of Heilbronn, *J. London Math. Soc.* (2), **4**(1971), 545-550.
12. Soifer, A., From problems of Mathematical Olympiads to open problems of mathematics, Preprint.
13. Yang Lu & Zhang Jingzhong, The problem of 6 points in a square, in "Lectures in Math. (2)", 151-175, Sichuan People's Publishing House 1980, pp. 151-175. (in Chinese).
14. Yang Lu & Zhang Jingzhong, A conjecture concerning six points in a square, in "Mathematical Olympiad in China", Hunan Education Publishing House 1990.
15. Yang Lu, Zhang Jingzhong & Zeng Zhenbing, On the conjecture and computing for exact values of the first several Heilbronn numbers, *Chin. Ann. Math.*, **13**:4(1992),503-515. (in Chinese).
16. Yang Lu, Zhang Jingzhong & Zeng Zhenbing, On the first several Heilbronn numbers of a triangle, *Acta Math. Sinica*, **37**:5(1994), 678-689. (in Chinese).
17. Yang Lu, Zhang Jingzhong & Zeng Zhenbing, Heilbronn problem for five points, Preprint, International Centre for Theoretical Physics, 1991, IC/91/252.
18. Yang Lu, Zhang Jingzhong & Zeng Zhenbing, On Goldberg's conjecture: computing the first several Heilbronn numbers, Preprint, Universität Bielefeld, 1991, ZiF-Nr.91/29, SFB-Nr.91/074.
19. Yang Lu, Zhang Jingzhong & Zeng Zhenbing, On exact values of Heilbronn numbers in triangular regions, Preprint.
20. Dress, A., Yang Lu & Zeng Zhenbing, Heilbronn problem for six points in a planar convex body, This volume.

ON THE COMPLEXITY OF MIN-MAX OPTIMIZATION PROBLEMS AND THEIR APPROXIMATION*

KER-I KO and CHIH-LONG LIN
Department of Computer Science, State University of New York at Stony Brook, Stony Brook, NY 11794.

Abstract. The computational complexity of optimization problems of the min-max form is naturally characterized by Π_2^P, the second level of the polynomial-time hierarchy. We present a number of optimization problems of this form and show that they are complete for the class Π_2^P. We also show that the constant-factor approximation versions of some of these optimization problems are also complete for Π_2^P.

1. Introduction

Consider an optimization problem of the following form:

MAX-A: for a given input x, find $\max_y\{|y| : (x,y) \in A\}$,

where A is a polynomial-time computable set such that $(x,y) \in A$ only if $|y| \le p(|x|)$ for some polynomial p (we say A is *polynomially related*). For instance, if $A = \{(G,Q)\colon G = (V,E)$ is a graph and $Q \subseteq V$ is a clique in $G\}$, then MAX-A is the well-known maximum clique problem.[1] It is immediate that the decision version of MAX-A, i.e., the problem of determining whether $\max_y\{|y| : (x,y) \in A\}$ is greater than or equal to a given constant K, is in *NP*. In the past twenty years, a great number of optimization problems of this type have been shown to be *NP*-complete [3].[2]

Assume that $A \in P$, A is polynomially related, and that we are given m input instances $x_1, \ldots, x_m$ and are asked to find

$$\min_{1\le i\le m} \max_y \{|y| : (x_i, y) \in A\}.$$

Then, (the decision version of) this problem is still in *NP*, if the instances $x_1, \ldots, x_m$ are given as input explicitly. However, if m is exponentially large relative to $|x|$ and the instances $x_1, \ldots, x_m$ have a succinct representation then the complexity of the problem may be higher than *NP*. For instance, consider the following problem MINMAX-CLIQUE: The input to the problem MINMAX-CLIQUE is a graph $G = (V,E)$ with its vertices V partitioned into subsets $V_{i,j}$, $1 \le i \le I$, $1 \le j \le J$. For any function $t : \{1, \ldots, I\} \to \{1, \ldots, J\}$, we let G_t denote the induced subgraph of G on the vertex set $V_t = \bigcup_{i=1}^{I} V_{i,t(i)}$.

* Research supported in part by NSF grant CCR 9121472.

1 A subsete $Q \subseteq V$ is a clique of a graph $G = (V,E)$ if $\{u,v\} \in E$ for all $u, v \in Q$.

2 Strictly speaking, their decision versions are shown to be *NP*-complete. In the rest of the paper, we will however use the term *NP*-complete for both the decision and the optimization versions of the problems.

D.-Z. Du and P. M. Pardalos (eds.), Minimax and Applications, 219–239.

MINMAX-CLIQUE: given a graph G with the substructures described above, find $f_{\text{CLIQUE}}(G) = \min_t \max_Q\{|Q| : Q \subseteq V \text{ is a clique in } G_t\}$.

Intuitively, the input G represents a network with I components, with each component V_i having J subcomponents $V_{i,1}, \ldots, V_{i,J}$. At any time t, only one subcomponent $V_{i,t(i)}$ of each V_i is *active*, and we are interested in the maximum clique size of G for all possible *active subgraphs* G_t of G. For people who are familiar with the *NP*-completeness theory, it is easy to see that the problem MINMAX-CLIQUE is in Π_2^P, the second level of the polynomial-time hierarchy; i.e., the decision problem of determining whether $f_{\text{CLIQUE}}(G) \leq K$, for a given constant K, is solvable by a polynomial-time nondeterministic machine with the help of an *NP*-complete set as the oracle. Therefore, it is probably not in *NP*. Indeed, we will show in Theorem 10 that this problem is complete for Π_2^P.

In general, if an input instance x contains an exponential number of subinstances $(x_1, \ldots, x_m)$ (called a *parameterized input*), then the problem of the form

MINMAX-A: for a given parametrized input x, find $f_{\text{MINMAX-A}}(x) = \min_{1 \leq t \leq m} \max_y\{|y| : (x_t, y) \in A\}$,

is a natural generalization of the problem MAX-A and its complexity is in Π_2^P. In this paper, we present a number of optimization problems of this type and show that they are complete for Π_2^P, and hence are not solvable in deterministic polynomial time even with the help of an *NP*-complete set as the oracle, assuming that the polynomial-time hierarchy does not collapse to $\Delta_2^P = P^{NP}$. These problems include the generalized versions of the maximum clique problem, the maximum 3-dimensional matching problem, the dynamic Hamiltonian circuit problem and the problem of computing the generalized Ramsey numbers.

We remark that although numerous optimization problems have been known to be *NP*-complete, there are relatively fewer natural problems known to be complete for Π_2^P (or, for Σ_2^P, the class of complements of sets in Π_2^P) (cf. [8, 12, 13, 14]). Our results here demonstrate a number of new Π_2^P-complete problems. We hope it could be a basis from which more Π_2^P-complete problems can be identified.

In the recent celebrated result of the *PCP* characterization of *NP*, Arora et al [1] showed that the constant-factor approximation versions of many optimization problems of the form MAX-A, including MAX-CLIQUE, are also *NP*-complete. It implies that if $P \neq NP$, then there is a constant $\epsilon > 0$ such that no polynomial-time algorithm can find for each input x a solution y of size $|y| \geq (1-\epsilon)|y^*|$ such that $(x, y) \in A$, where y^* is an optimum solution for x. Through a nontrivial generalization, the *PCP* characterization of *NP* has been successfully extended to Π_2^P [2, 5, 6]. We apply this characterization to show that some of the problems of the form MINMAX-A also have similar nonapproximability property. That is, if $\Pi_2^P \neq \Delta_2^P$, then there exists a constant $\epsilon > 0$ such that no polynomial-time oracle algorithm using an *NP*-complete set as the oracle can compute, for each x, a value k such that $k^*/(1+\epsilon) \leq k \leq (1+\epsilon)k^*$, where $k^* = f_{\text{MINMAX-A}}(x)$. These problems include the min-max versions of the maximum clique problem, the maximum 3-dimensional matching problem and the longest circuit problem.

2. Definitions

In this section, we review the notion of Π_2^P-completeness, and present the definition of the optimization problems. In the following, we assume that the reader is familiar with the complexity classes P, NP and the notion of NP-completeness. For the formal definitions and examples, the reader is referred to any standard text, for instance, [3].

For any string x in $\{0,1\}^*$, we denote by $|x|$ the length of x. Let $\langle x, y\rangle$ be any *pairing* function, i.e., a one-to-one mapping from strings x and y to a single string in polynomial time. A well-known characterization of the class NP is as follows: $A \in NP$ if and only if there exists a set $B \in P$ such that for all $x \in \{0,1\}^*$,

$$x \in A \iff (\exists y, |y| \le p(|x|))\langle x, y\rangle \in B,$$

where $p(n)$ is some polynomial depending only on A. The complexity class Π_2^P is a natural extension of the class NP: $A \in \Pi_2^P$ if and only if there exists a set $B \in P$ such that

$$x \in A \iff (\forall y, |y| \le p(|x|))(\exists z, |z| \le p(|x|))\langle x, \langle y, z\rangle\rangle \in B.$$

It is obvious that $NP \subseteq \Pi_2^P$. Between the complexity classes NP and Π_2^P lies the complexity class Δ_2^P (or, P^{NP}) that consists of all problems that are solvable in polynomial time with the help of an NP-complete problem as the oracle. Whether $NP = \Delta_2^P$ and/or $\Delta_2^P = \Pi_2^P$ are major open questions in complexity theory.

A decision problem A is Π_2^P*-complete*, if $A \in \Pi_2^P$, and for every $A' \in \Pi_2^P$, there is a polynomial-time computable function f such that for each $x \in \{0,1\}^*$, $x \in A' \iff f(x) \in A$ (f is called a *reduction* from A' to A). There are a few natural problems known to be complete for Π_2^P. A standard Π_2^P-complete problem that will be used in our proofs is the following generalization of the famous NP-complete problem SAT. Suppose that F is a 3-CNF boolean formula. We write $F(X,Y)$ to emphasize that its variables are partitioned into two sets X and Y. For a 3-CNF boolean formula $F(X,Y)$, and for any truth assignments $\tau_1 : X \to \{0,1\}$ and $\tau_2 : Y \to \{0,1\}$, we write $F(\tau_1, \tau_2)$ to denote the formula F with its variables taking the truth values defined by τ_1 and τ_2. We also write $tc(F(\tau_1,\tau_2))$ to denote the number of clauses of F that are true to the truth assignments τ_1 and τ_2.

> SAT$_2$: for a given 3-CNF boolean formula $F(X,Y)$, determine whether it is true that for all truth assignments $\tau_1 : X \to \{0,1\}$, there is a truth assignment $\tau_2 : Y \to \{0,1\}$ such that $F(\tau_1,\tau_2) = 1$.[3]

Proposition 1 SAT$_2$ *is* Π_2^P*-complete.*

The problem SAT$_2$ may be viewed as (a subproblem of) the decision version of the following optimization problem:

> MINMAX-SAT: for a given 3-CNF boolean formula $F(X,Y)$, find $f_{\text{SAT}}(F) = \min_{\tau_1 : X \to \{0,1\}} \max_{\tau_2 : Y \to \{0,1\}} tc(F(\tau_1,\tau_2))$.

[3] Throughout the paper, we identify 1 with *true* and 0 with *false*.

In the following, we introduce some new optimization problems of the min-max form. For each optimization problem, we also list its corresponding decision version. First, we consider the problem MINMAX-SAT in the restricted form.

MINMAX-SAT-B: for a given 3-CNF boolean formula $F(X,Y)$ in which each variable occurs in at most b clauses where b is a constant independent of the size of F, find $f_{\text{SAT}}(F)$. (Decision version: for an additional input $K > 0$, is $f_{\text{SAT}}(F) \geq K$?)

In addition to the problem MINMAX-CLIQUE defined in Section 1, we introduce a few more min-max optimization problems that are the generalizations of some famous *NP*-complete optimization problems based on the idea of parameterized inputs. Recall that for a graph $G = (V, E)$ with its vertex set V partitioned into subsets $V_{i,j}$, $1 \leq i \leq I$, $1 \leq j \leq J$, and for a function $t : \{1, \ldots, I\} \to \{1, \ldots, J\}$, we let $V_t = \bigcup_{i=1}^{I} V_{i,t(i)}$ and let G_t be the induced subgraph of G on the vertex set V_t. The following generalized vertex cover problem is a dual problem of MINMAX-CLIQUE. For a graph $G = (V, E)$, we say that a subset $V' \subseteq V$ is a *vertex cover* if $V' \cap \{u, v\} \neq \emptyset$ for all edges $\{u, v\} \in E$.

MAXMIN-VC: given a graph G with its vertex set V partitioned into subsets $\{V_{i,j}\}_{1 \leq i \leq I, 1 \leq j \leq J}$, find $f_{\text{VC}}(G) = \max_t \min_{V'} \{|V'| : V' \subseteq V_t \text{ is a vertex cover of } G_t\}$. (Decision version: Is $f_{\text{VC}}(G) \leq K$?)

The following problem is the generalization of the Hamiltonian circuit problem. For a graph $G = (V, E)$ and a subset $V' \subseteq V$, we say G *has a circuit on* V' if there is a cycle of G going through each vertex of V' exactly once. We say G is *Hamiltonian* if G has a circuit on V.

MINMAX-CIRCUIT: given a graph G with its vertex set V partitioned into subsets $\{V_{i,j}\}_{1 \leq i \leq I, 1 \leq j \leq J}$, find $f_{\text{CIRCUIT}}(G) = \min_t \max_{V'} \{|V'| : V' \subseteq V_t, G_t \text{ has a circuit on } V'\}$. (Decision version: Is $f_{\text{CIRCUIT}}(G) \geq K$?)

The next problem is the generalization of the maximum 3-dimensional matching problem. Let W be a finite set and S be a collection of 3-element subsets of W. Let W' be a subset of W. A subset $S' \subseteq S$ is called a *(3-dimensional) matching in* W' if all sets $s \in S'$ are mutually disjoint, and are contained in W'. In the following, if $W = \bigcup_{i=1}^{I} \bigcup_{j=1}^{J} W_{i,j}$, and t is a function from $\{1, \ldots, I\}$ to $\{1, \ldots, J\}$, we write $W(t)$ to denote the set $\bigcup_{i=1}^{I} W_{i,t(i)}$.

MINMAX-3DM: given mutually disjoint finite sets $W_{i,j}$, $1 \leq i \leq I$, $1 \leq j \leq J$, and a set S of 3-element subsets of $W = \bigcup_{i=1}^{I} \bigcup_{j=1}^{J} W_{i,j}$, find $f_{\text{3DM}}(W, S) = \min_t \max_{S'} \{|S'| : S' \subseteq S, S' \text{ is a matching in } W(t)\}$, where t ranges over all functions from $\{1, \ldots, I\}$ to $\{1, \ldots, J\}$. (Decision version: Is $f_{\text{3DM}}(W, S) \geq K$?)

In addition to the above problems based on the idea of parameterized inputs, we consider several natural problems in Π_2^P. First, we consider the problem of computing the Ramsey number. For any graph $G = (V, E)$, a function $c : E \to \{0, 1\}$ is called a *coloring* of G (with two colors). For any complete graph G with a two-color coloring c, a set $Q \subseteq V$ is a *monochromatic clique* if all edges between vertices in Q are of the same color. Ramsey theorem states that for any positive integer K, there exists an integer $n = R_K$ such that for all two-colored complete graph G of size n, there is a monochromatic clique Q of size K. To study the complexity of computing the Ramsey function mapping K to the minimum R_K, we consider the following generalized version. In the following, we say a function $c : E \to \{0, 1, *\}$ is a *partial coloring* of a graph $G = (V, E)$ ($c(e) = *$ means the edge e is not colored yet). A coloring $c' : E \to \{0, 1\}$ is a *restriction* of a partial coloring c, denoted by $c' \preceq c$, if $c'(e) = c(e)$ whenever $c(e) \neq *$.

GENERALIZED RAMSEY NUMBER(GRN): given a complete graph $G = (V, E)$ with a partial coloring c, find $f_{\text{GRN}}(G, c) = \min_{c' \preceq c} \max_Q\{|Q| : Q$ is a monochromatic clique under $c'\}$. (Decision version: Is $f_{\text{GRN}}(G, c) \geq K$?)

Notice that the Ramsey number R_K can be found by a binary search for the graph G of the minimum size that has $f_{\text{GRN}}(G, c_0) \geq K$ with respect to the empty coloring c_0 (i.e., $c_0(e) = *$ for all edges e).

The next two problems are the variations of the Hamiltonian circuit problem. The first problem is to find, from a given digraph and a subset of *alterable* edges, the length of the longest circuit in any alteration of those edges. Let $G = (V, E)$ be a digraph and D a subset of E. We let G_D denote the subgraph of G with vertex set V and edge set $(E - D) \cup \text{inv}(D)$, where $\text{inv}(D) = \{(s, t) : (t, s) \in D\}$.

LONGEST DIRECTED CIRCUIT (LDC): given a digraph $G = (V, E)$ and a subset E' of E, find $f_{\text{LDC}}(G, E') = \min_{D \subseteq E'} \max_{V'}\{|V'| : V' \subseteq V, G_D$ has a circuit on $V'\}$. (Decision version: Is $f_{\text{LDC}}(G, E') \geq K$?)

The next problem is similar to the above problem, but is about the longest circuits in undirected graphs. For simplicity, we formulate it as a special case of its decision version.

DYNAMIC HAMILTONIAN CIRCUIT (DHC): given a graph $G = (V, E)$, and a subset B of E, determine whether $G_D = (V, E - D)$ is Hamiltonian for all subsets D of B with $|D| \leq |B|/2$.

3. Π_2^P-Completeness Results

All the problems defined in Section 2 can be easily seen belonging to Π_2^P. In this section, we show that MINMAX-CIRCUIT, GRN and DHC are actually Π_2^P-complete. The problems MINMAX-CLIQUE and MINMAX-3DM will be shown to be Π_2^P-complete in Section 5 together with the stronger results that their contant-factor approximation versions are also Π_2^P-complete. The Π_2^P-completeness of MAXMIN-VC is a corollary of that of MINMAX-CLIQUE. The proofs for the Π_2^P-completeness of the problems MINMAX-SAT-B and LDC are much more involved; we prove them in a separate paper [7].

Theorem 2 MINMAX-CIRCUIT *is complete for* Π_2^P.

Proof. We construct a reduction from SAT_2 to (the decision version of) MINMAX-CIRCUIT. The construction is a modification of the reduction from SAT to the Hamiltonian circuit problem. Let F be a 3-CNF boolean formula over variables $X = \{x_1, \ldots, x_r\}$ and $Y = \{y_1, \ldots, y_s\}$. Assume that $F = C_1 \wedge C_2 \wedge \cdots \wedge C_n$, where each C_j is the OR of three literals. We will define a graph G over $18n + 4r + 2s$ vertices.

For each clause C_j, we define a subgraph H_j of 18 vertices as shown in Figure 1. This subgraph H_j will be connected to other parts of the graph G only through the vertices labeled $\alpha_k[j]$ and $\beta_k[j]$, $k = 1, 2, 3$. Thus, it has the following property: if a Hamiltonian circuit of G enters H_j through $\alpha_k[j]$ for some $k = 1, 2, 3$, then it must exit at $\beta_k[j]$, and visit either one or two or all three rows of H_j.

In addition to subgraphs H_j, we have some more vertices: For each variable x_i in X, we define four vertices: $u_{i,0}, u_{i,1}, \bar{u}_{i,0}, \bar{u}_{i,1}$. For each variable y_i in Y, we define two vertices: $v_{i,0}, v_{i,1}$.

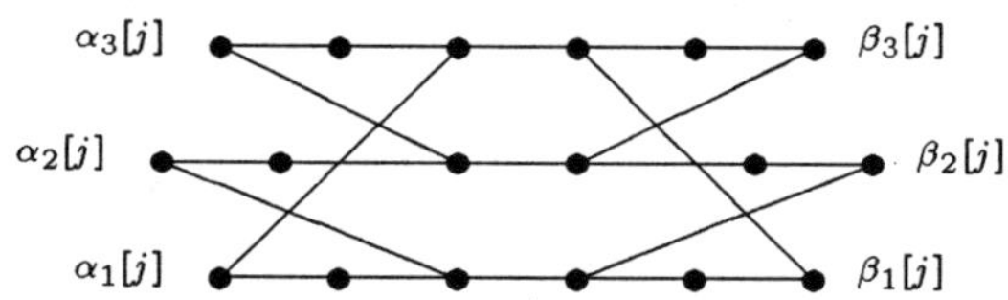

Fig. 1. The subgraph H_j for a clause C_j.

We define the edges between these components as follows.

(1) For each i, $1 \leq i < r$, we define edges $\{u_{i,1}, u_{i+1,0}\}$, $\{u_{i,1}, \bar{u}_{i+1,0}\}$, $\{\bar{u}_{i,1}, u_{i+1,0}\}$, $\{\bar{u}_{i,1}, \bar{u}_{i+1,0}\}$. For $i = r$, we have two edges $\{u_{r,1}, v_{1,0}\}$, $\{\bar{u}_{r,1}, v_{1,0}\}$.

(2) For each i, $1 \leq i < s$, we define an edge $\{v_{i,1}, v_{i+1,0}\}$. For $i = s$, we have two edges $\{v_{s,1}, u_{1,0}\}$, $\{v_{s,1}, \bar{u}_{1,0}\}$.

(3) For each literal z and for $k = 0, 1$, let

$$w(z)_k = \begin{cases} u_{i,k} & \text{if } z = x_i, \\ \bar{u}_{i,k} & \text{if } z = \bar{x}_i, \\ v_{i,k} & \text{if } z = y_i \text{ or } \bar{y}_i. \end{cases}$$

Then we define edges to form a path from $w(z)_0$ to $w(z)_1$: assume that z occurs as the k_1th, k_2th, ..., k_mth literal in clauses $C_{j_1}, C_{j_2}, \ldots, C_{j_m}$, respectively, with $j_1 < j_2 < \cdots < j_m$. Then we add edges $\{w(z)_0, \alpha_{k_1}[j_1]\}$, $\{\beta_{k_m}[j_m], w(z)_1\}$, and for each $\ell = 1, \ldots, m-1$, $\{\beta_{k_\ell}[j_\ell], \alpha_{k_{\ell+1}}[j_{\ell+1}]\}$. (Note that for each pair $(u_{i,0}, u_{i,1})$ or $(\bar{u}_{i,0}, \bar{u}_{i,1})$, there is a path between them, and for each pair $(v_{i,0}, v_{i,1})$ there are two paths between them, corresponding to the occurrences of two literals y_i and $\bar{y}_i$.)

The above completes the definition of all edges. To complete the reduction, we let $I = r+1$, $J = 2$, and for each $1 \leq i \leq r$, $V_{i,0} = \{\bar{u}_{i,0}, \bar{u}_{i,1}\}$ and $V_{i,1} = \{u_{i,0}, u_{i,1}\}$, and all other vertices are in $V_{r+1,0} = V_{r+1,1}$. (For convenience, we define $V_{i,0}$ and $V_{i,1}$ instead of $V_{i,1}$ and $V_{i,2}$.) Finally, we let $K = 18n + 2r + 2s$, which is equal to the size of $|V_t|$ for all functions $t : \{1, \ldots, I\} \to \{0, 1\}$.

The correctness of this reduction is very easy to see. We only give a short sketch here. First, assume that for each truth assignment τ_1 on X, there is a truth assignment τ_2 on Y satisfying all clauses C_j in F. Let $t : \{1, \ldots, r+1\} \to \{0, 1\}$ be any function. Then, t defines a truth assignment $\tau_1(x_i) = t(i)$, and all vertices in $V_{i,t(i)}$ corresponds to "true" literals under τ_1. From this τ_1, there is a truth assignment τ_2 on Y that satisfies all clauses. Now, for each "true" literal z under τ_1 and τ_2, we have a path from $w(z)_0$ to $w(z)_1$ in G_t. Connecting these paths together forms a Hamiltonian circuit for G_t, since each subgraph H_j is visited by at least one such paths.

Conversely, assume that for any $t : \{1, \ldots, I\} \to \{0, 1\}$, there is a Hamiltonian circuit Π_t in G_t. Then, by the basic property of subgraphs H_j, this circuit Π_t defines, for each pair of nodes $(w(z)_0, w(z)_1)$ in V_t, a path from $w(z)_0$ to $w(z)_1$. That is, for each τ_1 on X such that $\tau_1(x_i) = t(i)$, we can define a truth assignment τ_2 on Y by $\tau_2(y_i) = 1$ (or, $\tau_2(y_i) = 0$) if the path of Π_t from $v_{i,0}$ to $v_{i,1}$ visits nodes corresponding to the literal y_i (or, respectively, $\bar{y}_i$). Since each subgraph H_j is visited by at least one of such paths, the assignments τ_1 and τ_2 together must satisfy each clause C_j. □

Theorem 3 GRN *is complete for* Π_2^P.

Proof. We construct a reduction from SAT_2 to GRN. Let F be a 3-CNF formula over variables $X = \{x_1, \ldots, x_r\}$ and $Y = \{y_1, \ldots, y_s\}$. Assume that $F = C_1 \wedge C_2 \wedge \cdots \wedge C_n$, where each C_i is the OR of three literals. We further assume that $r \geq 2$ and $n \geq 3$. Let $K = 2r+n$. The graph G has $N = 6r+4n-4$ vertices. We divide them into three groups: $V_X = \{x_{i,j}, \bar{x}_{i,j} : 1 \leq i \leq r, 1 \leq j \leq 2\}$, $V_C = \{c_{i,j} : 1 \leq i \leq n, 1 \leq j \leq 3\}$ and $V_R = \{r_i : 1 \leq i \leq 2r + n - 4\}$. The partial coloring c on the edges of G is defined as follows (we use colors *blue* and *red* instead of 0 and 1):

(1) The edges among $x_{i,1}$, $x_{i,2}$, $\bar{x}_{i,1}$ and $\bar{x}_{i,2}$, for each i, $1 \leq i \leq r$, are colored by red, except that the edges $e_i = \{x_{i,1}, x_{i,2}\}$ and $\bar{e}_i = \{\bar{x}_{i,1}, \bar{x}_{i,2}\}$ are not colored (i.e., $c(e_i) = c(\bar{e}_i) = *$).

(2) All other edges between two vertices in V_X are colored by blue; i.e., $c(\{x_{i,j}, x_{i',j'}\}) = c(\{x_{i,j}, \bar{x}_{i',j'}\}) = \text{blue}$ if $i \neq i'$.

(3) All edges among vertices in V_R are colored by red.

(4) For each i, $1 \leq i \leq k$, the three edges among $c_{i,1}, c_{i,2}$ and $c_{i,3}$ are colored by red.

(5) The edge between two vertices $c_{i,j}$ and $c_{i',j'}$, where $i \neq i'$, is colored by red if the jth literal of C_i and the j'th literal of $C_{i'}$ are complementary (i.e., one is x_q and the other is $\bar{x}_q$, or one is y_q and the other is $\bar{y}_q$ for some q). Otherwise, it is colored by blue.

(6) The edge between any vertex in V_R and any vertex in V_X is colored by red, and the edge between any vertex in V_R and any vertex in V_C is colored by blue.

(7) For each vertex $c_{i,j}$ in V_C, if the jth literal of C_i is y_q or $\bar{y}_q$ for some q, then all edges between $c_{i,j}$ and any vertex in V_X are colored by blue. If the jth literal of C_i is x_q for some q, then all edges between $c_{i,j}$ and any vertex in V_X, except $\bar{x}_{q,1}$ and $\bar{x}_{q,2}$, are colored by blue, and $c(\{c_{i,j}, \bar{x}_{q,1}\}) = c(\{c_{i,j}, \bar{x}_{q,2}\}) = \text{red}$. The case where the jth literal of C_i is $\bar{x}_q$ for some q is symmetric; i.e., all edges between $c_{i,j}$ and any vertex in V_X, except $x_{q,1}$ and $x_{q,2}$, are colored by blue, and $c(\{c_{i,j}, x_{q,1}\}) = c(\{c_{i,j}, x_{q,2}\}) = \text{red}$.

The above completes the construction of the graph G and its partial coloring c. Notice that the partial coloring c has $c(e) \neq *$ for all edges e except e_i and $\bar{e}_i$, for $1 \leq i \leq r$. Now we prove that this construction is correct. First assume that for each assignment $\tau_1 : X \to \{0,1\}$, there is an assignment $\tau_2 : Y \to \{0,1\}$ such that $F(\tau_1, \tau_2) = 1$. We verify that for any two-coloring restriction c' of c, there must be a size-K monochromatic clique Q.

We note that if $c'(e_i) = c'(\bar{e}_i) =$ red for some $i \leq r$, then the vertices $x_{i,1}$, $x_{i,2}$, $\bar{x}_{i,1}$, $\bar{x}_{i,2}$, together with vertices in V_R, form a red clique of size $|V_R| + 4 = K$. Therefore, we may assume that for each i, $1 \leq i \leq r$, at least one of $c'(e_i)$ and $c'(\bar{e}_i)$ is blue. Now we define an assignment τ_1 on X by $\tau_1(x_i) = 1$ if and only if $c'(e_i) =$ blue. For this assignment τ_1, there is an assignment τ_2 on Y such that each clause C_i has a true literal. For each i, $1 \leq i \leq k$, let j_i be the least j, $1 \leq j \leq 3$, such that the jth literal of C_i is true to τ_1 and τ_2. Let $Q_C = \{c_{i,j_i} : 1 \leq i \leq n\}$ and $Q_X = \{x_{i,j} : c'(e_i) = \text{blue}, 1 \leq j \leq 2\} \cup \{\bar{x}_{i,j} : c'(e_i) = \text{red}, 1 \leq j \leq 2\}$. Let $Q = Q_C \cup Q_X$. It is clear that Q is of size $2r + n$. Furthermore, Q is a blue clique: (i) every two vertices in Q_C are connected by a blue edge because they both have value true under τ_1 and τ_2 and so are not complementary; (ii) every two vertices in Q_X are connected by a blue edge by the definition of Q_X, and (iii) if a vertex c_{i,j_i} in Q_C corresponds to a literal x_q, then $\tau_1(x_q) = 1$ and so $\bar{x}_{q,1}, \bar{x}_{q,2} \notin Q_X$ and hence all the edges between c_{i,j_i} and each of $x_{i',j'}$ or $\bar{x}_{i',j'} \in Q_X$ are colored blue.

Conversely, assume that there exists an assignment τ_1 on X such that for all assignments τ_2 on Y, $F(\tau_1, \tau_2) = 0$. Then, consider the following coloring c' on edges e_i and $\bar{e}_i$: $c'(e_i) =$ blue and $c'(\bar{e}_i) =$ red if $\tau_1(x_i) = 1$, and $c'(e_i) =$ red and $c'(\bar{e}_i) =$ blue if $\tau_1(x_i) = 0$. By the definition of c', the largest red clique in V_X is of size 3. Also, the largest red clique in V_C is of size 3, since every edge connecting two noncomplementary literals in two different clauses is colored by blue. Thus, the largest red clique containing V_R is of size $K - 1$, and the largest red clique containing at least one vertex of V_C is of size $\leq 6 < K$.

Next, assume by way of contradiction that there is a blue clique Q of G of size K. From our coloring, it is clear that, for each i, $1 \leq i \leq r$, Q contains exactly two vertices in $\{x_{i,1}, x_{i,2}, \bar{x}_{i,1}, \bar{x}_{i,2}\}$, and for each i, $1 \leq i \leq n$, Q contains exactly one c_{i,j_i}, for some $1 \leq j_i \leq 3$. Define $\tau_2 : Y \to \{0,1\}$ by $\tau_2(y_q) = 1$ if and only if the j_ith literal of C_i is y_q for some i, $1 \leq i \leq k$. Then, there is a clause C_i such that C_i is not satisfied by τ_1 and τ_2. In particular, the j_ith literal of C_i is false to τ_1 and τ_2.

Case 1. The j_ith literal of C_i is x_q for some q. Then, $\tau_1(x_q) = 0$, and so $c'(e_q) =$ red, and the edges between c_{i,j_i} and each of $\bar{x}_{q,1}$ and $\bar{x}_{q,2}$ are red. This contradicts the above observation that Q contains two vertices in $\{x_{q,1}, x_{q,2}, \bar{x}_{q,1}, \bar{x}_{q,2}\}$

Case 2. The j_ith literal of C_i is $\bar{x}_q$ for some q. This is symmetric to Case 1.

Case 3. The j_ith literal of C_i is y_q for some q. This is not possible, because by the definition of τ_2, $\tau_2(y_q) = 1$, but by the property that C_i is not satisfied by τ_1 and τ_2, $\tau_2(y_q) = 0$.

Case 4. The j_ith literal of C_i is $\bar{y}_q$ for some q. Then, $\tau_2(y_q) = 1$, and hence, by the definition of τ_2, there must be another $i' \leq n$, $i' \neq i$, such that $c_{i',j_{i'}}$ is in Q and the $j_{i'}$th literal of $C_{i'}$ is y_q. So, the edge between c_{i,j_i} and $c_{i',j_{i'}}$ is colored by red. This is again a contradiction.

The above case analysis shows that there is no blue clique in G of size K either. So the theorem is proven. □

Theorem 4 DHC *is* Π_2^{P}*-complete.*

Proof. We reduce SAT_2 to DHC. Let $F = C_1 \wedge C_2 \wedge \cdots \wedge C_n$ be a boolean formula over variables in $X = \{x_1, \ldots, x_r\}$ and $Y = \{y_1, \ldots, y_s\}$ where C_i's are three-literal clauses. We construct a graph G from F. A basic component of the graph G is a NOT device as shown in Figure 2(a). (It was first introduced in [9], and was called an exclusive-OR device in [11]). Schematically, the two horizontal line paths are represented by two *broadened* line segments, one designated as input and the other output to the device, and the four vertical paths are "condensed" into one *arrow*, running from input to output (see Figure 2(c)). As shown in [9], there are only two ways for a Hamiltonian circuit to traverse such a device, one of them indicated in Figure 2(a) by thicker line segments. For convenience, if a NOT device is traversed as shown in Figure 2(a), we say that the NOT device has input *true*.

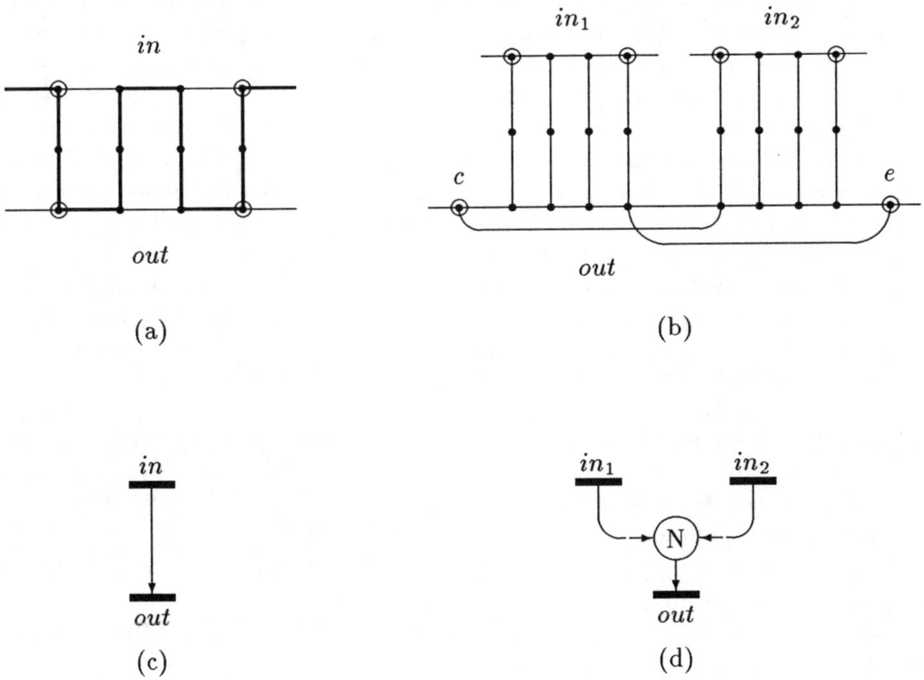

Fig. 2. The NOT and NAND devices.

Using two NOT devices, we also have the NAND device as shown in Figure 2(b) and 2(d). Here, in order for a Hamiltonian circuit to traverse from c to e or vice versa, *at least one* of the two input NOT devices have to be false. When using these devices, we require that connections to other parts of the graph G can only go through *circled* vertices.

The graph G we are going to construct consists of a set of interconnected subgraphs:

(1) For each $x_i \in X$, we have a *variable subgraph* $G_x(i)$ as shown in Figure 3(a), and for each $y_j \in Y$, a $G_y(j)$ as in Figure 3(b). The number of NOT devices

used in each variable subgraph will be defined in (3) below. In addition, we have a *component subgraph* G_w, which is a concatenation of $n+r$ NOT devices as shown in Figure 3(c). Note that each variable or component subgraph has some extra labeled vertices. They are not part of any NOT devices, except that c and e of $G_x(i)$ are precisely those in the NAND device shown in figure 2(b).

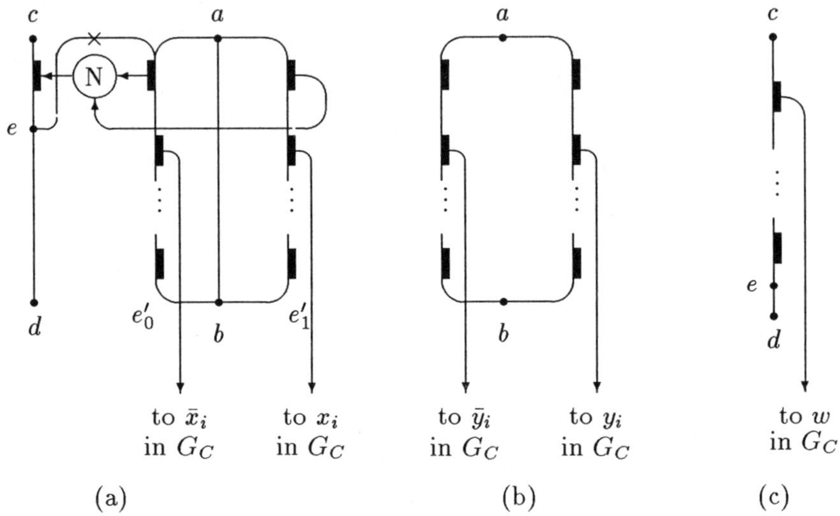

Fig. 3. The variable subgraphs: (a) $G_x(i)$, (b) $G_y(i)$ and (c) G_w.

(2) For each clause C_k, $1 \leq k \leq n$, we have a *clause subgraph* $G_C(k)$as shown in Figure 4. In addition, we have r extra clause subgraphs $G_C(n+1), \ldots, G_C(n+r)$, each of which is a *triangle* version of Figure 4. Note that if the four *conner vertices* are further connected as a clique, then any graph containing such a clause subgraph is Hamiltonian only if at least one of the four (or three) NOT devices has its input true.

(3) Arrows (of NOT devices) run from variable subgraphs to clause subgraphs as follows:

(a) If literal $\bar{x}_i$ (or x_i) occurs in m clauses, then there are $m+2$ NOT devices on the *negative* (or, respectively, *positive*) path, the path containing e'_0 (or, respectively, e'_1), of $G_x(i)$. The inputs of the first NOT devices on both paths serve as inputs to a NAND device as shown in Figure 3(a).

(b) The numbers of NOT devices in G_y's are defined analogously except that no NAND devices are involved.

(c) An arrow runs from the input of a NOT device in a variable subgraph to the output of a NOT device in a clause subgraph if and only if the corresponding variable, either positive or negative, occurs in that clause. For example, if $C_k = (x_i \vee \cdots)$, then there is an arrow running from the input of a NOT device on the positive path of $G_x(i)$ to the output of a NOT device in $G_C(k)$. Furthermore, for each $G_C(n+i)$, $1 \leq i \leq r$, we have two inputs from $G_x(i)$, one corresponding to x_i and the other $\bar{x}_i$, and another input from one of the NOT

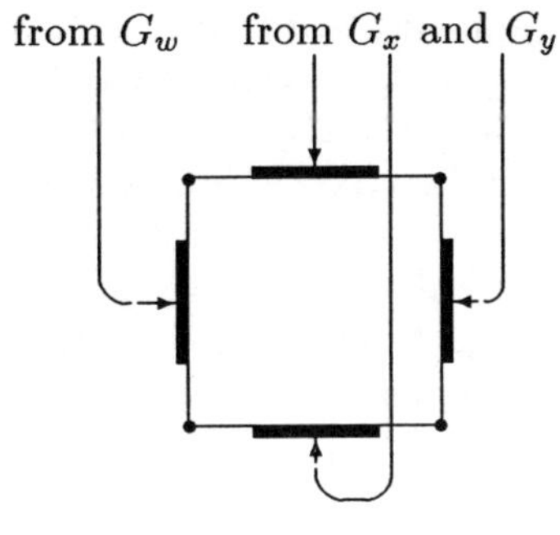

Fig. 4. The clause subgraph $G_C(i)$ that has two literals from X and one from Y.

devices in G_w.

(4) Let $a_x(i)$ denote the vertex of $G_x(i)$ labeled with a in Figure 3(a). Let $b_x(i)$, $c_x(i)$, $d_x(i)$, $e_x(i)$, $a_y(j)$, $b_y(j)$, $e_0'(i)$, $e_1'(i)$, c_w, e_w and d_w be defined analogously. Then, in addition to those arrows going to G_C from G_x, G_y and G_w, these subgraphs are further connected by the following edges (see Figure 5):

$\langle b_x(i), a_x(i+1)\rangle$, $\langle c_x(i), a_x(i+1)\rangle$ and $\langle d_x(i), c_x(i+1)\rangle$ for $1 \leq i \leq r-1$;
$\langle b_y(j), a_y(j+1)\rangle$ for $1 \leq j \leq s-1$;
$\langle b_x(r), a_y(1)\rangle$ and $\langle c_x(r), a_y(1)\rangle$;
$\langle d_x(r), c_w\rangle$, $\langle c_w, a_x(1)\rangle$ and $\langle d_w, c_x(1)\rangle$;
edges that connect all the corner vertices of all clause subgraphs into a clique;
$\langle b_y(s), a_C\rangle$, $\langle b_C, e_w\rangle$, and $\langle b_C, a_x(1)\rangle$, where a_C and b_C are two distinguished corner vertices in two different clause subgraphs.

The graph G is now completed. Finally we define B to be the set of edges $e_0'(i)$ and $e_1'(i)$ in $G_x(i)$, for all $1 \leq i \leq r$. This finishes the construction.

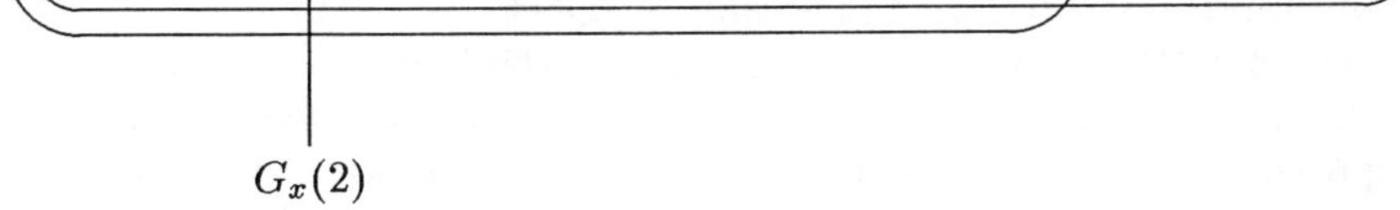

Fig. 5. The interconnection among subgraphs.

We now show that the above construction is a reduction from SAT_2 to DHC. First suppose that $F \in \text{SAT}_2$. We claim that G_D is Hamiltonian for any $D \subseteq B$ with

$|D| \leq |B|/2$. Consider the following two cases:

Case 1. For some i, $1 \leq i \leq r$, neither $e'_0(i)$ nor $e'_1(i)$ is in D. We show the existence of a Hamiltonian circuit H in this case. Let i be the smallest index affirming the case. Starting from $a_x(1)$, H visits all vertices in G_D in the following order:

(i) H first visits $a_x(1), b_x(1), a_x(2), b_x(2), \ldots, a_x(i-1)$, $b_x(i-1)$ and then $a_x(i)$.

(ii) It clockwisely visits all the NOT devices on the "loop" of $G_x(i)$, making the inputs to these NOT devices true, take the edge marked with $\times$ (see Figure 3(a)) to $e_x(i)$ and leaves at $d_x(i)$.

(iii) It then proceeds to visit $c_x(i+1)$, $d_x(i+1)$, $\ldots$, $c_x(r)$, $d_x(r)$, c_w, d_w, $c_x(1)$, $d_x(1)$, $\ldots$, $c_x(i-1)$, $d_x(i-1)$, $c_x(i)$. While going from each $c_x(k)$ to $d_x(k)$, $k \neq i$, it traverses the NAND device in between (and so sets both inputs false), and from c_w to d_w, sets all the NOT devices in between true.

(iv) It then visits $a_x(i+1), b_x(i+1), \ldots, a_x(r), b_x(r)$, and then sets all y_j to false (i.e., visits all the NOT devices on the left half of each $G_y(j)$).

(v) H finally visits G_C completely, starting at a_C and leaving via b_C, and returns to $a_x(1)$. Note that we are able to do so because all clause subgraphs contain the output of a NOT device whose input is in G_w, which had been set to true previously; further, all NOT devices in variable subgraphs not traversed before are visited here.

Case 2. Exactly one of $e'_0(i)$ and $e'_1(i)$ is in D for all i, $1 \leq i \leq r$. The Hamiltonian circuit H in this case is as follows. It starts from $a_x(1)$.

(i) It chooses the positive (negative) path of $G_x(i)$ if $e'_1(i)$ (respectively, $e'_0(i)$) is not in D, and visits all the NOT devices on the path.

(ii) Since the edges $e'_0(i)$ or $e'_1(i)$ not in D correspond to a truth assignment τ_1 on X (i.e., $\tau_1(x_i) = 1$ if and only if $e'_1(i)$ is not in D), there exists a truth assignment τ_2 on Y such that $F(\tau_1, \tau_2)$ is true. H sets each y_j accordingly by traversing the positive (or negative) path of $G_y(j)$ if y_j is set to true (or, respectively, false).

(iii) H then traverses G_C completely, leaving via b_C. We are able to do so because $F(\tau_1, \tau_2)$ is true and so each $G_C(j)$, $1 \leq j \leq n$, has at least one NOT device having the true input. Further, for each $G_C(n+i)$, $1 \leq i \leq r$, one of the NOT devices corresponding to x_i or $\bar{x}_i$ has its input true. All NOT devices in variable subgraphs not traversed previously are done at this stage.

(iv) It visits e_w, then visits d_w, $c_x(1)$, $d_x(1)$, $\ldots$, $c_x(r)$, and finally $d_x(r)$, traversing each NAND device between $c_x(k)$ and $e_x(k)$, since exactly one input has been set true. It leaves via c_w and finishes at $a_x(1)$.

Since $|D| \leq |B|/2$, we cannot have $e'_0(i)$ and $e'_1(i)$ both in D without having $e'_0(i')$ and $e'_1(i')$ both not in D for some i'. Therefore, the above two cases are exhaustive.

Suppose on the other hand that for some truth assignment τ_1 on X, $F(\tau_1, Y)$ is not satisfiable. We let $e'_0(i) \in D$ if $t(x_i) = 1$, and $e'_1(i) \in D$ otherwise. We need to show that G_D is not Hamiltonian. Suppose, for the sake of contradiction, that G_D has a Hamiltonian circuit H. First we can exclude the possibility that H enters and leaves a NOT device (including those within an NAND device) on different sides. Thus we can virtually "ignore" those arrows as far as H is concerned. As a consequence, either all the NOT devices in G_w are set to true by H or all set to false by H. We consider these two cases separately.

Case 1. All NOT devices in G_w are set to true by H. Orientation ignored, we assume that H starts from c_w, visiting all NOT devices, and then e_w. H then has to take d_w since taking b_C will leave d_w unvisited. Next it must visit $c_x(1)$, the only choice.

To visit other parts of G_D, H eventually has to leave the bottom loop which contains all the distinguished vertices labeled with c, e or d. It cannot leave from a vertex $e_x(j)$ since this would keep $d_x(j)$ out of the reach. Therefore, it must leave at $c_x(k)$ for some $1 \leq k \leq r$. It then has to visit $e_x(k)$ later. It means that the two inputs to the NAND device between $c_x(k)$ and $e_x(k)$ must be set to true, which is impossible because only one of $e'_0(k)$ and $e'_1(k)$ is in G_D.

Case 2. All of the NOT devices in G_w are set to false by H. Orientation ignored, we assume that H starts at $a_x(1)$, visiting each variable subgraph. There are two possibilities:

(a) If during traversing $G_x(i)$, H enters the bottom loop at either $c_x(i)$ or $e_x(i)$, then it has to leave the loop before visiting $d_x(r)$; otherwise, it would have to visit c_w and then either ends its journey prematurally or visits G_w, contradictary to our assumption. A contradiction can be derived as in Case 1.

(b) H visits all variable subgraphs in a normal manner (i.e., not going to the bottom loop from $G_x(i)$'s). Then H defines a truth assignment on X and Y which is consistent with τ_1 when restricted to X. By asumption, at least one of the clause C_k of F is not satisfied, meaning that none of the NOT devices of $G_C(k)$ has true input. As indicated previously, H then cannot traverse $G_C(k)$ completely, which is a contradiction. □

4. Approximation Problems and Their Hardness

We first formalize the notion of approximating optimization problems. Garey and Johnson [3] have defined the notion of approximating optimization problems of the form MAX-A. Since our min-max optimization problems have two parameters, their definition is not suitable to our case. In the following, we define a simple notion of *approximating a function*. Let $\mathbf{Q}^+$ be the set of positive rationals and $\mathbf{R}^+$ the set of positive reals.

Definition 5 *Let $f, g : \{0,1\}^* \to \mathbf{Q}^+$ and $c : \mathbf{N} \to \mathbf{R}^+$, $c(n) > 1$ for all n, be given. We say that g* approximates *f to* within a factor of c *(*c-approximates *f in short) if for all $x \in \{0,1\}^*$, we have $f(x)/c(|x|) < g(x) < c(|x|) \cdot f(x)$. The* c-approximation problem *of f is to compute a function g that c-approximates f.*

To define the notion of completeness of c-approximation problem of a function f, we generalize the notion of reductions between decision problems. In the following, we write $\langle A, B \rangle$ to denote a pair of sets over $\{0,1\}$ with $A \cap B = \emptyset$.

Definition 6 *(a) A pair $\langle A, B \rangle$ is* polynomial-time separable *(or, simply, $\langle A, B \rangle \in P$) if there exists a set $C \in P$ such that $A \subseteq C$ and $B \subseteq \overline{C}$.*

(b) For any two pairs $\langle A, B \rangle$ and $\langle A', B' \rangle$, we say that $\langle A, B \rangle$ is G-reducible *to $\langle A', B' \rangle$ if there is a polynomial-time computable function f such that $f(A) \subseteq A'$*

and $f(B) \subseteq B'$. Let $\mathcal{C}$ be a complexity class. We say that $\langle A, B\rangle$ is $\mathcal{C}$-hard if there exists a set C that is $\mathcal{C}$-hard and $\langle C, \overline{C}\rangle$ is G-reducible to $\langle A, B\rangle$.

It is clear that if $P \neq \mathcal{C}$ and that $\langle A, B\rangle$ is $\mathcal{C}$-hard, then $\langle A, B\rangle$ is not polynomial-time separable.

For any functions $s, l : \{0,1\} \to \mathbf{Q}^+$ such that $s(x) < l(x)$, we write $\langle f : l(x), s(x)\rangle$ to dentoe the pair of sets $\langle \{x \mid f(x) \geq l(x)\}, \{x \mid f(x) \leq s(x)\}\rangle$. The following proposition relates the hardness of approximating functions to that of pairs of decision problems.

Proposition 7 *Let $c : \mathbf{N} \to \mathbf{Q}^+$, $c(n) > 1$ for all $n \geq 0$, be polynomial-time computable. Let $s, l : \{0,1\}^* \to \mathbf{Q}^+$ be two polynomial-time computable functions satisfying $c(|x|)s(x) < l(x)/c(|x|)$. If $\langle f : l(x), s(x)\rangle$ is not polynomial-time separable, then the c-approximation problem of f is not computable in polynomial time.*

Proof. Assume that g is a function c-approximating f. Then, for any x, $|x| = n$, if $f(x) \geq l(x)$, then $g(x) > l(x)/c(n)$; if $f(x) \leq s(x)$, then $g(x) < c(n)s(x) < l(x)/c(n)$. Thus, from $g(x)$ and $l(x)/c(|x|)$, we can tell an instance in $\{x : f(x) \geq l(x)\}$ from an instance in $\{x : f(x) \leq s(x)\}$. □

Based on the above proposition, we define the notion of hardness of $\mathcal{C}$-approximation problems as follows:

Definition 8 *Let $f : \{0,1\}^* \to \mathbf{Q}^+$ be a given function and $c : \mathbf{N} \to \mathbf{Q}^+$, $c(n) > 1$ be a polynomial-time computable function. We say that the c-approximation problem of f is $\mathcal{C}$-hard if there exist polynomial-time computable functions $s, l : \{0,1\}^* \to \mathbf{Q}^+$, $s(x) < l(x)$, such that*

1. *for all x of length n, $c(n)s(x) < l(x)/c(n)$; and*
2. *$\langle f : l(x), s(x)\rangle$ is $\mathcal{C}$-hard.*

We say the c-approximation problem of MINMAX-A is Π_2^P-complete if the decision version of MINMAX-A is in Π_2^P and the c-approximation problem of $f_{\text{MINMAX-A}}$ is Π_2^P-hard.

Remark. In practice, we often prove $\mathcal{C}$-hardness of $\langle f : l(x), s(x)\rangle$, where $l(x) = l \cdot \text{size}(x)$, $s(x) = s \cdot \text{size}(x)$ for some simple function size(x) and constants $0 < s < l \leq 1$. From Proposition 7, it follows that the $(l/s)^{1/2}$-approximation problem for f is $\mathcal{C}$-hard. The function size(x) is not necessarily the natural size of the instance x. Rather, it is a measure designed to prove the hardness of approximation. In the case of MINMAX-SAT, a 3-CNF boolean formula F has size$(F) = |F|$, the number of clauses in F.

The recent breakthrough of Arora et al [1] on the *NP*-hardness of many optimixation problems in the form of MAX-A is based on a characterization of *NP* in terms of the notion of *probabilistically checkable proofs* (*PCP*). Through a generalization of the notion of *PCP*, this characterization has been extended to the class Π_2^P (it was implicit in [2], and explicit in [6] and [5]). A consequence of this characterization is that the c-approximation problem of MINMAX-SAT is Π_2^P-complete for some constant $c > 0$. This will be our basis for proving other Π_2^P-complete c-approximation problems.

Proposition 9 *There exists a constant* $0 < \epsilon < 1$ *such that* $\langle f_{\text{SAT}} : |F|, (1-\epsilon)|F|\rangle$ *is* Π_2^P*-hard. Therefore, there is a constant* $c > 1$ *such that the c-approximation problem for* f_{SAT} *is* Π_2^P*-complete.*

5. Nonapproximability Results

Our proofs of the Π_2^P-completeness results for c-approximation problems MINMAX-A will be done by G-reductions from MINMAX-SAT to MINMAX-A. More precisely, we will construct G-reductions from $\langle f_{\text{SAT}} : |F|, (1-\epsilon)|F|\rangle$ to a pair $\langle f_{\text{MINMAX-A}} : (1-\epsilon_2)\text{size}(x), (1-\epsilon_1)\text{size}(x)\rangle$, where $\epsilon_1 > \epsilon_2 \geq 0$. For the proofs below, ϵ_2 are always 0. However, in [7], the G-reductions from MINMAX-SAT to MINMAX-SAT-B and LDC have $\epsilon_2 > 0$.

We first present the proof that the decision version of the problem MINMAX-CLIQUE is Π_2^P-complete. The result on the approximation version follows as a corollary.

Theorem 10 MINMAX-CLIQUE *is* Π_2^P*-complete.*

Proof. We reduce the problem SAT_2 to MINMAX-CLIQUE. Let F be a 3-CNF formula over variables $X = \{x_1, \ldots, x_r\}$ and $Y = \{y_1, \ldots, y_s\}$. Let $F = C_1 \wedge \ldots \wedge C_n$, where each C_i is the OR of three literals. We may assume that each clause C_i has at most one literal in $\{x_1, \ldots, x_r, \bar{x}_1, \ldots, \bar{x}_r\}$ (called *X-literals*). Otherwise, we can convert a clause C_i of two X-literals to two clauses each with one X-literal without changing the membership in SAT_2. For instance, if $C_i = x_1 \vee x_2 \vee y_1$, then we replace C_i with $C_{i,1} = x_1 \vee y_1 \vee z$ and $C_{i,2} = x_2 \vee y_1 \vee \bar{z}$, where z is a new variable not in $X \cup Y$, and it can be checked that the resulting formula $F'(X, Y \cup \{z\})$ is in SAT_2 if and only if $F(X,Y)$ is in SAT_2. (A clause C_i with 3 X-literals is trivially false for some τ_1.) We now describe the construction of the graph G.

The vertex set V of G is partitioned into $2n$ subsets $V_{i,j}$, $1 \leq i \leq n$, $j = 0, 1$ (i.e., $I = n, J = 2$). For each $1 \leq i \leq n$ and $0 \leq j \leq 1$, $V_{i,j}$ has 3 vertices $a_{i,j}[k]$, $k = 1,2,3$, corresponding to the 3 literals of C_i, together with $n' = \lceil n/2 \rceil$ other vertices $b_{i,j}[k]$, $k = 1, \ldots, n'$. Let $B_{i,j} = \{b_{i,j}[k] : k = 1, \ldots, n'\}$. Within $B_{i,j}$, all vertices are connected (and so $B_{i,j}$ is a clique of size n'). There is no other edge within $V_{i,j}$, and there is no edge between $V_{i,0}$ and $V_{i,1}$.

To define the edges between the vertices in $V_{i,j}$ and vertices in $V_{i',j'}$ with $i \neq i'$, we associate an X-literal $xl(V_{i,j})$ to each $V_{i,j}$. Each C_i has at most one X-literal. Suppose C_i has an X-literal; then we let $xl(V_{i,1})$ be the literal in C_i and $xl(V_{i,0})$ be its complement. Suppose C_i has no X-literals; then we add a dummy variable x_{r+1} and let $xl(V_{i,0}) = xl(V_{i,1}) = x_{r+1}$. In addition, we define the following terms on vertices $a_{i,j}[k]$: A vertex $a_{i,j}[k]$ is *negative* if it corresponds to a X-literal in C_i and $j = 0$. Two vertices $a_{i,j}[k]$ and $a_{i',j'}[k']$ are *complementary* if they correspond to two complementary literals, i.e., one is x_k (or, y_k) and the other is $\bar{x}_k$ (or, respectively, $\bar{y}_k$) for some k.

Now, we define edges between $V_{i,j}$ and $V_{i',j'}$ with $i \neq i'$ as follows:

(1) If $xl(V_{i,j})$ and $xl(V_{i',j'})$ are complementary, then we connect all vertices between $B_{i,j}$ and $B_{i',j'}$. There is no other edge between $V_{i,j}$ and $V_{i',j'}$.

(2) If not (1), then there is no edge between $B_{i,j}$ and $V_{i',j'}$ and no edge between $B_{i',j'}$ and $V_{i,j}$. For any two vertices $a_{i,j}[k]$ and $a_{i',j'}[k']$, they are connected by an edge if and only if they are nonnegative and noncomplementary.

The above completes the graph G. Now we prove that this construction is correct with respect to the bound $K = n$. It suffices to prove the following stronger statement:

Claim. If $f_{\text{SAT}}(F) > \lceil |F|/2 \rceil$, then $f_{\text{CLIQUE}}(G) = f_{\text{SAT}}(F)$.

Proof of Claim. First assume that for each truth assignment τ_1 on X, there is a truth assignment τ_2 on Y that satisfies k^* clauses of F (i.e., $f_{SAT}(F) = k^*$). Let t be any mapping from $\{1, \ldots, n\}$ to $\{0,1\}$. First, if for some $i \neq i'$, $xl(V_{i,t(i)})$ and $xl(V_{i',t(i')})$ are complementary, then we get a clique $B_{i,t(i)} \cup B_{i',t(i')}$ that is of size $\geq n \geq k^*$. Second, if for all $i \neq i'$, $xl(V_{i,t(i)})$ and $xl(V_{i',t(i')})$ are noncomplementary, then there is a unique truth assignment τ_1 on X such that $\tau_1(xl(V_{i,t(i)})) = 1$ for all i, $1 \leq i \leq n$. (We always let τ_1 on the dummy variable x_{r+1} be 1.) For this assignment τ_1, there is a truth assignment τ_2 on Y that satisfies k^* clauses. Let $I_T = \{i : C_i(\tau_1, \tau_2) = 1\}$. For each $i \in I_T$, pick a vertex $a_{i,t(i)}[k]$ that corresponds to a *true* literal in C_i under τ_1 and τ_2. Note that if we picked a vertex $a_{i,t(i)}[k]$ that corresponds to the X-literal in C_i, then $t(i)$ must be equal to 1. Thus, it is easy to check that these vertices $a_{i,t(i)}[k]$ form a clique of size k^*: (i) if $a_{i,t(i)}[k]$ corresponds to an X-literal, then as observed above $t(i) = 1$ and it is nonnegative; (ii) no two selected vertices are complementary, since the corresponding literals must be noncomplementary to be satisfied by τ_1 and τ_2.

Conversely, assume that the maximum clique size of all G_t for all $t : \{1, \ldots, n\} \to \{0,1\}$ is at least k^*. Let τ_1 be any truth assignment on X. We need to show that there is a truth assignment τ_2 on Y that satisfies k^* clauses. Define a mapping $t : \{1, \ldots, n\} \to \{0,1\}$ by $t(i) = 1$ if and only if $\tau_1(xl(V_{i,1})) = 1$, i.e., $\tau_1(xl(V_{i,t(i)})) = 1$ for all $i \leq n$ (we assume that $\tau_1(x_{r+1}) = 1$). Thus, no two X-literals $xl(V_{i,t(i)})$ and $xl(V_{i',t(i')})$ are complementary, and so there is no edge between $B_{i,t(i)}$ and $V_{i',t(i')}$ if $i \neq i'$. It follows that the maximum clique Q of G_t must consist of a single vertex $a_{i,t(i)}[k]$ in $V_{i,t(i)}$, for k^* indices i. Let $I_Q = \{i : Q \cap V_{i,t(i)} \neq \emptyset\}$. Now, we define a truth assignment τ_2 on Y as follows: if y_ℓ ever occurs as a literal corresponding to some vertex in the clique Q, then assign $\tau_2(y_\ell) = 1$; otherwise, assign $\tau_2(y_\ell) = 0$. We check that τ_1 and τ_2 satisfy C_i for all $i \in I_Q$. In particular, for each $i \in I_Q$, the literal corresponding to the vertex $a_{i,t(i)}[k]$ in $V_{i,t(i)} \cap Q$ must be true to τ_1 and τ_2:

(1) If $a_{i,t(i)}[k]$ corresponds to an X-literal and it belongs to the clique Q, then it must be nonnegative and so $t(i) = 1$. That means the corresponding X-literal is the same as $xl(V_{i,1})$ and has the value 1 under τ_1.

(2) If $a_{i,t(i)}[k]$ corresponds to a Y-literal y_ℓ, then since y_ℓ occurs in Q, $\tau_2(y_\ell) = 1$.

(3) If $a_{i,t(i)}[k]$ corresponds to a Y-literal $\bar{y}_\ell$, then y_ℓ does not occur in the clique Q, because y_ℓ and $\bar{y}_\ell$ are complementary and so they cannot be connected. This implies that $\tau_2(\bar{y}_\ell) = 1$.

The above completes the proof of the claim and hence the correctness of the reduction. □

Corollary 11 *There exists a constant $c > 1$ such that the c-approximation problem of* Minmax-Clique *is Π_2^P-complete.*

Proof. Let g be the reduction of the above theorem. In the above proof, we showed that for any 3-CNF formula F, $f_{\text{SAT}}(F) = f_{\text{CLIQUE}}(g(F))$ as long as $f_{\text{SAT}}(F) > \lceil |F|/2 \rceil$. For any graph G whose vertex set V is partitioned into $V_{i,j}$, $1 \le i \le I$, $1 \le j \le J$. We let size$(G) = I$. Then the above observation implies that g is a G-reduction from $\langle f_{\text{SAT}} : |F|, (1-\epsilon)|F| \rangle$ to $\langle f_{\text{CLIQUE}} : \text{size}(G), (1-\epsilon)\text{size}(G) \rangle$. □

We note that the Π_2^P-completeness of MAXMIN-VC follows from that of MINMAX-CLIQUE since they are the dual problems to each other. However, the c-approximation results do not carry over.

Corollary 12 MAXMIN-VC *is* Π_2^P*-complete.*

Next, we prove that MINMAX-3DM and its c-approximation version are Π_2^P-complete. In order to do this, we need the Π_2^P-completeness result on a stronger version of the problem MINMAX-SAT.

> MINMAX-SAT-YB: given $F(X, Y)$, with the number of occurrences of each $y \in Y$ bounded by a constant b, find $f_{\text{SAT-YB}}(F) = f_{\text{SAT}}(F)$.

The more general case of MINMAX-SAT-B, in which the number of occurrences of all variables in X or Y are bounded, is also Π_2^P-complete. Its proof is more involved and is given in a separate paper [7]. Here we give a sketch for the Π_2^P-completeness of this simpler case MINMAX-SAT-YB.

Theorem 13 MINMAX-SAT-YB *is* Π_2^P*-complete.*

Proof. (Sketch) The proof is a simple modification of the reduction from the maximum satisfiability problem to the bounded-occurrence maximum satisfiability problem in [10]. It was shown in [10] that there exist a polynomial-time computable function f and two integers $\alpha, b > 0$ such that

(i) for each 3-CNF formula $F(X)$ with m clauses, $f(F(X)) = F'(X')$ is a 3-CNF boolean formula with $(\alpha + 1)m$ clauses in which each variable occurs at most b times, and

(ii) $\max_{\tau':X'\to\{0,1\}} tc(F'(\tau')) = \alpha m + \max_{\tau:X\to\{0,1\}} tc(F(\tau))$,

Now, for each $F(X, Y)$, we treat all $x \in X$ as constants and compute $f(F(X, Y)) = F'(X, Y')$. Then, we have

$$\begin{aligned}\min_{\tau_1:X\to\{0,1\}} \max_{\tau_2':Y'\to\{0,1\}} tc(F'(\tau_1, \tau_2')) &= \min_{\tau_1:X\to\{0,1\}} \left[\alpha m + \max_{\tau_2:Y\to\{0,1\}} tc(F(\tau_1, \tau_2))\right] \\ &= \alpha m + \min_{\tau_1:X\to\{0,1\}} \max_{\tau_2:Y\to\{0,1\}} tc(F(\tau_1, \tau_2)).\end{aligned}$$

Thus, if $\langle f_{\text{SAT}} : |F|, (1-\epsilon)|F| \rangle$ is Π_2^P-hard, then $\langle f_{\text{SAT-YB}} : |F|, (1-\epsilon/(\alpha+1))|F| \rangle$ is also Π_2^P-hard. □

Theorem 14 MINMAX-3DM *is* Π_2^P*-complete.*

Proof. We will construct a G-reduction from MINMAX-SAT-YB to MINMAX-3DM. It is a modification of the L-reduction from the maximum satisfiability problem with bounded occurrences of variables to the maximum 3DM problem [4]. We first give a brief review of that proof.

Let $F(X) = C_1 \wedge \ldots \wedge C_n$ be a 3-CNF boolean formula over variables $Y = \{y_1, \ldots, y_s\}$. Let d_i be the number of occurrences of y_i or $\bar{y}_i$ in F. (Assuming that each clause $C_\ell$has exactly 3 literals, we have $\sum_{i=1}^{s} d_i = 3n$.) We assume that $d_i \leq b$ for all $i = 1, \ldots, s$. Let M be the minimum number greater than $3b/2 + 1$ such that M is a power of 2. We describe below a collection S of 3-element subsets of a set W (called *triples*), without explicitly writing down all the names of elements in W.

(1) For each variable y_i, define M identical sets of *ring triples*. For each y_i and each k, $1 \leq k \leq M$, the ring $R_{i,k}$ contains two sets of triples:

$$R^1_{i,k} = \{\{\bar{y}_i[j,k], a_i[j,k], b_i[j,k]\} : 1 \leq j \leq d_i\},$$
$$R^0_{i,k} = \{\{y_i[j,k], b_i[j,k], a_i[j+1,k]\} : 1 \leq j \leq d_i\},$$

where $j+1$ in $a_i[j+1,k]$ is the addition modulo d_i. This ring is the basic component of the reduction from SAT to 3DM in, e.g., [3]. We show it in Figure 6(a). Here $a_i[j,k]$ and $b_i[j,k]$ are local elements and appear in no other triples. Thus, any complete matching must take all triples in $R^1_{i,k}$ or all triples in $R^0_{i,k}$, corresponding to setting all occurrences of y_i true or all false.

(2) For each variable y_i and each j, $1 \leq j \leq d_i$, construct two sets of *tree triples*: $T^0_{i,j}$ and $T^1_{i,j}$. Set $T^1_{i,j}$ forms a tree of size $2M-1$ with leaves $y_i[j,k]$, $1 \leq k \leq M$, and set $T^0_{i,j}$ with leaves $\bar{y}_i[j,k]$, $1 \leq k \leq M$. We show a tree $T^1_{i,j}$ in Figure 6(b). All the internal nodes of the trees, except the roots, are the local elements. The root of $T^1_{i,j}$ is called $u_i[j]$ and the root of $T^0_{i,j}$ is called $\bar{u}_i[j]$. It is shown in [4] that, a maximum matching must take, for any i and j, all $y_i[j,k]$'s by ring triples or all by tree triples. For instance, in Figure 6, with $d_i = 4$ and $M = 8$, a maximum matching must take all circled triples or all noncircled triples. In other words, for each i, $1 \leq i \leq s$, the maximum matching will match all ring triples and tree triples so that only all $u_i[j]$'s are left free or only all $\bar{u}_i[j]$'s are left free. If the matching leaves $u_i[j]$'s free, then we say it corresponds to the truth assignment that sets y_i true. For instance, the matching taking all circled triples in Figure 6 corresponds to assigning y_i false.

(3) Identify $u_i[j]$ with the jth occurrence of y_i, and identify $\bar{u}_i[j]$ with the jth occurrence of $\bar{y}_i$. For each clause C_ℓ, define 3 *clause triples*: $\{s_1[\ell], s_2[\ell], w\}$, where w ranges over the three roots of the trees $T^0_{i,j}$ and $T^1_{i,j}$ that are identified as above to the three literals in C_ℓ.

(4) Define *garbage triples* $\{g_{2q-1}, g_{2q}, u_i[j]\}$ and $\{g_{2q-1}, g_{2q}, \bar{u}_i[j]\}$ for all $q = 1, \ldots, 2n$ and all $i = 1, \ldots, s$ and all $j = 1, \ldots, d_i$.

The above are all triples. Some simple calculation shows that there are totally $18nM$ elements in W and the number of matching is at most $6nM$ that covers all elements. We let $K = 6nM$.

We now show that the reduction is correct. From the remarks in (1) and (2) above, we see that the maximum matchings on ring triples and tree triples correspond one-to-one with the truth assignments on Y. So, if F is satisfied by a truth assignment τ on Y, then we select disjoint triples from ring triples and tree triples so that for each y_i with $\tau(y_i) = 1$, only the nodes $u_i[j]$'s are left free, and for each y_i with $\tau(y_i) = 0$, only the nodes $\bar{u}_i[j]$'s are left free. It can be checked that there are $(6M-3)n$ such triples. For each clause C_ℓ, we select the clause triple that covers

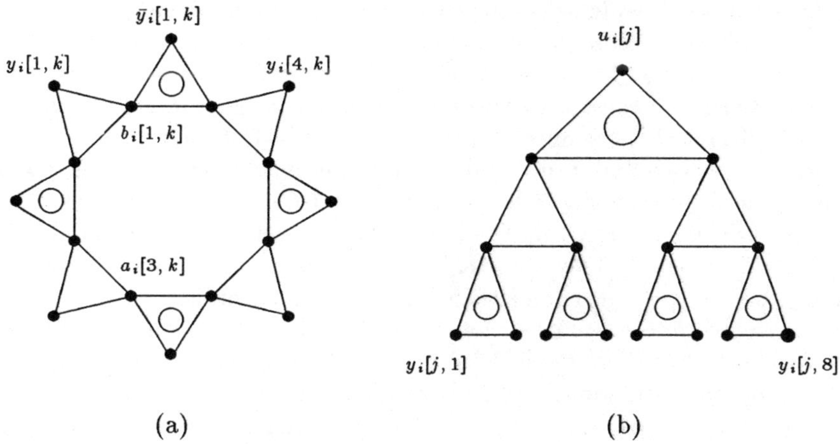

Fig. 6. (a) Ring triples with $d_i = 4$. (b) Tree triples with $M = 8$. Each triangle denotes a triple.

the root that corresponds to a true literal in C_ℓ. Finally, we cover all other roots by garbage clauses. This is a complete matching of size $6nM$.

Conversely, if there is a matching that covers every element, then it must contain for each clause C_ℓ a clause triple $\{s_1[\ell], s_2[\ell], w\}$, where w is a free root note and also corresponds to a literal in C_ℓ. By the property of the maximum matchings discussed in (2) above, we can define truth assignment τ on Y to make all such literals true and so to satisfy F.

Now we describe our modification for MINMAX-3DM. First, we divide W into n groups $W_1, \ldots, W_n$, with each W_ℓ containing all elements of W that occur in the ring triples and tree triples related to clause C_ℓ. More specifically, suppose the jth occurrence of y_i or $\bar{y}_i$ is in C_ℓ, then W_ℓ contains all elements in the trees $T^0_{i,j}$ and $T^1_{i,j}$, plus the internal elements $a_i[j,k]$ and $b_i[j,k]$ for all $k \le K$. In addition, W_ℓ contains $s_1[\ell], s_2[\ell]$ and $g_{4\ell-p}$, $p = 0,1,2,3$. For each ℓ, $1 \le \ell \le n$, we have some local triples that contain only elements in W_ℓ and inter-group triples that contain some elements in W_ℓ and some not in W_ℓ. For instance, all tree triples and clause triples are local, some ring triples are local and some ring triples and garbage triples are inter-group.

Now, suppose $F(X,Y) = C_1 \wedge \ldots \wedge C_n$ is a 3-CNF formula over two variable sets $X = \{x_1, \ldots, x_r\}$ and $Y = \{y_1, \ldots, y_s\}$, with each variable y_i of Y occurring in F at most b times. As explained in the proof of Theorem 10, we may assume that each clause C_ℓ contains at most one X-literal. We treat the variables in X as constants, and define the triples as above from F, and divide them into groups W_ℓ, $\ell = 1, \ldots, n$. (Note that for each clause C_ℓ with an X-literal, it has only 2 clause triples of the form $\{s_1[\ell], s_2[\ell], w\}$.) Next, for each $1 \le \ell \le n$ and each $m = 0, 1$, we define $W_{\ell,m}$ to be a copy of W_ℓ ; i.e., for each element in W_ℓ, attach an additional index m to it (so, e.g., $s_1[\ell]$ becomes $s_1[\ell, 0]$ in $W_{\ell,0}$). Then, for each group $W_{\ell,m}$, we add elements $\alpha_{\ell,m}[k], \beta_{\ell,m}[k], \gamma_{\ell,m}[k]$, for $k = 1, \ldots, n$. If C_ℓ has an X-literal which is positive, we add one more element σ_ℓ to $W_{\ell,1}$, else we add it to $W_{\ell,0}$. We define the set S' as follows:

(1) For each local triple in W_ℓ, we include its *copies* in both $W_{\ell,0}$ and $W_{\ell,1}$ in S'.

(2) For each inter-group triple between W_ℓ and $W_{\ell'}$, we include all its *copies* between $W_{\ell,m}$ and $W_{\ell',m'}$, for all $m, m' = 0, 1$, in S'.

(3) For each C_ℓ, if x_i is a literal of C_ℓ, then add a triple $\{s_1[\ell,1], s_2[\ell,1], \sigma_\ell\}$ to S'; if $\bar{x}_i$ is a literal of C_ℓ, then add a triple $\{s_1[\ell,0], s_2[\ell,0], \sigma_\ell\}$ to S';

(4) We say two pairs (ℓ, m) and (ℓ', m') are *inconsistent* if both C_ℓ and $C_{\ell'}$ have the same X-literal but $m \neq m'$ or if C_ℓ and $C_{\ell'}$ have the complementary X-literal but $m = m'$. If (ℓ, m) and (ℓ', m') are inconsistent, then we add the triples $\{\alpha_{\ell,m}[k], \beta_{\ell,m}[k], \gamma_{\ell',m'}[k]\}$ to S' for all $k = 1, \ldots, n$.

Finally, we let $K = 6nM$, and claim that the reduction is correct.

First, assume that $F(X,Y) \in \mathrm{SAT}_2$, and let t be a function from $\{1, \ldots, n\}$ to $\{0,1\}$. We check that there are at least $6nM$ matchings in $W_t = \bigcup_{\ell=1}^{n} W_{\ell,t(\ell)}$. First, as in the original reduction (from SAT to 3DM), we can select $(6M-1)n$ disjoint triples from ring triples, tree triples and garbage triples. Suppose for some ℓ, ℓ', $(\ell, t(\ell))$ and $(\ell', t(\ell'))$ are inconsistent. Then, we can get from (4) above at least n disjoint triples to make a matching of at least $6nM$ triples. Suppose t is consistent. Then, it defines a truth assignment τ_1 on X and for this τ_1 there is a truth assignment τ_2 on Y satisfying F. It follows from the analysis of the original reduction that there is a matching of $6nM$ triples. Note that for each clause C_ℓ if τ_1 satisfies C_ℓ, then the corresponding $W_{\ell,t(\ell)}$ must contain σ_ℓ and $\{s_1[\ell,t(\ell)], s_2[\ell,t(\ell)], \sigma_\ell\}$ must be in S'.

Conversely, if $F(X,Y) \notin \mathrm{SAT}_2$ then there exists a truth assignment τ_1 on X such that $F(\tau_1, Y)$ is not satisfiable. Choose the corresponding t, i.e., $t(\ell) = 1$ if and only if τ sets the X-literal in C_ℓ true. This function t must be consistent and so there is no triple from the extra elements such as $\alpha_{\ell,m}[k]$. The only triples are the *copies* of those in the original reduction, and there are less than $6nM$ disjoint triples. □

Corollary 15 *There exists a constant $c > 1$ such that the c-approximation problem for* MINMAX-3DM *is Π_2^P-complete.*

Proof. We observe that the original reduction (from SAT to 3DM), preserves the optimum solution in the following sense: if the maximum number of satisfiable clauses is β, then the maximum matching has $(6M-1)n + \beta n$ triples [4]. The main idea was that the design of the tree triples forces the maximum matching to make consistent truth assignments to the different occurrences of y_i. In the new reduction, this property is preserved if the function t is consistent. (If t is not consistent, then there are always at least $6nM$ disjoint triples.)

For each instance (W, S) of MINMAX-3DM with W partitioned into subsets $W_{\ell,m}$, with $1 \leq \ell \leq I$, $1 \leq m \leq J$, let $\mathrm{size}(W,S) = 6MI$. Then, the above observation shows that the new reduction is a G-reduction from $\langle f_{\mathrm{SAT-YB}} : |F|, (1-\epsilon)|F| \rangle$ to $\langle f_{\mathrm{3DM}} : \mathrm{size}(W,S), (1-\epsilon/6M)\mathrm{size}(W,S) \rangle$. □

6. Conclusion and Open Questions

We have demonstrated a number of min-max optimization problems to be Π_2^P-complete. Using the idea of parameterized inputs, there are apparently many more similar results on the generalization of *NP*-complete problems. For instance, the Π_2^P-completeness results also hold for the generalized knapsack problem and the generalized maximum set covering problem. It is hoped that these new Π_2^P-completeness results are useful for proving other natural problems such as GRN to be complete for Π_2^P or Σ_2^P.

Although the Π_2^P-completeness results for the min-max optimization problems appear easy to prove, the corresponding Π_2^P-completeness results for the *c*-approximation problems are harder to get. We were successful only for a few such problems. It would be interesting to develop techniques for classifying the complexity of the *c*-approximation problem of the min-max problems (like the class MAX SNP for problems of the form MAX-A). In particular, it would be interesting to know whether the *c*-approximation problems of f_{CIRCUIT} and f_{VC} are Π_2^P-complete.

References

1. A. Arora, C. Lund, R. Motwani, M. Sudan and M. Szegedy, Proof verification and hardness of approximation problems, *Proceedings, 33rd IEEE Symposium on Foundations of Computer Science* (1992), 14-23.
2. A. Condon, J. Feigenbaum, C. Lund and P. Shor, Probabilistically checkable debate systems and approximation algorithms for PSPACE-hard functions, *Proceedings, 25th ACM Symposium on Theory of Computing* (1993), 305-314.
3. M. R. Garey and D. S. Johnson, *Computers and Intractability: A Guide to the Theory of NP-Completeness*, Freeman, San Francisco, 1979.
4. V. Kann, Maximum bounded 3-dimensional matching is MAX SNP-complete, *Inform. Proc. Lett.* 37 (1991), 27-35.
5. M. Kiwi, C. Lund, A. Russell, D. Spielman and R. Sundaram, Alteration in Interaction, *Proceedings, 9th Structure in Complexity Theory Conference*, IEEE (1994), 294-303.
6. K. Ko and C.-L. Lin, Non-approximability in the polynomial-time hierarchy, *Tech. Report TR-94-2*, Department of Computer Science, State University of New York at Stony Brook, Stony Brook, 1994.
7. K. Ko and C.-L. Lin, On the longest circuit in an alterable digraph, preprint, 1994.
8. K. Ko and W.-G. Tzeng, Three Σ_2^P-complete problems in computational learning theory, *comput. complex.* 1 (1991), 269-301.
9. C. H. Papadimitriou and K. Steiglitz, *Combinatorial Optimization: Algorithms and Complexity*, Prentice-Hall, Englewood Cliffs, New Jersey, 1982.
10. C. H. Papadimitriou and M. Yannakakis, Optimization, approximation, and complexity classes, *J. Comput. System Sci.* 43 (1991), 425-440.
11. C. H. Papadimitriou and M. Yannakakis, The traveling salesman problem with distances one and two, *Math. Oper. Res.* 18 (1993), 1-11.
12. Y. Sagiv and M. Yannakakis, Equivalences among relational expressions with the union and difference operations, *J. Assoc. Comput. Mach.* 27 (1980), 633-655.
13. L. J. Stockmeyer, The polynomial-time hierarchy, *Theoret. Comput. Sci.* 3 (1977), 1-22.
14. K. W. Wagner, The complexity of combinatorial problems with succinct input representation, *Acta Inform.* 23 (1986), 325-356.

A COMPETITIVE ALGORITHM FOR THE COUNTERFEIT COIN PROBLEM

X. D. HU*
RUTCOR, Rutgers University, New Brunswick, NJ 08903

and

F. K. HWANG
AT&T Bell Laboratories, Murray Hill, NJ 07974

Abstract. The classical counterfeit coin problem asks for the minimum number of weighings on a balance scale to find the counterfeit coin among a set of n coins. The classical problem has been extended to more than one counterfeit coin, but the knowledge of the number of counterfeit coins was previously always assumed. In this paper we assume no such knowledge and propose an algorithm with uniform good performance.

1. Introduction

Let $S(n)$ denote the sample space of n coins where a coin is either *light* or *heavy* (all light coins have one weight and all heavy coins have another). We call the majority kind *regular* and the other kind *counterfeit*. Let $S(d,n)$ denote the sample space of n coins containing exactly d counterfeit coins. Denote by $M_G(d,n)$ the maximum number of weighings required by an algorithm G to find all counterfeit coins with a balance scale where the sample $s \in S(d,n)$. Define

$$M(d,n) = \min_G M_G(d,n) \ .$$

The determination of $M(1,n)$ is a well studied and solved problem (see [4] for a review) since Schell proposed it in 1945 as an Amer. Math. Monthly problem. The determination of $M(2,n)$ is much harder and still unsolved [1], [2], [3], [6]. Tošic [7], [8], [9], [10] gave near-optimal algorithms for $d = 2, 3, 4$ and 5. Pyber [5] devised an ingenious algorithm which requires at most $\lceil \log_3 \binom{n}{d} \rceil + 15d$ weighings.

All the above-mentioned algorithms assume that d is known (Pyber's algorithm assumes that d is an upper bound of the number of counterfeit coins). In practice, one may not know the value of d and has to devise algorithms assuming the sample is from $S(n)$. Let $M_G(n : d)$ denote the maximum number of weighings required by an algorithm G over all samples from $S(d,n)$ (but the algorithm is designed for $S(n)$). Define

$$M(n : d) = \min_G M_G(n : d) \ .$$

* This author thanks the Air Force Office of Scientific Research for its support under grants AFOSRI-90-0008 to Rutgers University.

D.-Z. Du and P. M. Pardalos (eds.), Minimax and Applications, 241–250.

Algorithm G is called *competitive* if

$$M_G(n:d) \leq cM(n:d) + b,$$

where c and b are constants (c is called the *competitive constant*). A competitive algorithm guarantees a uniformly bounded performance regardless of from which sample space the natural sample is taken from. In this paper we give a competitive algorithm for the counterfeit coin problem.

The counterfeit coin problem is perhaps one of the most popular mathematical problems, well-known to mathematicians and nonmathematicians alike. As such, it provides a readily recognizable example of discrete search problems. Amazingly, this easily understood problem is also extremely difficult to solve, except for the version of one counterfeit coin. It is hoped that the study and solution of the counterfeit coin problem may lead to a deeper understanding of information theory.

2. Some Lower Bounds of M(n:d)

Since $M(n:d)$ is unknown in general, we usually establish the competitiveness of an algorithm by comparing its required number of weighings with a lower bound of $M(n:d)$. Clearly, $M(d,n)$ is a lower bound of $M(n:d)$ since the knowledge of d can never hurt (we don't have to use it). However, $M(d,n)$ is also not known in general. Thus we will use the information-theoretic lower bound $\lceil \log_3 \binom{n}{d} \rceil$ of $M(d,n)$, where $\binom{n}{d}$ is the cardinality of $S(d,n)$ and 3 is the number of outcomes in each weighing, as a lower bound of $M(n:d)$. It turns out that using this lower bound, our proposed algorithm will have a large competitive constant for the two cases $d = 0, 1$. Thus we construct better lower bounds of $M(n:0)$ and $M(n:1)$, in fact, we solve them in this section.

We will call a set of coins *heavy-uniform (light-uniform)* if it contains only heavy (light) coins, and *heavy-unique (light-unique)* if it contains all except one heavy (light) coin. We use *uniform* and *unique* if we don't care about the weight class or it is understood. In the following we will use capital letters to denote sets of coins, and lower-case letters to denote single coins. An (A, B) weighing means weighing A against B. It is well known that if $|A| \neq |B|$, then the weighing does not provide any useful information. Thus we assume $|A| = |B|$ throughout. We write $W(A) = W(B)$ if A and B are found (could be from deduction) of equal weight, and $W(A) < W(B)$ if A is lighter than B. In the rest of this section, we assume without loss of generality that the heavy kind is regular.

Theorem 1 $M(n:0) = \lceil \log_2 n \rceil$.

Proof: Theorem 1 is trivially true for $n = 1$. We prove the general n case by induction. Suppose we first weigh A against B. Since the sample is from $S(0,n)$, the outcome of this weighing must be $W(A) = W(B)$. Let $|A|(= |B|) = k$. Then the problem is reduced to $M(n-k:0)$ since once we know A is uniform, we can deduce that B is also uniform. Therefore

$$\begin{aligned} M(n:0) &= 1 + \min_{1\le k\le \lfloor n/2\rfloor} M(n-k:0) \\ &= 1 + \min_{1\le k\le \lfloor n/2\rfloor} \lceil \log_2(n-k)\rceil \\ &= 1 + \lceil \log_2(n-\lfloor n/2\rfloor)\rceil = \lceil \log_2 n\rceil \ . \end{aligned}$$

The proof for $M(n:1)$ is more complicated. We first give a lemma.

Lemma 1 *A regular coin x can be identified if and only if one of the following conditions is met:*

(i) x is in a set A with $W(A) = W(B)$ where B is a heavy-uniform set.

(ii) x is in a set A with $W(A) > W(B)$ where B is either heavy-uniform or heavy-unique (of course from $W(A) > W(B)$ we can deduce B is heavy-unique).

(iii) x is in a set A, which contains a light coin other than x, with $W(A) = W(B)$ where B is a heavy-unique set.

Proof: Straightforward.

Corollary 1 *The first set of coins identified must consist of one heavy and one light coin.*

Proof: The only case meeting the conditions of Lemma 1 without requiring the presence of an identified coin is case (ii) with B consisting of a single coin. From $W(A) > W(B)$, clearly, A consists of a heavy coin and B a light one.

Theorem 2 $M(n:1) = \lceil \log_2 n\rceil$.

Proof: Since the sample is from $S(1,n)$, condition (iii) of Lemma 1 cannot be met since it requires two counterfeit coins. Consider the partial order of regular coins such that if the identification of a requires the presence of a set B of identified coins, then every regular coin in B precedes a. Note that the only regular coins identified without requiring any other identified regular coins are those compared to the counterfeit coin, one at a time. We will construct a linear extension of this partial order which minimizes the number of weighings required. This can be achieved by maximizing the number of newly identified coins at each weighing (luckily, we will show that these maximizations at different weighings do not conflict). With that goal in mind, we will put all regular coins say, k of them, identified through a comparison with the lone counterfeit coin at the first k weighings, since each of these weighings identifies a single regular coin no matter where they are, but their presence at the beginning builds up the number of identified coins to benefit other weighings which require them. It is now easily seen that the maximum number of the $(k+i)^{\text{th}}$ weighing can identify $i(k+1)$ new regular coins (using condition (ii) of Lemma 1) for $i = 1, 2, \ldots$, as long as there are enough regular coins. It is also easily verified that by setting $k = 1$, the number of newly identified regular coins is maximized at every weighing. This implies j weighings can identify a maximum of 2^j coins (including the counterfeit coin). So $M(n:1) \geq \lceil \log_2 n \rceil$.

We next give an algorithm for $S(n)$ which achieves $M(n:1) = \lceil \log_2 n \rceil$. Split the set evenly (with a left-over coin c if n is odd) into two sets A and B. Weigh A against B and assume $W(A) \geq W(B)$ without loss of generality. Replace the original set by $B \cup \{c\}$ and apply the same procedure. As soon as we know the original set contains more than one counterfeit coin, compare coins individually until every one is identified. It is easily verified that for an $S(1,n)$ sample, after $\lceil \log_2 n \rceil$ weighings we identify a counterfeit coin, and every other coin a belongs to a set S_a with either a weighing $W(S_a) > W(S)$ where S is unique, or a weighing $W(S_a) = W(S)$ where S is uniform. Thus we conclude that a is regular by Lemma 1. (It is also easily verified that this algorithm uses only $\lceil \log_2 n \rceil$ weighing for an $S(0,n)$ sample.)

3. A Competitive Algorithm

By comparing one fixed coin with every other coin, we can find all counterfeit coins in a set of n coins in $n-1$ weighings. When d is large, say, $d = fn$ where f is a fraction, $M(d,n)$ is lower bounded by $\lceil \log_3 \binom{n}{d} \rceil \cong -d \log_3 f = -fn \log_3 f$. So $n-1$ is a constant (given f) multiple of $M(n,d)$. However, this simple-minded algorithm is not competitive when d is small. The challenge of a competitive algorithm is to reduce the number of weighings from $O(n)$ to $O(\log n)$ when d is small. We will give such an algorithm in this section.

We first give the main idea of our algorithm. Suppose we partition the coins into g sets of equal size. When d is less than g, then some of the sets must be uniform, i.e., containing only regular coins. If several uniform sets have been weighed against each other and found of equal weight, then as soon as we find out the content of one uniform set, we know coins in all other sets of equal weight to this uniform set are regular without further ado (this is the same idea used in the proof of Theorem 1). The smaller d is, the more frequently this will happen. However, if g is too large, then too many weighings are required just to identify sets of equal weight.

We still need to take care of several technical points. The main difficulty is that we don't know d beforehand, and hence we don't know how to set g. This then calls for a sequential algorithm which first sets g small and then gradually increases it when not enough uniform sets are found. This is accomplished by introducing a binary tree structure for grouping. We also need a method to identify potential uniform sets at an early stage. Note that a uniform set is either the heaviest or the lightest (depending on whether heavy, or light, is regular) among all groups of the same size. Therefore we need to identify all such groups as candidates. Finally, we need to take care of the case when n is not divisible by g. Define $N = \lfloor \log_2 n \rfloor$. Suppose that

$$n = a_N 2^N + a_{N-1} \cdot 2^{N-1} + \cdots + a_1 \cdot 2 + a_0,$$

where $a_N = 1$ and $a_i \in \{0,1\}$ for $i = 0, 1, \ldots, N-1$. We will represent the *binary algorithm* by $\sum_{i=0}^{N} a_i$ binary trees $T_N, \ldots, T_0$, where T_i exists if and only if $a_i = 1$ and has leaves labeled by 2^i coins. Every internal node is labeled by the set of coins labeling the leaves of the subtree rooted at the internal node. Thus the root of T_i is labeled by its sets of 2^i coins. The *weight* of a node is simply the sum of weights of the labeling coin set. We will arrange the binary forest in such a way that a node at level i (leaves at level 0) is labeled by a 2^i-set. Then there are $n_i = \sum_{j=1}^{N} a_j \cdot 2^{j-i} = \lfloor n/2^i \rfloor$ nodes in level i. Fig. 1 illustrates such a forest with $7 = 1 \cdot 2^2 + 1 \cdot 2 + 1$ coins.

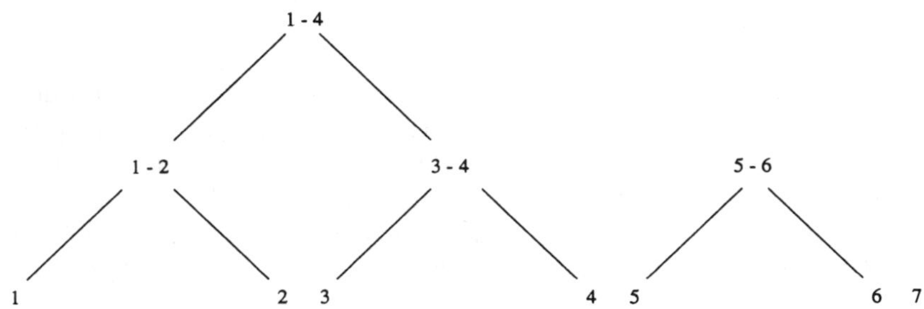

Fig. 1. A FOREST WITH 7 COINS.

Starting from level $N-1$ and moving downwards, the binary algorithm divides nodes at level i into four classes: H_i, L_i, M_i, U_i where H_i consists of heaviest nodes, L_i of lightest nodes among level i nodes which have been compared, M_i consists of other compared nodes and U_i uncompared nodes. The set C_i of compared nodes consists of the root of T_i, children of nodes in M_{i+1} and children of one node $h_{i+1} \in H_{i+1}$ and one node $l_{i+1} \in L_{i+1}$. Pyber [5] gave the following result.

Lemma 2 *A balance scale can find all heaviest coins and lightest coins in a set of n coins of arbitrary weights in $\lfloor 3(n-1)/2 \rfloor$ weighings.*

Proof: We will only compare two coins at a time. It is well known that a heaviest (lightest) coin among n coins can be found in $n-1$ paired comparisons. Suppose n is even. First divide the n coins into $n/2$ pairs and compare each pair. Find a heaviest (lightest) coin among the $n/2$ winners (losers) in $n/2-1$ comparisons. By keeping record of ties (a tie is broken arbitrarily to decide the winner of a comparison), all heaviest (lightest) coins are simultaneously found. The total number of comparisons (weighings) is

$$\frac{n}{2} + \frac{n}{2} - 1 + \frac{n}{2} - 1 = \lfloor \frac{3(n-1)}{2} \rfloor .$$

Suppose n is odd. Then we apply the above to any $n-1$ coins to find all heaviest and lightest coins among them. By comparing the last coin with one such heaviest coin and one such lightest coin, we find all heaviest and lightest coins for the original set of n coins. The number of weighings is $\lfloor 3(n-2)/2 \rfloor + 2 = \lfloor 3(n-1)/2 \rfloor$.

Corollary 2 *Suppose that the initial $\lfloor n/2 \rfloor$ paired comparisons produce t ties between either heaviest or lightest coins (that they are heaviest or lightist are unknown at this moment). Then the total number of required weighings is reduced by t.*

By treating each node as a super-coin, we can divide C_i into H_i, L_i and M_i in $\lfloor 3(|C_i|-1)/2 \rfloor$ weighings. The initial $\lfloor |C_i|/2 \rfloor$ weighings will always include the comparisons between the two h_{i+1} and the two l_{i+1} to maximize the possible saving from Corollary 2.

The following lemma shows that sometimes we can do better than paired comparisons.

Lemma 3 *Given three pairs of coins each consisting of a heavy and a light coin, and also given a known heavy and a known light coin, We can determine the nature of every coin in two weighings.*

Proof: Let (a_1, a_2), (b_1, b_2), (c_1, c_2) denote the three pairs and let h and l denote the known heavy and light coin, respectively. First weigh $\{a_1, b_1, c_1\}$ against $\{a_2, h, l\}$. If the first set is heavier, a_1 must be heavy while b_1 and c_1 cannot both be light. Weighing b_1 against c_1 clarifies the remaining ambiguity. The case the first set is lighter is analogous. When the two sets are of equal weight, either a_1 is light while both b_1 and c_1 are heavy, or a_1 is heavy while both b_1 and c_1 are light. Weighing a_1 against a_2 resolves the ambiguity.

After the comparisons at level 0 are done, we know the states of all coins in C_0. Inductively, we know the number of heavy (hence light) coins for each node in C_i for all i. Let k(or j) denote the highest level such that h_k(or l_j) is a heavy (light)-uniform set. Without loss of generality assume that $k \geq j$. Let $\bar{M}_i$ be obtained from U_i by deleting those nodes which have an ancestor known to be a uniform (heavy or light) set. Then for each level i, $k \geq i > j$, we identify the heaviest nodes in $h_i \cup \bar{M}_i$ in $|\bar{M}_i|$ weighings. Note that the contents of these heaviest nodes are the same as h_i, hence containing only heavy coins. At each level $i, j \geq i \geq 1$, we pair nodes in $\bar{M}_i$ for comparisons to determine winners and losers. We then identify all heaviest nodes by comparing $h_i \cup \{\text{winners}\}$ and all lightest nodes by comparing $l_i \cup \{\text{losers}\}$ in $2 + \lfloor 3(\bar{M}_i - 1)/2 \rfloor$ weighings. Since the number is larger than $|\bar{M}_i|$, we assume that we have to use that many weighings at every level.

For three integers a, b, c satisfying $a + b = c$, it is easily verified that $\lfloor 3(a - 1)/2 \rfloor + \lfloor 3(b - 1)/2 \rfloor + 2 = \lfloor 3(c - 1)/2 \rfloor$ if at least one of a and b is even. Define $m_i = |C_i| + |\bar{M}_i|$. Then we can count the weighings in C_i and $\bar{M}_i$ together as if they form a set of m_i coins as long as $|C_i|$ is even, this can be achieved by leaving out a node of C_i (to be picked up by $\bar{M}_i$) whenever $|C_i|$ is odd.

Let w_i denote the total number of weighings required to find all heaviest and lightest nodes at level i.

Lemma 4 *For $d \geq 2$, $w_i \leq 3d$ for $1 \leq i < N$.*

Proof: For a given level i we consider two cases.

(i) There exists a node in L_{i+1} containing at least two counterfeit coins. Then there exist at most $2(d-2)$ nonuniform sets at level i and $m_i \leq 2+2+2(d-2)+a_i \leq 2d+1$,
$$w_i = \lfloor 3(m_i - 1)/2 \rfloor = 3d\ .$$

(ii) Every node in L_{i+1} contains exactly one counterfeit coin. Then there exist d nodes in L_{i+1} but no node in M_{i+1}. Furthermore, h_{i+1} is uniform, hence both its two children are uniform and of equal weight. Therefore C_i contains at most

5 nodes with one tie in the initial two comparisons. By Corollary 2, 5 weighings suffice to resolve C_i. After the comparisons at level 0 are done, we know that each node in L_{i+1} contains exactly one counterfeit coin. Let l_{i+1} be such a node. Then one child of l_{i+1} must contain only heavy coins and the other contains exactly one counterfeit coin. By comparing the two children of each l_{i+1}, we find $d-1$ uniform sets and $d-1$ unique sets in $d-1$ weighings. So

$$w_i = 5 + d - 1 \le 3d \text{ for } d \ge 2 .$$

Lemma 5 $w_0 \le 2 + \lceil 2(d-1)/3 \rceil$.

Proof: If H_1, L_1 and M_1 all exist, then we know H_1 must be heavy-uniform, L_1, must be light-uniform and M_1 must consist of one heavy and one light. Since there are at most $d-2$ nodes in M_1 (and $\bar{M}_1$ is empty), it requires only $\lceil 2(d-2)/3 \rceil$ weighings by Lemma 3. If M_1 is empty, we weigh the two children of $h_1(l_1)$ against each other. If they are of equal weight, then $h_1(l_1)$ is heavy (light)-uniform. Otherwise, it consists of one heavy and one light. Since $w(h_1) > w(l_1)$, at most one of them is nonuniform, say it is l_1. There are at most d nodes in L_1. Therefore it requires $2 + \lceil 2(d-1)/3 \rceil$ weighings by Lemma 3.

4. Analysis of Competitiveness

Since the binary algorithm is symmetric with respect to heavy and light coins, we may assume that heavy is regular, i.e., there are at least as many heavy coins as light ones. Define $w = \sum_{i=0}^{N-1} w_i$.

Theorem 3 *The binary algorithm is a competitive algorithm with the competitive constant* $3\log_2 3$ *for the counterfeit coin problem.*

Proof: Case (i). $\frac{n}{2} \ge d \ge \frac{n}{4}$.

$$\begin{aligned} w &\le \textstyle\sum_{i=1}^{N} \lfloor \frac{3}{2}(n_i - 1) \rfloor + n - 1 \\ &\le \textstyle\frac{n}{2} \sum_{i=1}^{N} \frac{1}{2^i} - \frac{3}{2}N + n - 1 \\ &= \textstyle\frac{3}{2}n - \frac{n}{2}2^{-N} - \frac{3}{2}N - 1 \le \frac{3}{2}n - \frac{3}{2}\log_2 n - \frac{3}{2} . \end{aligned}$$

Note that

$$M(n:d) \ge \log_3 \binom{n}{d} \ge d \log_3 \frac{n}{d} \ge \frac{n}{2} \log_3 2 .$$

Hence $w \leq 3(\log_2 3)M(n;d) - \frac{3}{2}\log_2 n - \frac{3}{2}$.
Case (ii) $\frac{n}{4} > d \geq 2$. Define $k = \lceil \log_2 \frac{n}{d} \rceil - 1$.

$$
\begin{aligned}
w &\leq 2 + \lceil 2(d-1)/3 \rceil + \textstyle\sum_{i=1}^{k-1} 3d + \sum_{i=k}^{N} \lfloor \frac{3}{2}(n_i - 1) \rfloor \\
&\leq 2 + 2d/3 + 3d(k-1) + \textstyle\sum_{i=k}^{N} \left(\frac{3}{2} \lfloor \frac{n}{2^i} \rfloor - \frac{3}{2} \right) \\
&\leq -\left(\frac{3}{2}N + \frac{3n}{2^{N+1}} \right) + \left(3dk + \frac{3}{2}k + \frac{3n}{2^k} \right) - \left(\frac{7d}{3} - \frac{1}{2} \right) \\
&= -\frac{3}{2}\left(\lfloor \log_2 n \rfloor + n2^{-\lfloor \log_2 n \rfloor} \right) + 3\left(d\lceil \log_2 \frac{n}{d} \rceil + \frac{1}{2}\lceil \log_2 \frac{n}{d} \rceil + n2^{1-\lceil \log_2 \frac{n}{d} \rceil} \right) \\
&\quad - \left(\frac{16d}{3} + 1 \right) .
\end{aligned}
$$

Let $log_2 n = \lfloor \log_2 n \rfloor + u$ and $\lceil \log_2 \frac{n}{d} \rceil = \log_2 \frac{n}{d} + v$

$$
\begin{aligned}
w &\leq -\frac{3}{2}(\log_2 n - u + n \cdot 2^{u - \log_2 n}) \\
&\quad 3\left(d\left(\log_2 \frac{n}{d} + v \right) + \frac{1}{2}\left(\log_2 \frac{n}{d} + v \right) + n2^{1-\log_2 \frac{n}{d} - v} \right) - \left(\frac{16d}{3} + 1 \right) \\
&= 3d\log_2 \frac{n}{d} - \frac{3}{2}\log_2 d + \frac{3}{2}(u - 2^u) \\
&\quad + 3\left(dv + \frac{v}{2} + d2^{1-v} \right) - \left(\frac{16d}{3} + 1 \right)
\end{aligned}
$$

Note that $f(u) \equiv u - i^u$ is concave and achieves maximum at $u = \log_2 \log_2 e$, while $g(v) = dv + v/2 + d2^{1-v}$ is convex and achieves maximum at $g(1)$. Replacing $f(u)$ by $f(\log_2 \log_2 e)$ and $g(v)$ by $g(1)$, we obtain

$$
\begin{aligned}
w &\leq 3d\log_2 \frac{n}{d} - \frac{3}{2}\log_2 d + \frac{3}{2}(\log_2 \log_2 e - \log_2 e) \\
&\quad +3\left(2d + \frac{1}{2} \right) - \left(\frac{16d}{3} + 1 \right) \\
&= 3(\log_2 3)d\log_3 \frac{n}{d} - \frac{3}{2}\log_2 3 \cdot \log_3 d + \frac{3}{2}\log_2 \frac{\log_2 e}{e} + \frac{2d}{3} + \frac{1}{2} .
\end{aligned}
$$

Further note that

$$
\begin{aligned}
M(n:d) &\geq d\log_3 \frac{n}{d} + d\log_3 e\sqrt{\frac{3}{4}} - 0.5\log_3 d - 0.5\log_3 \frac{3}{4} - 1.567\log_3 2 . \\
&\geq d\log_3 \frac{n}{d} + d\log_3 e\sqrt{\frac{3}{4}} - \frac{1}{2}\log_3 d \\
&\quad -0.567\log_3 2 - \frac{1}{2} .
\end{aligned}
$$

Hence

$$
\begin{aligned}
w &\leq 3\log_2 3M(n:d) + \left(\frac{2}{3} - 3\log_2\left(\frac{e\sqrt{3}}{2} \right) \right) d \\
&\quad \frac{3}{2}\log_2 \frac{\log_2 e}{e} + 3 \times 0.567 + \frac{3}{2}\log_2 3 + 1/2 .
\end{aligned}
$$

Since

$$\tfrac{2}{3} - 3\log_2\left(\tfrac{e\sqrt{3}}{2}\right) = \tfrac{11}{3} - 3\log_2 e\sqrt{3} < -3,$$

$$\tfrac{3}{2}\log_2 \tfrac{\log_2 e}{e} = -1.3754,$$

$$\tfrac{3}{2}\log_2 3 = 2.3770,$$

we have

$$w = 3(\log_2 3)M(n:d) - 3d + 3.3 \ .$$

Case (iii). $d = 1$.

Suppose that the lone counterfeit coin is in T_j. Then $H_i = L_i$ for $j \le i \le N$. Hence $m_i = 2 + a_i \le 3$. By Corollary 2, $w_i \le \lfloor c(3-1)/2 \rfloor - 1 = 2$ for these levels. For $1 \le i < j$, $m_i = 4 + a_i$. If $a_i = 1$, then $|C_i| = m_i = 5$ is odd. So the root of T_i is transferred from C_i to $\bar{M}_i$. We add the following rule to the binary algorithm. If H_1 is uniform but not L_1, and if $|\bar{M}_i| = 1$ for any $i \ge 1$, then the lone node in $\bar{M}_i$ will be treated as the heaviest coin in $\bar{M}_i$ (rather than the lightest), i.e., it will be compared with h_i before l_i. The rule is also symmetric with respect to H_1 and L_1. This rule does not affect anything else except it helps the $d = 1$ case with $i < j$. Note that under the assumption that heavy is regular, the root of T_i for $i < j$ is heavy-uniform. This fact is ascertained after it is compared to h_i, hence the comparison to l_i can be skipped. Therefore

$$\begin{aligned} w_i &= \lfloor 3(5-1)/2 \rfloor - 1 - 1 = 4 \\ w &\le 3(N - j + 1) + 4j \le 4N + 3 \\ &\le 4M(n:1) + 3 \quad \text{by Theorem 2} \ . \end{aligned}$$

Case (iv). $d = 0$. $h_i = L_i$ for all $1 \le i \le N$. Therefore $m_i = 2 + a_i$ and

$$\begin{aligned} w_i &\le \lfloor 3(3-1)/2 \rfloor - 1 = 2 \ . \\ w &\le 2N \le 2M(n:0) \quad \text{by Theorem 1} \ . \end{aligned}$$

5. Conclusion

We give the first competitive algorithm for the popular counterfeit coin problem. The competitive constant is $3\log_2 3 < 5$, though we suspect that a more careful analysis of lower bounds will reduce the constant significantly, i.e., the algorithm is better than the analysis shows. Instead of the binary algorithm we can also have a t-nary algorithm by considering t-nary forests. Then m_i has a leading term td, but the forest has only $\lfloor \log_t n \rfloor$ levels. Since $t \log_t n$ is minimized over integers at $t = 3$, in theory, a ternary algorithm should be best. However, we obtain a competitive constant slightly greater than 5 for the ternary case. This can be explained by the fact that the competitive constant is the maximum over three intervals of d while the theoretical advantage for ternary is valid only in one such interval, and that the binary algorithm uses some special subroutines like those given in Lemma 2 and 3.

Other counterfeit coin problems studied in the literature include the average number of weighings and a model where weighing outcomes can be erroneous. We hope that the competitive algorithm idea of this paper can be extended to these models.

6. Acknowledgment

The authors thank Prof. D. Z. Du for bringing the problem to their attention.

References

1. R. Bellman and B. Gluss, On various versions of the defective coin problem, Information and Control 4 (1961), 118–131.
2. S. S. Cairns, Balanced scale sorting, Amer. Math. Monthly 70 (1963), 136–148.
3. G. O. H. Katona, Combinatorial search problem, in J. N. Srivastava, ed., A Survey of Combinatorial Theory (North-Holland, Amsterdam, 1973), 285–305.
4. B. Manvel, Counterfeit coin problems, Math. Mag. 50 (1977), 90–92.
5. L. Pyber, How to find many counterfeit coins? Graphs and Combinatorics 2 (1986), 173–177.
6. C. A. B. Smith, The counterfeit coin problem, Math. Gazette 31 (1947), 31–39.
7. R. Tošic, Two counterfeit coins, Disc. Math. 46 (1983), 295–298.
8. R. Tošic, Three counterfeit coins, Rev. Res. Sci. Univ. Novi. Sad 15 (1985), 225–233.
9. R. Tošic, Four counterfeit coins, Rev. Res. Sci. Univ. Novi. Sad 14 (1984), 99–108.
10. R. Tošic, Five counterfeit coins, J. Statist. Plan. Infer. 22 (1989), 197–202.

A MINIMAX $\alpha\beta$ RELAXATION FOR GLOBAL OPTIMIZATION

JUN GU

Dept. of Electrical and Computer Engineering,
University of Calgary,
Calgary, Alberta T2N 1N4, Canada
gu@enel.ucalgary.ca

Abstract. Local minima make search and optimization harder. In this paper, we give a new global optimization approach, minimax $\alpha\beta$ relaxation, to cope with the pathological behavior of local minima. The minimax $\alpha\beta$ relaxation interplays a dual step minimax local to global optimization, an iterative local to global information propagation, and an adaptive local to global algorithm transition, within a parallel processing framework. In minimax $\alpha\beta$ relaxation, α controls the rate of local to global information propagation and β controls the rate of algorithm transition from local to global optimization. Compared to the existing optimization approaches such as simulated annealing and local search, the minimax $\alpha\beta$ relaxation demonstrates much better convergence performance for certain classes of constrained optimization problems.[1]

1. Introduction

A wide range of problems in science, engineering, and management demands optimization solutions. The *goal* of optimization is to find a value assignment to the variables such that all the constraints are satisfied and the performance objective is optimized. A constrained optimization problem can be represented using a constraint network with nodes representing the variables and arcs representing the constraints. Solving a constrained optimization problem requires the generation of an explicit *global* solution to problem expressed in a network of implicit *local* constraints.

Many search and optimization methods have been developed to solve the constrained optimization problems. One important class of algorithms, which is particularly suitable for optimization problems represented in a constraint network, is the relaxation methods. A number of relaxation techniques have been proposed. They complement rather than exclude each others by being effective for particular optimization problems. A significant complication in the constrained optimization is the pathological behavior of local minima.

Recently we have developed a new global optimization approach, minimax $\alpha\beta$ relaxation, for constrained optimization problems [5]. A number of strategies in minimax $\alpha\beta$ relaxation, e.g., a dual step minimax local to global optimization, an iterative local to global propagation, and an adaptive local to global algorithm transi-

---

[1] This work was present in part in Technical Report UCECE-TR-91-003 [5] and in [6]. Jun Gu is presently on leave at Dept. of Computer Science, Hong Kong University of Science and Technology, Clear Water Bay, Kowloon, Hong Kong. E-mail: gu@cs.ust.hk

D.-Z. Du and P. M. Pardalos (eds.), Minimax and Applications, 251–268.

tion, are used to cope with the local minimum problem. In minimax $\alpha\beta$ relaxation, α controls the rate of local to global information propagation and β controls the rate of algorithm transition from local to global optimization. Within a parallel and distributed processing framework, the minimax $\alpha\beta$ relaxation tailors the dynamic progression of the optimization process and makes a balanced use of both local and global information. Compared to the existing optimization approaches such as the simulated annealing and local search, the minimax $\alpha\beta$ relaxation algorithm has shown much better convergence performance for certain classes of constrained optimization problems.

The rest of this paper is organized as follows. In the next section, we give a constraint network model representing the optimization problem. In Section 3, we briefly review the existing relaxation methods which give a basis of the minimax $\alpha\beta$ relaxation. In Section 4, we describe a general $\alpha\beta$ relaxation algorithm. A parallel, minimax $\alpha\beta$ relaxation algorithm, i.e., $\alpha\beta4$, is given in Section 5. In Section 6, we show some experimental results of the $\alpha\beta4$ algorithm. We will also compare the performance of $\alpha\beta4$ to a number of existing optimization algorithms.

2. Problem Model

In this work, we focus our discussion on the following Constrained Optimization Problem (COP) model. The COP model consists of the following several components [5, 8]:

- n distinct <u>*variables*</u>: $\mathbf{X} = \{x_1, x_2, \ldots, x_n\}$. An <u>*assignment*</u> is a tuple of n values assigned to n variables.
- n <u>*domains*</u>: $\mathbf{D} = \{D_1, D_2, \ldots, D_n\}$. Domain D_i defines the possible values that variable x_i may be instantiated. The domain of the variables may be continuous or discrete. For discrete domain, for $i \neq j$, domain D_i may have the same elements as those in domain D_j. The size of D_i is denoted as $|D_i|$ $(1 \leq i \leq n)$.
- a set of <u>*constraints*</u>, $\mathbf{C}$, imposed on variables $x_1, x_2, \ldots, x_n$. An order-l constraint relation $(l \leq n)$ imposed on variables $x_{i_1}, x_{i_2}, \ldots, x_{i_l}$ $(\in \{x_1, x_2, \ldots, x_n\})$ is denoted as

$$R_{i_1,i_2,\ldots,i_l}(x_{i_1}, x_{i_2}, \ldots, x_{i_l}).$$

The order-l constraint relations may be represented as an order-l constraint:

$$C_{i_1,i_2,\ldots,i_l}(x_{i_1}, x_{i_2}, \ldots, x_{i_l}).$$

The constraint measures the *compatibility* or the *conflicting level* among the values of tuple $(x_{i_1}, x_{i_2}, \ldots, x_{i_l})$:

$$C_{i_1,i_2,\ldots,i_l}(x_{i_1}, x_{i_2}, \ldots, x_{i_l}) = \begin{cases} 0, & \text{if } x_i\text{s' values are \textit{compatible};} \\ 1, & \text{if } x_i\text{s' values are \textit{not compatible}.} \end{cases}$$

The above compatibility measure is defined based on the *negative constraints* [8]. A smaller value of the constraint means less conflicts or more compatibility

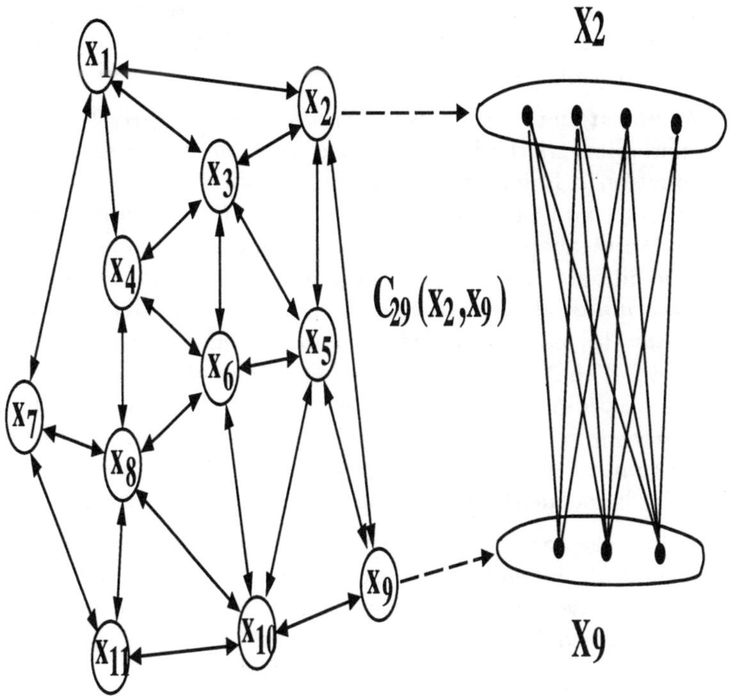

Fig. 1. A constrained optimization problem (COP) can be represented in a constraint network with *nodes* representing the variables and *arcs* representing the constraints (a). Constraints are defined among the values that may be instantiated to the variables. An example of constraints among the values for variables x_2 and x_9 is illustrated in (b).

among the values assigned to the variables. A larger constraint value indicates more conflicts or less compatibility among the values assigned to the variables. In practice, the *unary constraint*, $C_i(x_i)$, which is an order-l constraint, and the *binary constraint*, $C_{ij}(x_i, y_j)$, which is an order-2 constraint, are frequently used.

– an *objective function* indicating the performance criterion for optimization.

This COP model presents a general model for constraint satisfaction problem (CSP) [8, 14, 11, 15] and the satisfiability (SAT) problem models [2, 3].

A constrained optimization problem can be represented in a *constraint network* with nodes representing the variables and arcs representing the constraints (see Figure 1). The *goal* of COP is to find a value assignment to the variables such that all the constraints are satisfied and the performance objective is optimized.

3. Relaxation Approach

Solving a constrained optimization problem requires the generation of an explicit *global* solution to knowledge expressed in a network of implicit *local* constraints.

Many search and optimization methods can be used to solve COPs. One important class of algorithms, which is particularly suitable for COPs represented in a network of implicit *local* constraints, is the relaxation methods.

The *classical relaxation* technique was introduced in the early 1940s by Southwell [16, 17]. This numerical method can produce solutions for large systems of linear simultaneous equations. It has many practical applications in computing stresses [16], stereo disparity, and optical flow [12]. *Symbolic relaxation*, also known as Huffman-Clowes-Waltz labeling [1, 13, 18], is a technique searching for an assignment while enforcing symbolic constraints among a set of object variables. Symbolic relaxation has two major classes, stochastic relaxation and discrete relaxation. In *stochastic relaxation* [15], each variable has an associated vector which assigns to each value the likelihood of that value for the variable. The algorithm then modifies the likelihood vectors according to the local constraints of the variable. The solution to be found in the *discrete relaxation* involves the assignment of a set of discrete values to variables such that all the local constraints are satisfied [9, 10, 8]. The algorithm deletes a value if it is inconsistent with the assignment and the constraints.

Constraints play a key role in the relaxation methods [8]. In classical relaxation, the local constraints are introduced by expanding at each point the partial differential equation which gives the global constraint. Likewise, constraints in symbolic relaxation can be derived from the physical world or can be the result of purely symbolic considerations.

A major difficulty with the existing COP algorithms is the presence of local minima. Accordingly, a major issue of relaxation methods is the convergence property. Although convergence is easily proven for most formulations of discrete relaxation, the convergence properties for other relaxation methods are difficult to determine.

Before we give a minimax $\alpha\beta$ relaxation algorithm (in Section 5), in the next section, we will describe a general $\alpha\beta$ relaxation algorithm.

4. A General $\alpha\beta$ Relaxation Algorithm

The $\alpha\beta$ relaxation is developed to cope with the pathological behavior of local minima. A number of methods were incorporated in the $\alpha\beta$ relaxation algorithm which facilitate the algorithm to tailor the dynamic progression of the optimization process and to make a balanced use of both local and global information.

In the following, we first discuss several basic methods behind the $\alpha\beta$ relaxation and then we describe a general $\alpha\beta$ relaxation algorithm.

4.1. Basic Methods

In a constrained optimization problem, a great deal of local information is expressed in the local constraints and local minima. Solving a COP amounts to the derivation of a global solution from the local structures. A proper use of local information would make the optimization process more informed. It is our experience that a mixed local and global optimization, a periodical exchanging of local to global information, and a gradual transition from local to global optimization are crucial to the design a good optimization algorithm for the constraint satisfaction problems.

Dual Step Local to Global Optimization

Local structures, e.g., constraints, local variables, and local terrain structure of the search space, are the background that we derive a local solution and, furthermore, a global solution.

In $\alpha\beta$ relaxation, we incorporate a local optimization procedure inside each step of the global optimization procedure. This results in a *dual step optimization* procedure. In the first step, based on the local structure, a local optimization is performed that produces a locally optimal solution. In the second step, based on the locally optimal solution from local optimization, a global optimization is performed over the entire problem range, resulting in a "dually optimized" solution.

Depending on the formulation of problem constraints and the objective function, the dual step optimization can be carried out by a minimax optimization procedure. Through numerous experiments (see Section 6), we have observed that the solution produced from the dual step optimization procedure is much better than solutions from a single local optimization or a single global optimization [5].

Iterative Local to Global Propagation

Local minima are hard to deal with since they are unpredictable and intractable. Recently a number of multispace search techniques have been developed which handle the local minima through structured multispace operations [4, 7]. Previous work suggested that even a limited amount of local information exchange would significantly improve the performance of an optimization algorithm for COP.

In the $\alpha\beta$ relaxation, an iterative procedure for local information propagation is developed. The procedure spreads local information of each local variable to other variables in the constraint network. Depending on the problem and algorithm variations, the distribution of local information may take a moderate rate or in a more speedy way. This is controlled by the local information *propagation rate*, α.

In general, at the beginning of optimization, it would be advantageous to distribute the local information much more quickly, allowing it to inform other variables in the network at an early stage of optimization. So, the initial value of α is set to a large value. A large α value, however, may cause an nonuniform distribution of local information. As the iteration progresses, α is gradually decreased such that the local information in the network can be spread more uniformly and throughly in the network.

Adaptive Local to Global Algorithm Transition

In optimization, *phase transition* phenomena have been observed by many researchers. Initially, the objective function is reduced rapidly showing an *easy phase* of progression (although the initial optimization process may take different descent paths in the search space, as affected by the optimization algorithm and the terrain surface structure of the search space). In this stage, most optimization algorithms perform well, indicating minor differences in terms of convergence behavior. The *difficult phase* of an optimization process appears at the later stage of locating the global optimum. Due to the complication of numerous local minima, the difficulty of finding the global optimum is increased considerably. For hard optimization problems, many optimization algorithms that exhibit satisfactory performance at the earlier stage perform poorly at the later stage of optimization, showing inferior convergent behavior.

This suggests that different phases of an optimization process should be treated separately and an optimization algorithm should be designed to adapt to the different phases of the optimization process. Furthermore, if the phases are distinct, we may use different algorithms in different phases of optimization (i.e., *algorithm switching*); otherwise, we may handle the phase transition and the corresponding algorithm transition in a gradual manner (i.e., *algorithm transition*). This is a key idea behind the $\alpha\beta$ relaxation.

In $\alpha\beta$ relaxation, an adaptive local-to-global algorithm transition is built that changes the composition of local and global information and performs a dynamic algorithm transition under the control of *transition rate*, β. In a basic $\alpha\beta$ relaxation algorithm, the β value is initially set to 1 and the algorithm corresponds to a local optimization algorithm. This makes $\alpha\beta$ relaxation a full use of the local information.

As the optimization process evolves, a global solution is gradually derived from the local information. The local information becomes less important. Accordingly, the β value is gradually reduced to minimize the contribution of the local information and to maximize the effect of global information. Eventually, $\alpha\beta$ relaxation gradually approaches an optimization algorithm utilizing global information.

Parallel and Distributed Processing

Applying parallel processing techniques to $\alpha\beta$ relaxation is a direct consequence of several earlier parallel discrete relaxation VLSI architectures [9, 10]. Both discrete relaxation and $\alpha\beta$ relaxation share the same constraint network structure. They differ in the variable domain, the constraint domain, and the local to global transition techniques. The $\alpha\beta$ relaxation is an extension of the discrete relaxation algorithms.

Most operations in the $\alpha\beta$ relaxation can be executed in parallel which result in a number of parallel $\alpha\beta$ relaxation algorithms [5].

4.2. A General $\alpha\beta$ Relaxation Algorithm

A general $\alpha\beta$ relaxation algorithm is shown in Figure 2 [5]. The algorithm consists of initialization and relaxation two stages.

```
procedure αβ1 ()
begin
    /* initialization */
    given a COP problem instance;
    initialize α to a larger positive value;
    initialize β to be 1;
    initialize the constraint network structure;
    initialize the objective functions;
    initialize the information functions;

    /* relaxation */
    k := 0;
    while the objective function is not zero do
    begin
        /* dual step local to global optimization */
        for each of the variables do
        begin
            /* optimizing the objective with local structures */
            local optimization of the objective function;
            update local and global information;
            /* optimizing the objective with global network */
            global optimization of the objective function;
            update global information;
        end;
        /* local to global information propagation */
        updating propagation rate α;
        /* local to global algorithm transition */
        updating transition rate β;
        k := k + 1;
    end;
end;
```

Fig. 2. **$\alpha\beta$1: A general $\alpha\beta$ relaxation algorithm in the real space.**

Initialization. During the initialization stage, a COP problem instance is given. It is represented/initialized as a constraint network with nodes representing the variables and arcs representing the constraints. The initial values of the objective function is the *sum* of all unary and binary constraints. Initially, the propagation rate, α, is set to a larger positive value, ensuring an earlier, fast distribution of local information. And the transition rate, β, is set to be 1.

Relaxation. The relaxation step is an iterative procedure. It interplays a dual step local to global optimization with local to global information propagation and local to global algorithm transition. For each variable of the optimization problem, a local optimization to the objective function within the local neighboring structure is performed, followed by a local and global information updating. A global optimization for the entire network is then conducted, followed by a global information updating. During each iteration, the propagation rate, α, the transition rate, β, and the iteration number, k, are updated accordingly.

Theoretically, in the real space, the relaxation process is terminated if the objective function is reduced to zero. For the discrete variable and constraint domain there are quantization errors. The relaxation process may be terminated if, in practice, the objective function is sufficiently small.

In the next section, we will give a minimax $\alpha\beta$ relaxation algorithm for global optimization.

5. A Minimax $\alpha\beta$ Relaxation Algorithm for COP

In this section, we give a minimax $\alpha\beta$ formulation and then a minimax $\alpha\beta$ relaxation algorithm for the discrete constrained optimization problems.

5.1. A Minimax $\alpha\beta$ Formulation of COP

We use a constraint network to represent the constrained optimization problem with nodes representing the variables and arcs representing the constraints. In the following discussion, let:

- n be the total number of nodes in the network,
- i be the ith node representing variable x_i,
- n_i be the degree of node i,
- N_i be a set of nodes directly connected to node i, $|N_i| = n_i$,
- $\vec{y}$ be a set of variables represented by nodes in N_i,
- $\mathcal{N}_i$ be a local network consisting of node i and its directly connected nodes,
- c_i be the local information function of node i,
- f_i be the objective function of the local network $\mathcal{N}_i$, and
- f be the objective function of the constraint network.

If no confusion arises, we will use i and x_i interchangeably, and $j \in N_i$ ($j = 1, 2, ..., |N_i|$) and $\vec{y} \in N_i$ interchangeably.

Each node and its directly connected nodes in the constraint network forms a local network. For n distinct variables, a constraint network can be divided into n small local networks. In Figure 3, for node 6, N_6 consists of nodes 3, 4, 5, 8, and 10. The degree of node 6, n_6, is 5. The local network $\mathcal{N}_6$ is formed by node 6 and its directly connected nodes in N_6.

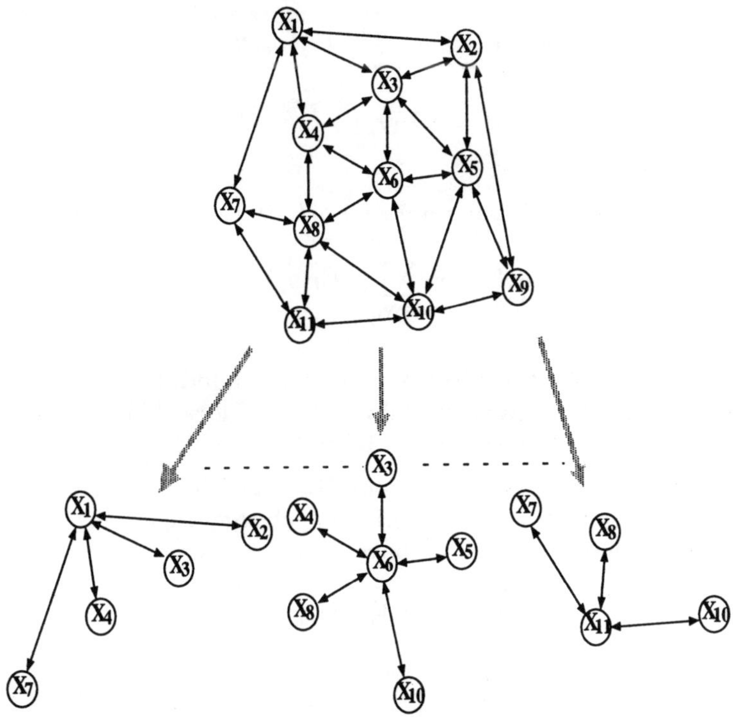

Fig. 3. Each node in the constraint network forms a local network. For n distinct variables, a constraint network can be divided into n simple local networks.

The *objective function of local network* $\mathcal{N}_i$, $f_i(.)$, is a function of unary and binary constraints defined on x_i and $\vec{y}$:

$$f_i(x_i, \vec{y}) = e^{-\left(R_i(x_i)+\sum_{y\in N_i}(R_j(y_j)+2R_{ij}(x_i,y_j))\right)}. \tag{1}$$

The first step of dual step minimax optimization is to optimize the local network. This can be done by fixing x_i to be a constant in D_i and, for each variable y_j, optimize the local network objective function. This produces a locally optimal solution for the neighboring variable:

$$y_j^* = \max_{j\in N_i, y_j\in D_j} f_i(x_i, y_j)|_{x_i=c} = \max_{j\in N_i, y_j\in D_j} f_i(y_j). \tag{2}$$

The *compatibility function*, $c_i(x_i)$, is a real function defined on variable x_i. For a value in D_i, c_i gives its value compatibility when the value is instantiated to variable x_i. The smallest value of $c_i(x_i)$ indicates that the value is the best solution to x_i. The compatibility function gives local and global information of the variables and the constraint network.

The *transition rate*, β, controls algorithm's functionality transition along with the optimization process progress. In this formulation, the unary and binary constraints are weighted by β as:

$$R'_{ij}(x_i, y_j) = \beta R_{ij}(x_i, y_j) \tag{3}$$

and

$$R'_i(x_i) = \beta R_i(x_i) + (1-\beta)c_i(x_i) \tag{4}$$

where $0 \le \beta \le 1$ and $0 \le i, j \le n$.

The *objective function of constraint network*, $f(.)$, is defined as:

$$f(x_i, \vec{y}) = \beta R_i(x_i) + (1-\beta)c_i(x_i) + \sum_{j \in N_i} (\beta R_j(y_j) + 2\beta R_{ij}(x_i, y_j) + (1-\beta)c_j(y_j)) . \tag{5}$$

In (5) the first and the third terms indicate the information of local unary constraints. The fourth term gives the information of local binary constraints. The second and the fifth terms show compatibility function.

Based on the locally optimal solutions (Eq. 2), the second step of dual step minimax optimization is to optimize $f(.)$ while all the neighboring variables remain as constants. Considering Eq. (2) we have:

$$\begin{aligned} x_i^* &= \min_{x_i} f(x_i, \vec{y}_j)|_{\vec{y}_j = \vec{y}_j^*} \\ &= \min_{x_i} f(x_i, \vec{y}_j^*) \\ &= \min_{x_i} f\left(x_i, \max_{j \in N_i, \vec{y}_j \in D_j} f_i(\vec{y}_j)\right) . \end{aligned} \tag{6}$$

This constitutes the dual step minimax optimization.

The local variable information, both local and global information, is propagated by iteratively updating the compatibility functions $c_i(x_i)$ in the following way:

$$c_i^{(k+1)}(x_i) \leftarrow (1-\alpha) \sum_j w_{ij} c_j^{(k)}(y_j^*) + \alpha f(x_i, y_j^*). \tag{7}$$

where $0 \le \alpha \le 1$ and w_{ij} are the weights for normalization purpose which satisfies:

$$\sum_j w_{ij} = 1 \text{ and } w_{ij} \ge 0. \tag{8}$$

One frequently used weight w_{ij}, which takes the average of the $n_i + 1$ variables in the local network, is:

$$w_{ij} = \begin{cases} \frac{1}{1+n_i}, & \text{for } j = i \text{ or } j \in N_i; \\ 0, & \text{otherwise.} \end{cases} \tag{9}$$

This selection of weight deals comfortably with the individual, small local network structures. It shows better algorithm performance than using weight $w_{ij} = \frac{1}{n}$ (which considers the entire network as a whole without reviewing the individual local network structures).

Parameter α controls the information propagation rate. For smaller α, the optimization information of $f(\vec{x})$ is spread more uniformly into the network but this would take more iteration times. For large α, the optimization information of $f(\vec{x})$ can be spread into the network quickly but the information is distributed in a less uniform manner. So it is desirable to start from a larger α at the beginning and gradually reduce it to a smaller value as the iteration progresses.

Parameter β weights the importance between the local constraints, i.e., $R_i(x_i)$ and $R_{ij}(x_i, y_j)$, and the global information, e.g., $c_i(x_i)$. This lends itself well to control the composition of local and global information and, furthermore, a dynamic algorithm transition. In $\alpha\beta 4$, β value is initially set to 1. Since the last term in Eq. (5) (i.e., $(1-\beta)c_i(x_i)$) disappears when β is 1, there is no global information available. Therefore, the optimization process is initially dependent on local constraints $R_i(x_i)$ and $R_{ij}(x_i, y_j)$. The $\alpha\beta 4$ is initially equivalent to a local consistency algorithm. As the optimization process progresses, much global information has been derived from local consistency resolution. The local information has become less important. So β is gradually reduced to minimize the contribution of the local information and to maximize the effect of global information. When α is small and $\beta \rightarrow 0$, local information is ignored. Eventually $\alpha\beta 4$ becomes an unconstrained optimization algorithm.

5.2. A Minimax $\alpha\beta$ Relaxation Algorithm: $\alpha\beta 4$

A parallel, discrete $\alpha\beta$ relaxation algorithm, $\alpha\beta 4$, is shown in Figure 4 [5]. It was developed to solve the discrete COPs. In Figure 4, we use a concurrent *for* construct, i.e., ||**for**, to describe processes running concurrently. For example, the statement

||**for** i := 1 **to** m **do** $f(i)$

creates m tasks which execute concurrently (at least in theory) the ith task computing $f(i)$. The statement after ||**for** is executed only after all m tasks are completed.

In $\alpha\beta 1$ (Figure 2), the domain of the variables is real. In each iteration a dual step optimization is performed to optimize the objective function. With the discrete variable domain, the $\alpha\beta 4$ is restricted to perform local search. For each variable x_i, its local neighboring variables (i.e., $y_j \in N_i$) are optimized first. During each iteration, a periodical exchanging of local and global information (i.e., updating $c_i(x_i)$ functions and α) and a gradual transition from local to global optimization (i.e., updating β in (3), (4), and (5)) are performed.

Apparently, a number of the statements in $\alpha\beta 4$ can be evaluated in parallel, resulting in a number of parallel $\alpha\beta$ relaxation algorithms.

5.3. Time Complexity

We briefly estimate the run time of a sequential version and then a parallel version of the $\alpha\beta 4$ relaxation algorithm.

In the sequential $\alpha\beta 4$ algorithm, the initialization stage takes $O(n^2)$ time. The *while* loop takes k iterations to terminate where the number of k is dependent on

```
procedure αβ4 ()
begin
    /* initialization */
    given_a_COP_instance;
    α(0) := α_0;
    β(0) := 1;
    x⃗(0) := select_an_initial_point();
    ||for i := 1, 2, ..., n do
    begin
        c_i^(0)(0) := evaluate_local_info_function(x_0);
        f_i^(0)(0,0) := evaluate_local_objective_function(x_{i,0}, y⃗_{j,0});
        f(0,0) := evaluate_objective_function(x_{i,0}, y⃗_{i,0}, β_0);
    end;

    /* relaxation */
    k := 0;
    while f(.) > 0 do
    begin
        ||for each variable i := 1, 2, ..., n do
        begin
            ||for each x_i value ∈ D_i do
            begin
                ||for each local variable j := 1, 2, ...|N_i| do
                    y_j^* := max_{j∈N_i, y_j∈D_j} f_i(x_i, y_j)|_{x_i=c};
                c_i^(k+1)(x_i) := (1 − α) Σ_j w_ij c_j^(k)(y_j^*) + α f(x_i, y⃗_j^*);
            end;
            x_i^* := min_{x_i} f(x_i, y⃗_j^*);
        end;
        if |f(.)| < E then return solution and quit;
        reduce α;
        reduce β;
        k := k + 1;
    end;
end;
```

Fig. 4. $\alpha\beta4$: A parallel minimax $\alpha\beta$ relaxation algorithm.

the specific problem instance. There are three *for* loops inside the *while* loop. For n variables the first *for* loop takes $O(n)$ time. Assume the maximum number of values that a variable may be instantiated is $m = max\{|D_1|, |D_2|, ..., |D_n|\}$. The second *for* loop will take $O(m)$ time. The third *for* loop takes $O(n)$ time since, in a complete graph, a node may have $O(n)$ directly connected nodes. For each variable $y_j \in N_i$, the first step of minimax optimization (Eq. 2) will check m values in D_j to determine the maximum value which takes O(m) time. Updating $c_i(.)$ takes O(n) time. The second step of dual optimization (Eq. 6) is inside the first *for* loop which takes, for m values in D_i, O(m) time. Summarizing the above, the sequential $\alpha\beta4$ will take O(kn^2m^2) time. In practical applications, the maximum number of values, m, may be a small constant so the time complexity of $\alpha\beta4$ may be reduced to O(kn^2).

In the parallel $\alpha\beta4$ algorithm, the initialization stage takes O(1) time. The *while* loop takes k iterations to terminate. There are three *for* loops inside the *while* loop they each take O(1) time. The first step of minimax optimization (Eq. 2), which determines the maximum value of x_j, can be done by a parallel sorting in O($logm$) time. Similarly, the second step of dual step optimization (Eq. 6) will take O($logm$) time. Summarizing the above, in theory, the parallel $\alpha\beta4$ will take O($klog^2m$) time.

6. Experimental Results

We have implemented a number of $\alpha\beta4$ relaxation algorithms and have tested their performance with large number of simulated and practical COP problem instances [5, 6]. Figures 5, 6, and 7 show convergence performance of an $\alpha\beta4$ algorithm for three tested problem instances. For the same problem instances, we also compare the performance of the $\alpha\beta4$ with those of the local search and the simulated annealing algorithms.

The first problem instance is a COP with 10 variables. The domain size, i.e., the number of possible values assigned to the variables, varies from 70 to 130. The second COP problem instance has 20 variables and the domain size varies from 20 to 80. In these two COPs, the constraints among variable assignments are generated randomly and uniformly with a probability of 0.5. The third COP problem instance was drawn from a practical industrial object recognition problem, i.e., recovering object orientation from an image intensity photo [6]. Since there is only one correct object orientation (which corresponds to the global minimum of the objective function), the object recognition problem is itself NP-hard.

In these three COP problem instances, the initial objective value is the *sum* of all the unary and binary constraints among all the possible assignments to the variables. Our goal is to reduce the objective function to as close to zero as possible, producing a consistent (i.e., conflict free) value assignment to the variables.

The convergence profiles shown in Figures 5, 6, and 7 indicate that, compared to the local search and simulated annealing algorithms, the $\alpha\beta4$ algorithm converges to the minimum value of the objective function in much less number of iterations, for the same problem instances.

Acknowledgements

Ding-Zhu Du and Panos Pardalos provided important comments for an early version of this chapter. This work was supported in part by 1987 and 1988 ACM/IEEE Academic Scholarship Awards, NSERC Strategic Grant MEF0045793, and NSERC Research Grant OGP0046423 and is presently supported in part by NSERC Strategic Grant STR0167029 and the Federal Micronet Research Grant.

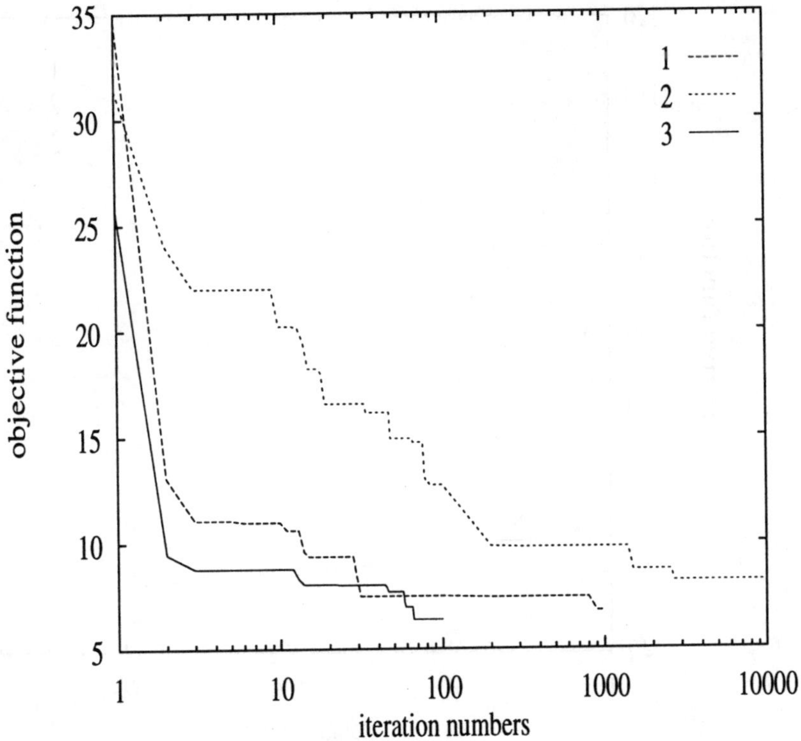

Fig. 5. Convergence performance of (1) local search, (2) simulated annealing, and (3) an $\alpha\beta 4$ algorithm.

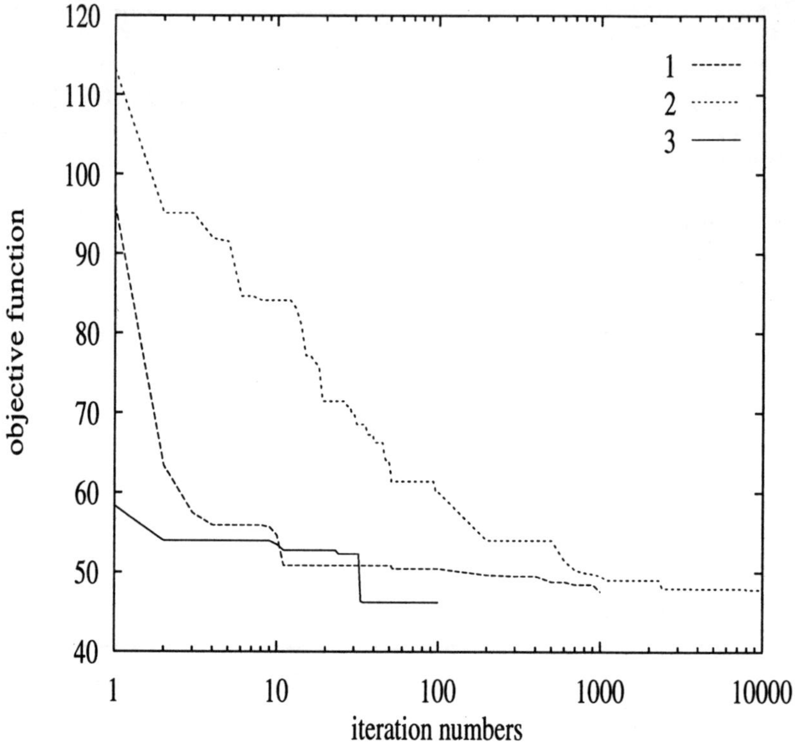

Fig. 6. Convergence performance of (1) local search, (2) simulated annealing, and (3) an $\alpha\beta 4$ algorithm.

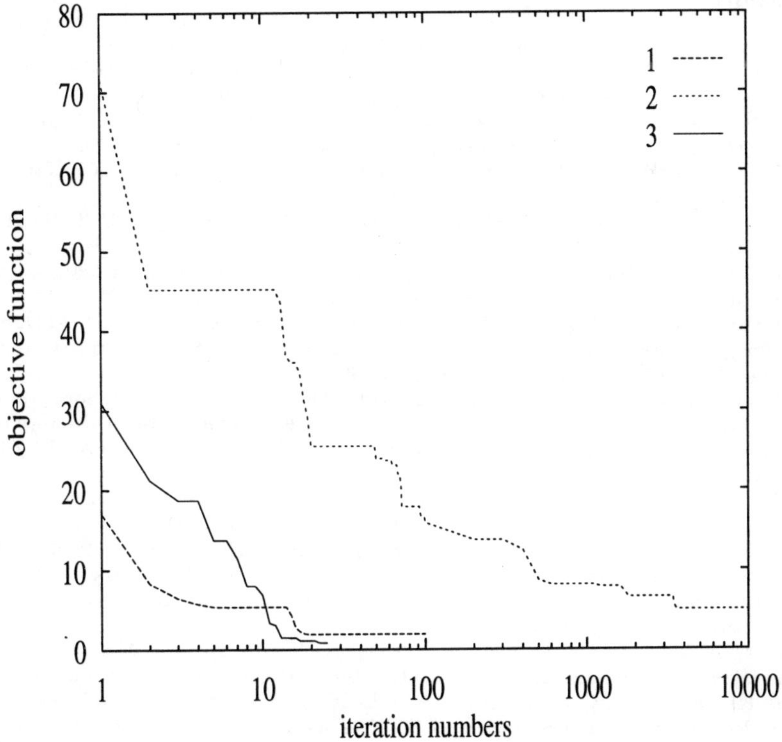

Fig. 7. Convergence performance of (1) local search, (2) simulated annealing, and (3) an $\alpha\beta4$ algorithm.

References

1. M.B. Clowes. On seeing things. *Artificial Intelligence*, 2:79–116, 1971.
2. S.A. Cook. The complexity of theorem-proving procedures. In *Proceedings of the Third ACM Symposium on Theory of Computing*, pages 151–158, 1971.
3. M.R. Garey and D.S. Johnson. *Computers and Intractability: A Guide to the Theory of NP-Completeness*. W.H. Freeman and Company, San Francisco, 1979.
4. J. Gu. Optimization by multispace search. Technical Report UCECE-TR-90-001, Dept. of Electrical and Computer Engineering, Univ. of Calgary, Jan. 1990.
5. J. Gu. An $\alpha\beta$-relaxation for global optimization. Technical Report UCECE-TR-91-003, Dept. of ECE, Univ. of Calgary, Apr. 1991.
6. J. Gu and X. Huang. A constraint network approach to a shape from shading analysis of a polyhedron. In *Proceedings of IJCNN'92*, pages 441–446, Beijing, Nov. 1992.
7. J. Gu. *Multispace Search: A New Optimization Approach (Summary). In* D.-Z. Du and X.-S. Zhang, editors, *Lecture Notes in Computer Science, Vol. 834: Algorithms and Computation*, pages 252–260. Springer-Verlag, Berlin, 1994.
8. J. Gu. *Constraint-Based Search*. Cambridge University Press, New York, 1995.
9. J. Gu, W. Wang, and T. C. Henderson. A parallel architecture for discrete relaxation algorithm. *IEEE Trans. on Pattern Analysis and Machine Intelligence*, PAMI-9(6):816–831, Nov. 1987.
10. J. Gu and W. Wang. A novel discrete relaxation architecture. *IEEE Trans. on Pattern Analysis and Machine Intelligence*, 14(8):857–865, Aug. 1992.
11. R. M. Haralick and L. G. Shapiro. The consistent labeling problem: Part 1. *IEEE Trans. on Pattern Analysis and Machine Intelligence*, PAMI-1(2):173–184, Apr. 1979.
12. B.K.P. Horn and B. G. Schunck. Determining optical flow. Technical Report AI Memo 572, AI Lab, MIT, Apr. 1978.
13. D.A. Huffman. *Impossible Objects as Nonsense Sentences. In* B. Meltzer and D. Michie, Eds., *Machine Intelligence*, pages 295–323. Edinburgh University Press, Edinburgh, Scotland, 1971.
14. A. K. Mackworth. Consistency in networks of relations. *Artificial Intelligence*, 8:99–119, 1977.
15. A. Rosenfeld, R. A. Hummel, and S. W. Zucker. Scene labeling by relaxation operations. *IEEE Trans. on Systems, Man, and Cybernetics*, SMC-6(6):420–433, June 1976.
16. R. V. Southwell. *Relaxation Methods in Engineering Science*. Oxford University Press, London, 1940.
17. R. V. Southwell. *Relaxation Methods in Theoretical Physics*. Oxford University Press, London, 1946.
18. D. Waltz. Generating semantic descriptions from drawings of scenes with shadows. Technical Report AI271, MIT, Nov. 1972.

MINIMAX PROBLEMS IN COMBINATORIAL OPTIMIZATION

FENG CAO, DING-ZHU DU, BIAO GAO and PENG-JUN WAN
Department of Computer Science, University of Minnesota, Minneapolis, MN 55455, U.S.A.

and

PANOS M. PARDALOS
Center for Applied Optimization and ISE Department, University of Florida, Gainesville, FL 32611, U.S.A.

1. Introduction

Classical minimax theory initiated by Von Neumann, together with duality and saddle point analysis, has played a critical role in optimization and game theory. However, minimax problems and techniques appear in a very wide area of disciplines. There are many interesting and sophisticated problems formulated as minimax problems. For example, many combinatorial optimization problems, including scheduling, location, allocation, packing, searching, and triangulation, can be represented as a minimax problem. Many of these problems have nothing to do with duality and saddle points, and they have not been considered in any general uniform treatment. Furthermore, many minimax problems have deep mathematical background, nice generalizations, and lead to new areas of research in combinatorial optimization. In this survey, we discuss a small but diverse collection of minimax problems, and we present some results (with a few key references) and open questions.

References

1. V.F. Demyanov and V.N. Malozemov, *Introduction to Minimax*, (Dover Publications, New York, 1974).
2. S. Gabler, *Minimax Solutions in Sampling from Finite Populations*, (Lecture notes in statistics, Vol. 64, Springer-Verlag, 1990).
3. R. Horst and P.M. Pardalos (Editors), *Handbook of Global Optimization*, (Nonconvex optimization and its applications ; vol. 2, Kluwer Academic Publishers, 1995).
4. A.P. Korostelev, *Minimax Theory of Image Reconstruction*, (Lecture notes in statistics, Springer-Verlag, vol. 82, 1993).
5. J. Von Neumann, *Collected works*, General editor, A. H. Taub, (6 vol., Oxford, New York, Pergamon Press, 1961-63)
6. J. Von Neumann and O. Morgenstern, *Theory of Games and Economic Behavior*, (New York, Science Editions, J. Wiley, 1964).
7. P.H. Rabinowitz, *Minimax Methods in Critical Point Theory With Applications to Differential Equations*, (Regional conference series in mathematics, no. 65, American Mathematical Society, 1986)
8. S. Trybula, *Some Investigations in Minimax Estimation Theory*, (Warszawa : Panstwowe Wydawn. Nauk., 1985).

D.-Z. Du and P. M. Pardalos (eds.), Minimax and Applications, 269–292.

9. A.G. Sukharev, *Minimax Models in the Theory of Numerical Methods*, (Dordrecht; Boston, Kluwer Academic Publishers, 1992).
10. M. Walk, *Theory of Duality in Mathematical Programming*, (Akademie-Verlag, Berlin, 1993).

2. Algorithmic Problems

The worst-case time complexity can be treated as a minimax problem of the following form:

$$\min_{M} \max_{x} \mathrm{TIME}_M(x)$$

where M belongs to a class of Turing machines and x denotes the input strings of length n. The problems in this section are basically having the same flavor, that is, the worst-case complexity is studied although it is measured in various different ways.

2.1. Group Testing

Combinatorial Group Testing: Consider a set of n items in which d items are *defective* and others are *good*. These items can identified by a sequence of tests. Each test is on a subset of items, which tells whether the subset contains a defective item or not. For an algorithm α, let $N_\alpha(s \mid d, n)$ be the number of tests that algorithm α spends on a sample s of n items with exactly d defective ones. The problem is to compute

$$M(d,n) = \min_{\alpha} \max_{s \in \mathcal{A}(d,n)} N_\alpha(s \mid d, n)$$

where $\mathcal{A}(d,n)$ is the set of samples of n items containing d defective items.

Group Testing was first proposed by R. Dorfman [1], however, in a different model. The above combinatorial version was first proposed by M. Sobel and P. A. Groll [4]. The computational complexity of combinatorial group testing is still an open problem [2]. It is well-known that $M(1,n) = \lceil \log_2 n \rceil$. But, for $d \geq 2$, it seems that $M(d,n)$ is very hard to determine. More information is contained in [3].

References

1. R. Dorfman, The detection of defective members of large populations, *Ann. Math. Statist.* 14 (1943) 436-440.
2. D.-Z. Du and Ker-I Ko, Some completeness results on decision trees and group testing, *SIAM Algebraic and Discrete Methods* 8 (1987) 762-777.
3. D.-Z. Du and F.K. Hwang, *Combinatorial Group Testing*, (World Scientific, Singapore, 1993).
4. M. Sobel and P.A. Groll, Group testing to eliminate efficiently all defectives from a binomial sample, *Bell System Tech. J.* 28 (1959) 1179-1252.

Symmetric Group Testing: This is a variation of the oraginary group testing problem. Given n items with d defectives, the problem is to identify them by a sequence of tests each of which is on a subset of items and has ternary outcomes; all good, all defective, and mixed (i.e., the subset contains at least one good and at least one defective item). This problem was first proposed by Sobel, Kummar, and Blumenthal [2] and extensively studied by Hwang [1].

References

1. F.K. Hwang, Three versions of a group testing game, *SIAM J. Alg. Disc. Methods* 5 (1984) 145-153.
2. M.S. Sobel, S. Kummar, and S. Blumenthal, Symmetric binomial group-testing with three outcomes, in *Statist. Decision Theory and Related Topics*, S.S. Gupta and J. Yackel (eds.), (Academic, 1971) 119-160.

Parametric Models: Suppose item i has a quality parameter q_i, where $q_i = 0$ if i is good and $q_i > 0$ if i is defective. For mathematical convenience, assume that all linear combinations of nonzero q_i are distinct. The feedback of the test on a subset G of items is the sum of q_i over all i in G. An item can be identified in many ways. For example, the following table shows how to identify all items in a sample of eight items with at most two defectives by using the feedback on the three tests on $\{1,2,3,4\}$, $\{1,2,5,6\}$, and $\{1,3,5,7\}$. (Assume the feedback from the test on the whole set is 1.)

defectives	1 2 3 4 5 6 7 8	1 1 1 1 1 1 1 2 2 2 2 2 2 3 3 3 3 3 4 4 4 4 5 5 5 6 6 7
		2 3 4 5 6 7 8 3 4 5 6 7 8 4 5 6 7 8 5 6 7 8 6 7 8 7 8 8
1234	1 1 1 1 0 0 0 0	1 1 1 a a a a 1 1 a a a a 1 a a a a b a a a 0 0 0 0 0 0
1256	1 1 0 0 1 1 0 0	1 a a 1 1 a a a a 1 1 a a 0 b b 0 0 a b 0 0 1 a a a a 0
1357	1 0 1 0 1 0 1 0	a 1 a 1 a 1 a b 0 b 0 b 0 a 1 a 1 a a 0 b 0 a 1 a b 0 a

where a and b are two unequal numbers between 0 and 1. Note that all feedback patterns are different. Also note that this model provides more information than the quantitative model. In [1], an algorithm for the above parametric model with up to two defectives was studied.

References

1. J.J. Metzner, Efficient replicated remote file comparison, *IEEE Trans Comput.* 40 (1991) 651-660.

Group Testing on Graphs:

Items are represented as edges of a graph G. Each test must be on an edge set of a subgraph. A conjecture on this model is that if there exists exactly one defective item, then the minimax number of tests to identify it is $\lceil \log_2 |E(G)| \rceil + 1$ where $|E(G)|$ is the number of edges in G [1, 3]. This conjecture has been proved for bipartite graph by Althöfer and Triesch [2]. For general graphs, they proved an upper bound $\lceil \log_2 |E(G)| \rceil + 3$.

References

1. M. Aigner, *Combinatorial Search*, (Wiley-Teubner, 1988).
2. I. Althöfer and E. Triesch, Edge search in graphs and hypergraphs of bounded rank, manuscript.
3. D.-Z. Du and F.K. Hwang, *Combinatorial Group Testing*, (World Scientific, Singapore, 1993).

2.2. Mutiaccess Channels

Multiaccess Channels: Consider a communication network where many users share a single multiaccess channel. A user is called an *active* user if he has a message to be transmitted. Assume that each message has unit length, which can be transmitted in one time slot. The transmission is done by broadcasting over the channel so that every user can receive. If at any time slot, more than one active users transmit their messages, then the messages conflict with each other and noise results. The problem is to find an algorithm which schedules the transmissions of active users into different time slots.

The multiaccess channel problem can be represented as a variation of group testing problem. Each test is done in one time slot, that is, the channel is open to a certain group of users at each time slot. The test has three outcomes: no active user, one active user, and more than one active user. They correspond to the channel receive no messages, one message, and noise. The optimal algorithm is to minimize the maximum number of time slots for every active user to transmit successfully.

Hayes [1] first proposed an algorithm using group query. Recent results can be found in [2, 3, 4].

References

1. J.F. Hayes, An adaptive technique for local distribution, *IEEE Transaction on Communication* 26 (1978) 1178-1186.
2. J.I. Capetanakis, Tree algorithms for packet broadcast channels, *IEEE Transactions on Information Theory* 25 (1979) 505-515.
3. B.S. Tsybakov and V.A. Mikhailov, Free synchronous packet access in a broadcast channel with feedback, *Probl. Inform. Transm.* 14 (1978) 259-280.
4. P.J. Wan and D.-Z. Du, An algorithm for the multi-access channel problem, TR 95-003, Department of Computer Science, University of Minnesota.

Multiaccess Channels with Additional Noise: This is a variation of the multiaccess channel problem. The situation is that there may be another source of noise so that the two cases, no transmission and more than one transmission, may result in the same outcome, noise. This variation was studied in [1].

References

1. T. Berger, N. Mehravari, D. Towsley, and J. Wolf, Random multiple-access communications and group testing, *IEEE Trans. Commun.* 32 (1984) 769-778.

The k-Channel: The k-channel is a generalization of multiaccess channel problem such that any fewer than k active users can transmit their messages successfully in one time slot. It was first studied by Tsybakov, Mikhailov, and Likhanov [3]. One may find more information in [2, 1].

References

1. X.M. Chang and F.K. Hwang, The minimax number of calls for finite population multi-access channels, in *Computer Networking and Performance Evaluation*, T. Hasegawa, H. Takagi,

and Y. Takahashi (eds.), (Elsevier, Amsterdam, 1986) 381-388.

2. R.W. Chen and F.K. Hwang, K-definite group testing and its application to polling in computer networks, *Congress Numerantium* 47 (1985) 145-159.
3. B.S. Tsybakov, V.A. Mikhailov, and N.B. Likhanov, Bounds for packet transmissions rate in a random-multiple-access system, *Prob. Inform. Transm.* 19 (1983) 61-81.

Quantitative Channels: In this model, the successful transmission is still for one transmission per time slot. However, the feedback from each time slot is the exact number of active users in the opening group [1].

References

1. B.S. Tsybakov, Resolution of a conflict with known multiplicity, *Probl. Inform. Transm.* 16 (1980) 65-79.

2.3. MICELLANEOUS

Ulam's Problem: Stanislaw M. Ulam (1909-1984) is one of the great mathematicians in the twentieth century. In his autobiography [3], he wrote the following.

" Someone thinks of a number between one and one million (which is just less than 2^{20}). Another person is allowed to ask up to twenty questions, to each of which the first person is supposed to answer only yes or no. Obviously the number can be guessed by asking first: Is the number in the first half-million? and then again reduce the reservoir of numbers in the next question by one-half, and so on. Finally the number is obtained in less than $\log_2(1,000,000)$ questions. Now suppose one was allowed to lie once or twice, then how many questions would one need to get the right answer? One clearly needs more than n questions for guessing one of the 2^n objects because one does not know when the lie was told. This problem is not solved in general."

Ulam's problem is a group testing problem with one defective and at most one or two erroneous tests. In general, more defectives and more errors may be considered. Suppose there are n items with d defectives and at most r errors are allowed. For any algorithm α identifying all defectives with such unreliable tests, let $N^r_\alpha(\sigma \mid d,n)$ denote the number of unreliable tests performed by α on sample σ and let

$$M^r_\alpha(d,n) = \max_{\sigma\in S(d,n)} N^r_\alpha(\sigma \mid d,n)$$
$$M^r(d,n) = \min_\alpha M^r_\alpha(d,n).$$

Ulam's problem is equivalent to finding $M^1(1,10^6)$ and $M^2(1,10^6)$. Pelc [2] determined $M^1(1,10^6) = 25$. Czyzowicz, Mundici, and Pelc [1] determined $M^2(1,2^m) = 29$.

References

1. J. Czyzowicz, D. Mundici, and A. Pelc, Solution of Ulam's problem on binary search with two lies, *J. Combin. Theory* A49 (1988) 384-388.
2. A. Pelc, Solution of Ulam's problem on searching with a lie, *J. Combin. Theory* A44 (1987) 129-140.

3. S.M. Ulam, *Adventures of a Mathematician*, (Scribner's New York 1976).

Counterfeit Coins: Consider a set of coins where each coin is either of the heavy type or the light type. The problem is to identify the type of each coin with minimal number of weighings on a balanced scale. The case that only one coin, called a counterfeit, has a different weight from others, is a classic mathematic puzzle [2]. Later works study the case of more than one counterfeits [1].

References

1. X.-D. Hu and F.K. Hwang, A competitive algorithm for the counterfeit coin problem, in this book.
2. S.S. Cairns, Balance scale sorting, *Amer. Math. Monthly* 70 (1963) 136-148.

Alphabetic Minimax Trees: Given vertices $v_1, ..., v_n$, with weights $w_1, ..., w_n$, construct a t-ary tree with leaves $v_1, ..., v_n$ in left to right order, such that if l_i denotes the length of the path from v_i to the root for each i, the maximum of $w_i + l_i$ is minimized. For the case where all the weights are integers, the problem can be solved in $O(n \log n)$ time [1].

References

1. D.G. Kirkpatrick and M.M. Klawe, Alphabetic minimax trees, *SIAM Journal of Computing*, vol. 14, No. 3, 1985, 514-526.

3. Geometric Problems

3.1. Packing and Spreading

The equivalence of the following two problems may explain the relationship between packing and spreading. It also explains why some packing problem can be formulated as a minimax problem.

Packing Circles in a Square: The problem is to determine the maximal radius of n equal circles that can be packed into a unit square.

Spreading Points in a Square: How should n points be arranged into a unit square such that the minimal distance between them is maximized?

The following problems are either transformed from packing or spreading with various goal functions.

Spreading Points in a Circle: How large can the least distance between a pair chosen from n points in the unit circle be? The exact value has been determined for $n \leq 10$. For $n \geq 11$, the value is unknown [1].

References

1. H.T. Croft, K.J. Falconer, and R.K. Guy, *Unsolved Problems in Geometry*, (Springer-Verlag, New York, 1991).
2. D.-Z. Du and P.M. Pardalos, A Minimax Approach and Its Applications, *TR 92-51*, 1992, Univ. Of Minnesota.
3. C. Maranas, C.A. Floudas and P.M. Pardalos, New results in the packing of equal circles in a square, Discrete Mathematics (1995).

Heilbronn Numbers: Given a planar convex body K (i.e. a compact convex set with non-empty interior) in the plane, find n points $r_1, r_2, \cdots, r_n$ inside K, to achieve the Heilbronn number which is defined by

$$H_n(K) := \max_{\{(r_1,r_2,\cdots,r_n)|r_i\in K, 1\le i\le n\}} \min_{1\le i<j<k\le n} \frac{area(r_ir_jr_k)}{area(K)}$$

So far there is no result about the exact lower bound of $H_n(K)$ for general K, even for $n = 4, 5$. For the upper bound, it was proved that $H_6(K) \le \frac{1}{6}$ and in [1] and $H_7(K) \le \frac{1}{9}$ and in [3]. Some other results consider some special cases of K, such as triangle[4], square[5] and disk[2]. However the exact values for only small n are known.

References

1. A. Dress, L. Yang and Z.B. Zeng, Heinlbronn problem for six points in a planar convex body, in this book.
2. K.F. Roth, Estimate of the area of the smallest triangle obtained by selecting three out of n points in a disc of unit area, *Amer. Math. Soc. Proc. Symp. Pure Math.*, 24: 251-262, 1973.
3. L. Yang and Z.B. Zeng, Heinlbronn problem for seven points in a planar convex body, in this book.
4. L. Yang, J.Z. Zhang and Z.B. Zeng, On exact values of Heinlbronn numbers in triangular regions, Preprint.
5. L. Yang, J.Z. Zhang and Z.B. Zeng, On Goldberg's conjecture: computing the first several Heinlbronn numbers, Preprint, Univ. Bielefeld, 1991, ZiF-Nr. 91/29, SFB-Nr. 91/074. seven points in a planar convex body, in this book.

3.2. TRIANGULATION

Given a set P of n points in the Euclidean plane, a *triangulation* $T(P)$ is a collection of non-intersecting line segments, each joining two points in the point set, that divides the interior of the convex hull into triangles. In the following problem, various minimax goal functions are considered.

Minimax Edge Length Triangulation: The problem is to minimize the maximum edge length in a triangulation, that is, to compute

$$\min_{T(P)} \max_{e\in Edge(T(P))} length(e).$$

where $Edge(T(P))$ consists of all three edges of all triangles in $T(P)$.

The problem can be solved in $O(n^2)$ time [1].

References

1. H. Edelsbrunner and T.S. Tan, A quadratic time algorithm for the minmax length triangulation, *Proceedings of 32nd Annual Symposium on Foundations of Computer Science*, (1991, San Jun, Purrto Rico) 414-423.

Maximin Angle Triangulation: The problem is to maximize the minimum angle in a triangulation, that is, to compute

$$\max_{T(P)} \min_{\alpha \in Angle(T(P))} \alpha.$$

where $Angle(T(P))$ consists of all three interior angles of all triangles in $T(P)$.

It is known that the Delaunay triangulation maximizes the minimum angle over all triangulations of the same pint set [5]. This result can be extended to a similar statement about the sorted angle vector of the Delaunay triangulation [2] and to the constrained case [2]. The Delaunay triangulation of n points in the plane can be constructed in time $O(n \log n)$ [3, 2], and even if some edges are prescribed its constrained version can be constructed in the same amount of time [4].

References

1. H. Edelsbrunner, *Algorithms in Combinatorial Geometry.* Springer-Verlag, Heidelberg, Germany, 1987.
2. D.T. Lee and A.K. Lin, Gernalized Delaunay triangulations for planar graphs. *Discrete Comput. Geom.* 1 (1986), 161-194.
3. F.P. Preparata and M.L. Shamos, *Computational Geometry - an Introduction.* Springer-Verlag, New York, 1985.
4. R. Siedel, Constrained Delaunay triangulations and Voronoi diagrams with obstacles. In *"1978-1988, 10-Years IIG"* a report of the Inst. Informat. Process¿, Techn. Univ. Graz, Austria, 1988, 178-191.
5. R. Sibson, Locally equiangular triangulation, *Comput. J.* 21 (1978) 243-245.

Minimax Angle Triangulation: The problem is to minimize the maximum angle in a triangulation, that is, to compute

$$\min_{T(P)} \max_{\alpha \in Angle(T(P))} \alpha.$$

where $Angle(T(P))$ consists of all three interior angles of all triangles in $T(P)$.

This problem, with or without prescribed edges, can be solved in time $O(n \log n)$ and space $O(n)$ in [1].

References

1. H. Edelsbrunner, T.S. Tan, and R. Waupotitsch, An $O(n^2 \log n)$ time algorithm for the minmax angle triangulation, *Proc. 6th Ann. Sympos. Compu. Geom.* 1990, 44-52.

Minimax Smallest Circumscibing Circle Triangulation: The problem is to minimize the maximum triangle circumradii (the radius of the smallest circle containing the triangle) in a triangulation, that is, to compute

$$\min_{T(P)} \max_{\delta \in T(P)} circumradii(\delta).$$

It is known [1, 2, 6] that the Delaunay triangulation of a set of points minimizes the maximum circumradii. Optimal $O(n \log n)$ time divide-and-conquer and plane sweep algorithms are known and elegant data structures to support their implementations exist [3, 4, 6].

Rajan [5] generalizes this result to d-dimensional Euclidean space R^d and shows that the Delaunay triangulation of a set of points in R^d minimizes the maximum circumradii. He also gives an incremental algorithm for the Delaunay triangulation with time complexity $O(n^{\lceil \frac{d+1}{2} \rceil})$.

References

1. F. Aurenhammer, *Voronoi diagrams -survey.* Inst. for Inf. Proc., Graz Tech. Univ. report, (1989).
2. H. Edelsbrunner, *Algorithms in Combinatorial Geometry.* Springer-Verlag, Heidelberg, Germany, 1987.
   Voronoi diagrams
3. S. Fortune, A sweepline algorithm for Voronoi diagrams. *Algorithmica,* 2(2) (1987), 153-174.
4. L. Guibas and J. Stolfi, Primitives form the manipulation of general subdivisions and the computations of Voronoi diagrams, *ACM Trans. Graphics* 4 (1985) 74-123.
5. V.T. Rajan, Optimality of the Delaunay triangulation in R^d, *Proceedings of 7th Ann. Sympos. Comput. Geom.,* 1991, 357-363.
6. M.I. Shamos and D. Hoey, Closest-point problems, em Proc. 16th Ann. IEEE Sypm. on Found. of Comp. Sci., (1975), 151-162.

3.3. Clustering

Euclidean Central Clustering: Given a set S of points on the Euclidean plane, find a partition $S = S_1 \cup S_2 \cup \cdots \cup S_k$ to achieve

$$\min_{S=S_1 \cup S_2 \cup \cdots \cup S_k} \max_{1 \le i \le k} radius(S_i).$$

The problem is NP-hard. Feder and Greene [1] shows that it is NP-hard to approximate the Euclidean central clustering with an approximation ratio smaller than 1.822. Gonzalez [2] proposed a polynomial-time approximation algorithm working in any metric space, called *farthest-point clustering,* which has approximation ratio 2.

References

1. T. Feder and D.H. Greene. Optimal algorithms for approximate clustering. *Proc. 20th ACM Symp. Theory of Computing,* 1988, 434-444.
2. T. Gonzalez. Clustering to minimize the maximum intercluster distances. *Therotical Computer Sciecne,* 38:293-306, 1985.

Euclidean Pairwise Clustering: Given a set S of points on the Euclidean plane, find partition $S = S_1 \cup S_2 \cup \cdots \cup S_k$ to achieve

$$\min_{S=S_1 \cup S_2 \cup \cdots \cup S_k} \max_{1 \le i \le k} diameter(S_i).$$

The problem is NP-hard. Feder and Greene [1] shows that it is NP-hard to approximate the Euclidean pairwise clustering with an approximation ratio smaller than

1.969. Gonzalez [2] proposed a polynomial-time approximation algorithm working in any metric space, called *farthest-point clustering*, which has approximation ratio 2.

References

1. T. Feder and D.H. Greene. Optimal algorithms for approximate clustering. *Proc. 20th ACM Symp. Theory of Computing*, 1988, 434-444.
2. T. Gonzalez. Clustering to minimize the maximum intercluster distances. *Therotical Computer Sciecne*, 38:293-306, 1985.

Euclidean Minimum Radius k-point Clustering: Given a set S of points on the Euclidean plane, seek the best single cluster C containing k points of S such that the radius of C achieves the minimum over all k-point clusters. The problem can be represented as a minimax problem in the following way:

$$\min_{C\subset S,|C|=k} \min_{c\in C} \max_{x\in C,x\neq c} d(x,c).$$

This problem is polynomially solvable [1, 2].

References

1. A. Datta, H.P. Lenhof, C. Schwarz, and M.Smid, Static and dynamic algorithms for $k$point clustering problems. *Proc. 3rd Workshop on Algorithms and Data Structures,* 265-276, LNCS 709, Springer-Verlag, 1993.
2. D. Eppstein and J. Erikson, Iterated nearest neighbors and finding minimal polytopes, *Disc. and Comp. Geometry,* 11:321-350, 1994.

Euclidean Minimum Diameter k-point Clustering: Given a set S of points on the Euclidean plane, seek the best single cluster C containing k points of S such that the diameter of C achieves the minimum over all k-point clusters. The problem can be represented as a minimax problem in the following way:

$$\min_{C\subseteq S,|C|=k} \max_{x,y\in C} d(x,y).$$

This problem is polynomially solvable [1, 2].

References

1. A. Datta, H.P. Lenhof, C. Schwarz, and M.Smid, Static and dynamic algorithms for $k$point clustering problems. *Proc. 3rd Workshop on Algorithms and Data Structures,* 265-276, LNCS 709, Springer-Verlag, 1993.
2. D. Eppstein and J. Erikson, Iterated nearest neighbors and finding minimal polytopes, *Disc. and Comp. Geometry,* 11:321-350, 1994.

p-center problem: Given n demand points on the plane, the p-center problem is the problem of locating p points, called *centers* on the plane so that the maximum distance from any demand point to its respective closest center is minimal. This is an *NP-Complete* problem for both the Euclidean and rectilinear metrics [4, 5]. Ko et. al. [3] present an $O(n^{p-2}\log e)$ algorithm for the rectilinear case. The analogous problem on networks (using network distances) is also *NP-Hard* [2]. For the practical solution of large-scale problems, Francis and Lowe [3] have addressed the problem of aggregation for location problems, including the p-center problem analytically; they showed that doing aggregation well is provably difficult.

References

1. R. L. Francis and T. J. Lowe, "On Worst-Case Aggregation Analysis for Network Location Problems," *Annals of Operations Research* (1992) Vol. 40, 229-246.
2. O. Kariv and S. L. Hakimi, "An Algorithmic Approach to Network Location Problems. Part 1: The p-Centers," *SIAM J. Appl. Math.* (1979) Vol. 37, 513-538.
3. M. T. Ko, R. C. T. Lee and J. S. Chang, "Rectilinear m-Center Problem," *Naval Research Logistics* (1990) Vol. 37, 419-427.
4. S. Masuyama, T. Ibaraki and T. Hasegawa, "The Computational Complexity of the m-Center Problems on the Plane", *The Transactions of the IECE of Japan* (1981) Vol E 64, No. 2, 57-64.
5. N. Meggido, and K. Supowit, "On the Complexity of Some Common Geometric Location Problems," *SIAM J. Computing* (1984) Vol. 13, 182-196.

Bounded Fan-Out Minimum-Radius Partition: Given a set $Q = \{q_i \mid 1 \leq i \leq k\}$ of k points in the plane, called the clock driver vertices, a set $P = \{p_i \mid 1 \leq i \leq n\}$ of n points, called the clock pin vertices, and a constant $B > 0$ for fan-out limit with $kB \geq n$, the problem is to find an optimal partition that divides the set P into k disjoint subsets $P_1, \cdots, P_k$ such that $\cup_{i=1}^k P_i = P$, $|P_i| \leq B$ for all $1 \leq i \leq k$, and the following value is achieved

$$\min_{(P_1,\cdots,P_k)} \max_{1\leq i\leq k} \max_{p\in P_i} d(q_i, p)$$

where $d(q_i, p)$ is the distance between two points q_i and p.

The problem can be solved in $O(n^{2.5}\sqrt{\log n})$ time [1].

References

1. Jan-ming Ho and Ren-Song Tsay, Clock tree regeneration, *IEEE International Conference on Computer-Aided Desigh, Degist of Technical Papers*, 1992, pp. 198-203.

3.4. Network Designs

Maximum Remote Minimum Spanning Tree: Given a set V of n points in the Euclidean plane and an integer $k \leq n$, find a subset $S \subset V$ of size k such that the minimal spanning tree for S achieves the maximum over all minimal spanning trees of all subsets of size k.

It is unknown whether this problem is NP-complete. However it is polynomial-time approximable within a factor of 2.252 from optimal [1].

References

1. M.M. Halldorsson, K. Iwano, N. Katoh and T. Tokuyama, Finding Subsets Maximizing Minimal Structures, *Proc. 6th Ann. ACM-SIAM Symp. on Disc. Algo.*, ACM-SIAM, 1995, to appear.

Maximum Remote Minimum Steiner Tree: Given a set V of n points in the Euclidean plane and an integer $k \leq n$, find a subset $S \subset V$ of size k such that the minimal Steiner tree for S achieves the maximum over all minimal Steiner trees of subsets of size k.

This problem is NP-complete and polynomial-time approximable within a factor of 2.16 from optimal [1].

References

1. M.M. Halldorsson, K. Iwano, N. Katoh and T. Tokuyama, Finding Subsets Maximizing Minimal Structures, *Proc. 6th Ann. ACM-SIAM Symp. on Disc. Algo.*, ACM-SIAM, 1995,to appear.

Maximum Degree of Minimum-Degree Minimum Spanning Trees: Given a metric space and a point set S, a minimum-degree minimal spanning tree of S is minimal spanning tree of S whose degree is the smallest among all the minimal spanning trees of S. The problem is how large can the degree of the minimum-degree minimal spanning tree over any point set be in a given metric space, that is to compute

$$\max_{|S|<\infty} \min_{T \in MST(S)} degree(T)$$

where $MST(S)$ is the set of minimal spanning trees of S.

It is shown in [1] that

1. In the rectilinear plane, the maximum degree of any minimum-degree MST over any finite point set is 4.
2. In the 3-dimensional rectilinear space, the maximum degree of any minimum-degree MST over any finite point set is either 13 or 14.
3. In the d-dimensional L_∞ space, the maximum degree of any minimum-degree MST over any finite point set is 2^d.
4. In the d-dimensional L_p space, the maximum degree of any minimum-degree MST over any finite point set is at least $\Omega(\sqrt{d}2^{d(1-E(\alpha))})$, where $\alpha = \frac{1}{2^p}$ and $E(x) = x \lg \frac{1}{x} + (1-x) \lg \frac{1}{1-x}$.
5. For each d, there is a p such that in the d-dimensional L_p space, the maximum degree of any minimum-degree MST over any finite point set exceeds 2^d.

References

1. G. Robins and J.S. Salowe, On the maximum degree of minimum spanning trees, *Proceedings of the 10th Annual Symposium on Computational Geometry*, 1994, 250-258.

4. Graph Problems

4.1. Diameter, and Path

Weighted Diameter: Consider a graph $G = (V, E)$ and a collection C of not necessarily distinct, $|E|$ nonnegative numbers. The problem is to find an optimal assignment for those numbers to edges to minimize the diameter of the resulting weighted graph. Since the diameter is the maximum distance between every two vertices, the problem can be represented as

$$\min_{c:E\to C} \max_{x,y\in V} d_c(x, y)$$

where c is a one-to-one mapping from E to C and $d_c(x, y)$ is the distance between vertices x and y under assignment c. This problem is NP-hard proved by Y. Perl and S. Zaks [1].

References

1. M.R. Garey and D.S. Johnson, *Computers and Intractability, A Guide to the Theory of NP-Completeness*, (Freeman, 1978).

Oriented Diameter: Given a graph, assign to each edge a direction to minimize the diameter of the resulting directed graph. This problem can be interpreted as a minimax problem in the same way as above. Its NP-hardness was established by Chvátal and Thomassen [1].

References

1. V. Chvátal and G. Thomassen, Distances in orientations of graphs, *Journal of Combinatorial Theory Ser. B* 24 (1978) 61-75.

The Route Covering Number of a Graph: Suppose $G = (V, E)$ is a connected graph on n vertices. For a permutation π, we consider a *route set* P, which is just some set of paths P_i joining each vertex v_i to its destination vertex $\pi(v_i)$, for $i = 1, ..., n$. For each edge e of G, we consider the number $rc(e, G, \pi, P)$ of paths P_i in P which contain e. The *route covering number* $rc(G)$ of G is defined as

$$\max_{\pi} \min_{P} \max_{e\in E} rc(e, G, \pi, P)$$

In other words, for each permutation, we want to choose the route set so that the maximum number of occurrences of any edge in the paths of the route set is minimized. The problem was proposed by N. Alon, F.R.K.Chung and R.L. Graham [1], they also prove that for the n-cube Q^n, $2 \leq rc(Q^n) \leq 4$. However, the problem of determining the exact value of $rc(Q^n)$ for general n remains unresolved.

References

1. N. Alon, F.R.K. Chung and R.L. Graham, Routing permutations on graphs via matchings, *SIAM Journal of Discrete Mathematics, vol. 7, No. 3* , 1994, 513-530.

Forwarding Index Problem: A routing R of a graph G of order n is a set of $n(n-1)$ elementary paths specified for all ordered pairs of vertices of G. The load of a vertex v for a given routing R of G, denoted by $\xi_G(R, v)$, is the number of paths P of R passing through v, such that v is not an end vertex of P.The problem is to compute the forwarding index of G which is defined to be

$$\min_{R} \max_{v \in V(G)} \xi_G(R, V)$$

The problem was raised by F.R.K. Chung, E. Coffman, M. Reiman and B. Simon [1] and was proved to be NP-hard by R. Saad [2].

References

1. F.R.K.Chung, E. Coffman, M. Reiman and B. Simon, The forwarding index of communication networks, *IEEE Trans. Inform. Theory*, 33, (1987), 224-232.
2. R. Saad, Complexity of the forwarding index problem, *SIAM Journal of Discrete Mathematics, vol. 6 No. 3*, 1993, 418-427.

The Cutwidth and Bandwidth of Graphs: Suppose G is a graph with vertex set $V(G)$ and edge set $E(G)$. A numbering π of G is a one-to-one mapping from $V(G)$ to the set of positive integers. The bandwidth problem of G is to compute

$$b(G) = \min_{\pi} \max_{\{u,v\} \in E(G)} \{|\pi(u) - \pi(v)|\}$$

The cutwidth problem of G is to compute

$$f(G) = \min_{\pi} \max_{i} |\{\{u, v\} \in E(G) \mid \pi(u) \leq i < \pi(v)\}|$$

These problems are related to the quadratic assignment problem [1]. The bandwidth problem for graphs is known to be NP-hard [1, 3], as is the bandwidth problem for trees [4]. The cutwidth problem for graphs is also NP-hard [5].

References

1. P.M. Pardalos and H. Wolkowicz (Editors), *Quadratic Assignment and Related Problems*, DIMACS Series Vol. 16, American Mathematical Society (1994).
2. M.R. Garey and D.S. Johnson, *Computers and Intractability, A Guide to the Theory of NP-Completeness*, (Freeman, 1978).
3. C.H. Papadimitriou, The NP-completeness of the bandwidth minimization problem, *SIAM Journal of Computing*, 16, (1976), 263-270.
4. M.R Garey, R.L. Graham, D.S. Johnson and D.E. Knuth, Complexity results for bandwidth minimization, *SIAM Journal of Applied Mathematics*, 34, (1978), 477-495.
5. M.R. Garey, D.S. Johnson and L. Stockmeyer, Some simplified NP-complete graph problems, *Theoretical Computer Science*, 1, (1976), 237-267.

4.2. Location and Partition

Min-max Multicenter: Given a weighted graph, find k vertices as centers to minimize the maximum distance from the centers to all vertices, that is,

$$\min_{v_1,\ldots,v_k \in V} \max_{v \in V} \min_{1 \leq i \leq k} d(v, v_i)$$

where $d(v, v_i)$ is the distance between vertices v and v_i.

References

1. M.R. Garey and D.S. Johnson, *Computers and Intractability, A Guide to the Theory of NP-Completeness*, (Freeman, 1978).

The p-Center Problem: Let $G = (V, E)$ be an undirected graph with node set $V = \{v_1, ..., v_n\}$, and edge set E with $|E| = m$. Assume that each node v in V is associated with a positive weight w_v. Each edge has a positive length and is assumed to be rectifiable. We refer to interior points on an edge by their distances (along the edge) from the two nodes of the edge. Let $A(G)$ denote the continuum set of points on the edges of G. The edge lengths induce a distance function on $A(G)$. For any x, y in $A(G)$, $d(x, y)$ will denote the length of the shortest path connecting x and y. Let $X = \{x_1, ..., x_p\}$ be a finite subset of points in $A(G)$. The p-center problem is to compute

$$\min_{X \subseteq A(G), |X|=p} \max_{v \in V} w_v d(v, X)$$

where $d(v, X) = \min_{x \in X} d(v, x)$.

When p is arbitrary, the problem is NP-hard [1, 2]. For fixed p, the problem can be solved in $O(m^p n^p \log^2 n)$ and $O(m^p n^p \log n)$ for the weighted and unweighted cases, respectively [3].

References

1. O. Kariv and S.L. Hakimi, An algorithmic approach to network location problems, I: The p-centers, *SIAM Journal of Applied Mathematics*, 37, 1979, 513-537.
2. Meggido, N. and K. Supowit, On the complexity of some common geometric location problems, SIAM J. Computing (1984) Vol. 13, 182-196.
3. Francis, R. L. and T. J. Lowe, On worst-case aggregation analysis for network location problems, Annals of Operations Research (1992) Vol. 40, 229-246.
4. A. Tamir, Improved complexity bounds for center location problems on networks by using dynamic data structures, *SIAM Journal of Discrete Mathematics*, vol. 1, No. 3, 1988, 377-396.

The p-Maximin Problem: The notation is the same as above, except that we have a set of weights, α_{ij}, for $1 \leq i \leq n$, $1 \leq j \leq p$, and β_{ij}, for $1 \leq i \neq j \leq p$. The p-maximin is defined as

$$\max_{X \subseteq A(G), |X|=p} \min \left\{ \begin{array}{l} \min\{\alpha_{ij} d(v_i, x_j) \mid 1 \leq i \leq n, 1 \leq j \leq p\}, \\ \min\{\beta_{ij} d(x_i, x_j) \mid 1 \leq i \neq j \leq p\} \end{array} \right\}$$

The problem was motivated from the problem of locating undesirable facilities [1, 2]. Even for the homogeneous case, where $\alpha_{ij} = \infty, 1 \leq i \leq n, 1 \leq j \leq p$, and $\beta_{ij} = 1, 1 \leq i \neq j \leq p$, the problem of finding a $\frac{2}{3}$-approximation solution is NP-hard, and there is a polynomial-time approximation algorithm with performance ratio of $\frac{1}{2}$ [3].

References

1. E. Erkut and S. Neuman, Analytical models for locating undesirable facilities, *European Journal of Operations Research*, 40 (1989), 275-291.

2. D. Moon and S.S. Chaudhry, An analysis of network location problems with distance constraints, *Management Science*, 30 (1984), 290-397.
3. A. Tamir, Obnoxious Facility location on graphs, *SIAM Journal of Discrete Mathematics*, vol. 4, No. 4, 1991, 550-567.

Tree-Decomposition and Treewidth: Let $G = (V, E)$ be a graph. A tree-decomposition D of G is a pair $(\{X_i \mid i \in I\}, T = (I, F))$, with $\{X_i \mid i \in I\}$ a family of subsets of V, and T a tree with $V(T) = I$ such that:

1. $\cup_{i \in I} X_i = V$;
2. $\forall (u, v) \in E, \exists i \in I, (u \in X_i) \wedge (v \in X_i)$;
3. For $i, j, k \in I$, if j lies on the path of T from i to k, then $X_i \cap X_k \subseteq X_j$.

The treewidth of a tree-decomposition $D = (\{X_i \mid i \in I\}, T = (I, F))$ is $\max_{i \in I}(|X_i| - 1)$. The problem is to compute

$$\min_D \max_{i \in I}(|X_i| - 1)$$

The problem was introduced by Robertson and Seymour [1]. Computing the treewidth and the corresponding tree-decomposition is known to be NP-hard for general graphs [2]. For fixed k, the problem of determining whether the treewidth of a given graph is at most k and building the corresponding tree-decomposition can be solved in $O(n \log n)$ time [3, 4].

References

1. N. Robertson and P. Seymour, Graph minors II. Algorithmic aspects of treewidth, *Journal of Algorithms*, 7 (1986), 309-322.
2. S. Arnborg, D. G. Corneil and A. Proskurowski, Complexity of finding embeddings in a k-tree, *SIAM Journal of Algebraic Discrete Methods*, 8 (1987), 277-284.
3. B. Reed, Finding approximate separators and computing treewidth quickly, *Proceedings of 24th Annual Symposium on Theory of Computing*, 1992, 221-228.
4. H. L. Bodlaender and T. Kloks, Better algorithms for the pathwidth and treewidth of graphs, *Proceedings of 18th International Collquium on Automata, Languages and Programming, Lecture Notes in Computer Science, vol. 510, Springer-Verlag, Berlin, New York*, 1988, 544-555.

Path-Decomposition and Pathwidth: Let $G = (V, E)$ be a graph. A path-decomposition D of G is a pair $(\{X_i \mid i \in I\}, I)$ with $\{X_i \mid i \in I\}$ a family of subsets of V, and $I = \{1, 2, ..., r\}$, such that

1. $\cup_{i \in I} X_i = V$;
2. $\forall (u, v) \in E, \exists i \in I, (u \in X_i) \wedge (v \in X_i)$;
3. For all i, j, k, with $1 \leq i < j < k \leq r$, $X_i \cap X_k \subseteq X_j$.

The pathwidth of a path-decomposition $D = (\{X_i \mid i \in I\}, I)$ is $\max_{i \in I} |X_i|$. The problem is to compute

$$\min_D \max_{i \in I} |X_i|$$

The problem was introduced by Robertson and Seymour [1]. Determining the pathwidth of a given graph is NP-hard [2]. For fixed k, the problem of determining whether the pathwidth of a given graph is at most k and building the corresponding path-decomposition can be solved in $O(n \log n)$ time [3, 4].

References

1. N. Robertson and P. Seymour, Graph minors II. Algorithmic aspects of treewidth, *Journal of Algorithms*, 7 (1986), 309-322.
2. S. Arnborg, D. G. Corneil and A. Proskurowski, Complexity of finding embeddings in a k-tree, *SIAM Journal of Algebraic Discrete Methods*, 8 (1987), 277-284.
3. B. Reed, Finding approximate separators and computing treewidth quickly, *Proceedings of 24th Annual Symposium on Theory of Computing*, 1992, 221-228.
4. H. L. Bodlaender and T. Kloks, Better algorithms for the pathwidth and treewidth of graphs, *Proceedings of 18th International Collquium on Automata, Languages and Programming, Lecture Notes in Computer Science, vol. 510, Springer-Verlag, Berlin, New York*, 1988, 544-555.

T-Colorings of Graphs: Given a finite set T of positive integers containing $\{0\}$, a T-coloring of a simple graph is a nonnegative integer function f_T defined on the vertex set of G, such that if $\{u, v\} \in E(G)$, then $|f_T(u) - f_T(v)| \notin T$. The T-span of a T-coloring f_T is defined as the difference of the largest and smallest colors used. The problem is to compute

$$\min_{f_T} \max_{\{u,v\} \subseteq V(G)} \{|f_T(u) - f_T(v)|\}$$

T-coloring was first introduced by W.K. Hale [1], who formulated the most general form of the channel assignment in graph-theoretic terms. A survey can be found in [3]. The problem of finding the T-coloring with minimum span is NP-hard by a trivial reduction from k-colorability problem by setting $T = \{0\}$.

References

1. W.K. Hale, Frequency assignment: Theory and applications, *Proc. IEEE*, 68 (1980), 1497-1514.
2. M.B. Cozzens and F.S. Robert, T-colorings of graphs and the channel assignment problem, *Congr. Numer.*, 35, (1982), 191-208.
3. F.S. Roberts, From garbage to rainbows: Generalization of graph coloring and their applications, *Proc. Sixth International Conference on the Theory and Applications of Graphs*, Y. Alari, G. Chartrand, O.R. Ollermann and A.J. Schwenk (eds.), John Wiley, New York, 1989.

Cycle-Bandwidth and Cycle-Separation for Graphs: Given a graph $G = (V, E)$, a cyclic layout f of G is a bijection $f : V \to Z_n$, where $|V| = n$, and Z_n is the set $\{0, ..., n-1\}$. The cycle-bandwidth problem is to compute

$$\min_{f} \max_{\{u,v\} \in E} \min\{|f(u) - f(v)|, n - |f(u) - f(v)|\}$$

The cycle-separation problem is to compute

$$\max_{f} \min_{\{u,v\} \in E} \min\{|f(u) - f(v)|, n - |f(u) - f(v)|\}$$

Both the cycle-bandwidth and cycle-separation problems are known to be NP-hard [1].

References

1. J. Y-T. Leung, O. Vornberger and J.D. Witthoff, On some variants of the bandwidth minimization problem, *SIAM Journal of Computing*, vol. 13, No. 3, 1984, 650-667.

4.3. Miscellaneous

Minimum Maxinal Matching: Given a graph, find a maximal matching of minimum cardinality. The problem is NP-hard [1].

References

1. M. Yannakakis and F. Gavril, Edge dominating sets in graphs, unpublished manuscript, 1978.

Bottleneck Traveling Salesman: Given m cities and a table of distances between cities, find a tour to minimize the longest edge in the tour. The problem is NP-hard [1].

References

1. M.R. Garey and D.S. Johnson, *Computers and Intractability, A Guide to the Theory of NP-Completeness*, (Freeman, 1978).

Min-max Spanning Tree: Consider an undirected graph $G = (V, E)$ and a set S of scenarios. We associate a nonnegative cost c_e^s to each edge $e \in E$ under each scenario $s \in S$. The min-max spanning tree problem is defined by

$$\min_T \max_{s \in S} \sum_{e \in T} c_e^s$$

where T is a spanning tree. The problem is NP-hard [2].

References

1. G. Yu and P. Kouvelis, On min-max optimization of a collection of classical discrete optimization problems, in this book.

Unit Capacity Concurrent Flow Problem Given an undirected graph $G = (V, E)$, and k commodities, each consisting of a source-sink pair $s_i, t_i \in V$ and a nonnegative integer demand $d(i)$. A flow f_i in G from node s_i to node t_i is defined as a collection of paths P_i from s_i to t_i with each $P \in P_i$ associated a nonnegative value $f_i(P)$. The value of the flow is defined by $\sum_{P \in P_i} f_i(P)$. The amount of flow f_i through an edge $\{v, w\}$ is$f_i(\{v, w\}) = \sum \{f_i(P) \mid P \in P_i \text{ and } \{v, w\} \in P\}$. The total amount of flows on edge $\{v, w\}$ is $f(\{v, w\}) = \sum_{i=1}^k f_i(\{v, w\})$. The goal is to find

$$\min_f \max_{\{v,w\} \in E} f(\{v, w\})$$

such that

$$\sum_{P \in P_i} f_i(P) = d(i)$$

The problem can be solved in $O(k^{3.5}n^3\sqrt{m}\log nD)$ time [1, 2] (where D denotes the sum of the demands).

References

1. S. Kapoor and P.M. Vaidya, Fast algorithms for convex quadratic programming and multicommodity flows, *Proceedings of the 18th Annual Symposium on Theory of Computing*, 1986, 147-159.
2. P.M. Vaidya, Speeding up linear programming using fast matrix multiplication, *Proceedings of the 30th Annual Symposium on Foundation of Computer Science*, 1989, 332-337.

5. Management Problems

5.1. Scheduling

Usually, a scheduling problem on more than one machines has a minimax objective function. The following are some examples of scheduling problems.

Multiprocessor Scheduling: Given m processors and a set J of nonpreemptive jobs, find a schedule that minimizes the maximum time over all processors.

The problem is NP-hard for $m \geq 2$ [1], but can be solved in pseudo-polynomial time for any fixed m. Some approximation algorithms are presented in [2].

References

1. J. Ullman, NP-complete scheduling problems, *J. Comput. System Sci.* 10 (1975), 384-393.
2. V. Tanaev, V. Gordon and Y. Shafransky, Scheduling theory. Single systems, *Kluwer Academic Publishers*, 1994, 312-323.

Precedence Constrained Scheduling: There are m processors and a set J of independent nonpreemptive jobs with a partial order, the problem is to find a schedule satisfying the partial order to minimize the maximum time over all processors.

It is NP-hard [1]. If all the tasks have the same length, there are some cases that can be solved in polynomial time [3]. Some approximation algorithms are presented in [2].

References

1. J. Ullman, NP-complete scheduling problems, *J. Comput. System Sci.* 10 (1975), 384-393.
2. V. Tanaev, V. Gordon and Y. Shafransky, Scheduling theory. Single systems, *Kluwer Academic Publishers*, 1994, 312-323.
3. E. Coffman and R. Graham, Optimal scheduling for two-processor system, *Acta Informat.*, 1 (1972), 200-213.

Preemptive Scheduling: There are m processors and a set J of preemptive jobs with a partial order, the problem is to find a schedule to minimize the maximum time over all processors.

It is NP-hard [1]. But there are a lot of special cases which can be solved in polynomial time [4]. Approximate algorithms for scheduling parallelizable jobs can be found in [2] and [3].

References

1. J. Ullman, NP-complete scheduling problems, *J. Comput. System Sci.* 10 (1975), 384-393.
2. J. Turek, J. Wolf and P. Yu, Approximate algorithms for scheduling parallelizable tasks, *4th Annual ACM Symposium on Parallel Algorithms and Architectures*, 1992, 323-332.
3. W. Ludwig and P. Tiwari, Scheduling malleable and nonmalleable parallel tasks, *Proc. 5th ACM-SIAM Symposium on Discrete Algorithms*, 1994, 167-176.
4. R. Muntz and E. Coffman, Preemptive scheduling on two-processor system, *J. Assoc. Comput. Math.*, 17 (1970), 324-338.

Resource Constrained Scheduling: There are m processors, i resources and a set J of jobs. Each job t requires $R_i(t)$ of resource i and each resource has a bound of B_i. The problem is to find a schedule such that the sum of $R_i(t)$ over all the jobs is at most B_i for all i, and this schedule minimizes the maximum time over all processors.

The problem has been studied by Ullman [1]. In general, it is NP-hard, but there are some cases that can be solved in polynomial time [1]. Some approximation algorithms are presented in [2].

References

1. J. Ullman, NP-complete scheduling problems, *J. Comput. System Sci.* 10 (1975), 384-393.
2. V. Tanaev, V. Gordon and Y. Shafransky, Scheduling theory. Single systems, *Kluwer Academic Publishers*, 1994, 312-323.

Open-Shop Scheduling: There are m processors and a set J of jobs. Each job $j \in J$ consists of m tasks $t_1[j], t_2[j], \ldots, t_m[j]$ with $t_i[j]$ to be executed by processor i. Given length $l(t)$ for each such task t, the problem is to find a schedule to minimize the maximum time over all processors. (A schedule should satisfy the condition that every job cannot be processed by two processors at the same time.)

It is NP-hard [1]. It can be solved in polynomial time if $m = 2$ or preemptive schedules are allowed. A lot of approximation algorithms are mentioned in [2].

References

1. T. Gonzalez and S. Sahni, Open shop scheduling to minimize finish time, *J. Assoc. Comput. March.*, 23 (1976), 665-679.
2. V. Tanaev, Y. Sotskov and V. Strusevich, Scheduling theory. Multi-stage system, *Kluwer Academic Publishers*, 1994, 271-279.

Flow-Shop Scheduling: There are m processors and a set J of jobs. Each job $j \in J$ consists of m ordered tasks $t_1[j]$, $t_2[j]$, ..., $t_m[j]$ with $t_i[j]$ to be executed by processor i. Given length $l(t)$ for each such task t, the problem is to find a schedule to minimize the maximum time over all processors. (A schedule should satisfy the condition that every job cannot be processed by two processors at the same time.)

It is NP-hard [1]. It can be solved in polynomial time if $m = 2$ or preemptive schedules are allowed [3]. Many polynomial approximation algorithms are described in [2].

References

1. M. Garey, D. Johnson and R. Sethi, The complexity of flowshop and jobshop scheduling, *Math. Oper. Res.*, 1 (1976), 117-129.
2. V. Tanaev, Y. Sotskov and V. Strusevich, Scheduling theory. Multi-stage system, *Kluwer Academic Publishers*, 1994, 104-113.
3. T. Gonzalez and S. Sahni, Flowshop and jobshop schedules: complexity and approximation, *Operations Res.*, 26 (1978), 36-52.
4. P.M. Pardalos (Editor), *Complexity in Numerical Optimization*, (World Scientific, Singapore, 1993).

No-Wait Flow-Shop Scheduling: The description is the same as Flow-Shop Scheduling. A schedule is defined to be a no-wait flow-shop schedule if for each job $j \in J$, the starting time of j on processor $i + 1$ is the sum of the starting time on processor i and the length of $t_i[j]$. The problem is to find a no-wait flow-shop schedule to minimize the maximum time over all processors.

It is NP-hard [1]. It can be solved in polynomial time if $m = 2$ [2]. One can find a comprehensive review on no-wait shop scheduling including flow-shop scheduling in [3] and [4].

References

1. J. Lenstra, A.H.G. Rinnooy Kan and P. Brucker, Complexity of machine scheduling problems, *Ann. Discrete Math.*, 1 (1977), 343-362.
2. P. Gilmore and R. Gomory, Sequencing a one state-variable machine: a solvable case of the traveling salesman problem, *Operations Res.*, 12 (1964), 655-679.
3. S. Goyal and C. Sriskandarajah, No-wait shop scheduling: computational complexity and approximation algorithms. *Opsearch*, 25, 1988, 220-244.
4. N. Hall and C. Sriskandarajah, Machine scheduling problems with no-wait in process, Working paper 91-05, 1991, Department of Industrial Engineering, University of Toronto, Toronto, Canada.

Two-Processor Flow-Shop With Bounded Buffer The description is the same as Flow-Shop Scheduling with m=2, with the addition of a buffer bound B. A schedule is said to satisfy the buffer bound if for all positive number u, the number of jobs in J for which the finishing time on processor 1 is no more than u and the starting time on processor 2 is later than u is at most B. The problem is to find a schedule satisfying the buffer bound to minimize the maximum time over all processors.

It is NP-hard [1]. It can be solved in polynomial time if $B = 0$ or $B \geq |J| - 1$ [2]. One can an algorithm of $O(n^{2(m-|J|-2)})$ where $m - |J| - 2 > 0$ in [3].

References

1. C. Papadimitriou and K. Steiglitz, Flowshop scheduling with limited temporary storage, 1978,unpublished manuscript.
2. P. Gilmore and R. Gomory, Sequencing a one state-variable machine: a solvable case of the traveling salesman problem, *Operations Res.*, 12 (1964), 655-679.
3. S. Redddi, Sequencing with finite intermediate storage, *Manag. Sci.*, 23, 1976, 216-217.

Job-Shop Scheduling There are m processors and a set J of jobs. Each job $j \in J$ consists of an ordered collection of tasks $t_k[j], 1 \leq k \leq n_j$ and t is on processor $p(t) \in \{1, 2, ..., m\}$. We consider the kind of schedules which satisfy the order of all tasks. The problem is to find a schedule which satisfies the order of all tasks to minimize the maximum time over all processors.

It is NP-hard [1]. One can find the first randomized and deterministic algorithms in [2] which yield polylogarithmic approximations to the optimal length schedule.

References

1. M. Garey, D. Johnson and R. Sethi, The complexity of flowshop and jobshop scheduling, *Math. Oper. Res.*, 1 (1976), 117-129.
2. D. Shmoys, C. Stein and J. Wein, Improved approximation algorithms for shop scheduling, *SIAM J. Computing*, vol 23, 6 (1994), 617-632.

Scheduling on a Chain-Like Task System: A chain-like system consists of m modules scheduled on n identical processors. The m modules are in a linear order; each is associated with an execution time and communicates with its neighbor with a communication cost. Each processor accepts only consecutive modules. If adjacent modules are assigned to the same processor, their communication time is assumed to be zero. The complete time for each processor consists of execution times of modules and communication costs to adjacent modules in outside. The problem is to schedule m modules to n processors in order to minimize the maximum completion time of processors.

This problem is polynomial-time solvable. One can find algorithms in [1] and [2] running in $0(m^3 n)$ time and $O(m^2 - n^2)$ time, respectively.

References

1. S.H. Bokhari, Partitioning problems in parallel, pipelined, and distributed computing, *IEEE Transactions on Computers* 37 (1988) 48-57.
2. C.-L. Chan and G. Young, Scheduling algorithms for a chain-like task system, *Lecture Notes in Computer Science* 762 (1993) 496-505.

5.2. MISCELLANEOUS

Min-max Product Control: Consider a finite time horizon. In each time period, there is a deterministic demand on a single product. Let K_t be the product capacity in period t. Let c_t^s and h_t^s be the product unit cost and the inventory holding cost in period t under scenario s, respectively. The problem is to find the product quantity

x_t and inventory quantity y_t in period t to optimize

$$\min_{x,y} \max_{s \in S} \sum_{t=1}^{T} (c_t^s x_t + h_t^s y_t)$$

subject to

$$\begin{aligned} y_t &\geq y_{t-1} + x_t & t &= 2, \cdots, T \\ 0 \leq x_t &\leq K_t & t &= 1, \cdots, T \\ y_t &\geq 0, & t &= 1, \cdots, T \\ x_t, y_t &\text{ are integers} & t &= 1, \cdots, T. \end{aligned}$$

When $|S| = 1$, the problem can be solved in $O(T^2)$ [1]. In general, the problem is NP-hard [2].

References

1. H.M. Wagner and T. Whitin, Dynamic problem in the theory of the firm, in *Theory of Inventory Management* T. Whitin (ed.) (Princeton University Press, Princeton, N.J., 1957).
2. G. Yu and P. Kouvelis, On min-max optimization of a collection of classical discrete optimization problems, in this book.

6. Miscellaneous

Multiquadratic and Semidefinite Programming: The Multiquadratic Programming Problem (MQP) is defined to be of the form: $f^* = \inf\{f_0(x) | f_i(x) \leq 0, i = 1, ..m\}$, where $f_i(x) = x^T Q_i x + b_i^T x + c_i, i = 0, .., m$ are quadratic functions. Define the Lagrangean function:

$$L(x, u) := f_0(x) + \sum_{i=1}^{m} u_i f_i(x).$$

Then $f^* = \inf_x \sup_{u \geq 0} L(x, u)$. The MQP is NP-Hard, so one considers the relaxation

$$\psi^* = \sup_{u \geq 0} \inf_x L(x, u),$$

and this relaxation reduces [2] to a Semidefinite Programming Problem(SDP) [1], which can be solved approximately in polynomial time. The important issue here is to identify classes of MQPs, for which the above relaxation is exact, i.e. $f^* = \psi^*$. Some work has already been done in this regard, see for instance [3] and [4].

References

1. F. Alizadeh, Optimization Over Positive Semi-Definite Cone; Interior-Point Methods and Combinatorial Applications, In "Advances in Optimization and Parallel Computing", P.M. Pardalos, editor, North–Holland, 1992.

2. M.V. Ramana, *An Algorithmic Analysis of Multiquadratic and Semidefinite Programming Problems*, Ph.D. Thesis, The Johns Hopkins University, Baltimore, MD 21218, 1993.
3. M.V. Ramana, A.J. Goldman, Quadratic Maps with Convex Images, RUTCOR Research Report, RRR 36-94, Rutgers University, New Brunswick, NJ 08903.
4. N.Z. Shor, Dual Estimates in Multiextremal Problems, *Journal of Global Optimization*, Vol. 2, No. 4 (1992), pp. 411-418.

AUTHOR INDEX

Nonconvex Optimization and Its Applications

1. D.-Z. Du and J. Sun (eds.): *Advances in Optimization and Approximation.* 1994. ISBN 0-7923-2785-3
2. R. Horst and P.M. Pardalos (eds.): *Handbook of Global Optimization.* 1995 ISBN 0-7923-3120-6
3. R. Horst, P.M. Pardalos and N.V. Thoai: *Introduction to Global Optimization* 1995 ISBN 0-7923-3556-2; Pb 0-7923-3557-0
4. D.-Z. Du and P.M. Pardalos (eds.): *Minimax and Applications.* 1995 ISBN 0-7923-3615-1

KLUWER ACADEMIC PUBLISHERS – DORDRECHT / BOSTON / LONDON